Mechanical Engineering Series

Series Editor

Francis A. Kulacki, Department of Mechanical Engineering, University of Minnesota, Minneapolis, MN, USA

The Mechanical Engineering Series presents advanced level treatment of topics on the cutting edge of mechanical engineering. Designed for use by students, researchers and practicing engineers, the series presents modern developments in mechanical engineering and its innovative applications in applied mechanics, bio-engineering, dynamic systems and control, energy, energy conversion and energy systems, fluid mechanics and fluid machinery, heat and mass transfer, manufacturing science and technology, mechanical design, mechanics of materials, micro- and nano-science technology, thermal physics, tribology, and vibration and acoustics. The series features graduate-level texts, professional books, and research monographs in key engineering science concentrations.

More information about this series at http://www.springer.com/series/1161

Hamid Arastoopour • Dimitri Gidaspow •
Robert W. Lyczkowski

Transport Phenomena in Multiphase Systems

 Springer

Hamid Arastoopour
Department of Chemical
and Biological Engineering
and Department of Mechanical,
Materials, and Aerospace Engineering
Wanger Institute for Sustainable Energy
Research (WISER)
Illinois Institute of Technology
Chicago, IL, USA

Dimitri Gidaspow
Department of Chemical and Biological
Engineering
Illinois Institute of Technology
Chicago, IL, USA

Robert W. Lyczkowski
Darien, IL, USA

Editor
Margaret M. Murphy
Wanger Institute for Sustainable Energy
Research (WISER)
Illinois Institute of Technology
Chicago, IL, USA

ISSN 0941-5122   ISSN 2192-063X   (electronic)
Mechanical Engineering Series
ISBN 978-3-030-68580-5   ISBN 978-3-030-68578-2   (eBook)
https://doi.org/10.1007/978-3-030-68578-2

This Springer imprint is published by the registered company Springer Nature Switzerland AG
The registered company address is: Gewerbestrasse 11, 6330 Cham, Switzerland

Preface

The most widely used transport phenomena textbook by Bird, Stewart, and Lightfoot, published 60 years ago in 1960, with a revised second edition published in 2007, is limited to the subject of single-phase flow heat and mass transfer. However, most chemical, pharmaceutical, energy, food, and biological processes are multiphase systems. There is a great need, therefore, for a book that is focused on fundamentals and application of multiphase transport phenomena. Furthermore, the widespread application of transport phenomena of multiphase systems and fluidization in industry demands an increase in efficiency and the development of predictive numerical simulations of industrial processes for design and scale-up purposes. This book is intended to address some of these issues and provide fundamental multiphase transport phenomena knowledge and needed design and scale-up tools for graduate students in energy, chemical, biological, mechanical, nuclear, and environmental engineering, as well as professional scientists and engineers working in areas related to multiphase flow systems.

In the past, research in multiphase flow systems was focused almost exclusively on the development of overall flow measurements and empirical correlations. However, during the last four decades, with the rise of high-speed computers and commercial software, notable advances have been made in analyzing multiphase flow systems that have significantly impacted the approach in the design and scale-up of processes based on multiphase transport phenomena.

There are significant reserves of unconventional natural gas in the world residing in low-permeability shales and hydrates. The manufacturing, utility, transportation, and chemicals industries are rapidly changing to use gas in place of oil and coal. Consequently, this book places heavy emphasis on processes that include solid particles suspended by gases or liquids and the application of Eulerian/Eulerian transport phenomena models for such systems.

Attempts to develop Eulerian/Eulerian models for fluid/particle systems and fluidization began in the 1960s and 1970s by Davidson, Jackson, Soo, and Crowe, for example. Starting in the 1980s, progress began to be made by Illinois Institute of Technology (IIT) researchers and others to obtain numerical solutions for one-dimensional fluid/solid flow equations to simulate flow in vertical pneumatic conveying systems. Subsequently, IIT researchers modeled each particle size as a phase using separate continuity and momentum equations, developed experimentally verified one-dimensional particle/particle collision models, and successfully compared computed flow parameters with experimental data for dilute gas/particle systems. This probably motivated several investigators to develop a theory of particle interactions and collisions based on the kinetic theory approach, which uses a granular temperature equation to determine the fluctuating turbulent kinetic energy of the particles. In fact, the kinetic energy theory approach for granular flow allows the determination of, for example, particle phase stress, pressure, and viscosity in place of empirical correlations. After this developmental stage of granular theory, there have been several modifications to the constitutive equations such as extension of the kinetic theory for granular flow for mixtures of multi-type particles. To account for continuous variation in particle property distribution due to phenomena such as chemical reaction, agglomeration, breakage, attrition, and growth, population balance equations (PBE) linked with computational fluid dynamics (CFD) were developed.

This book is aimed to address a need, which has existed for decades, for a text with basic governing and constitutive equations followed by several engineering problems on multiphase flow and transport that are not provided in current advanced texts, monographs, and handbooks. The unique emphasis of this book is on the sound formulation of the basic equations describing multiphase transport and how they can be used to design processes in selected industrially important fields. The clear underlying mathematical and physical bases of the interdisciplinary description of multiphase flow and transport are the main themes.

The multiphase kinetic theory approach in developing constitutive equations for fluid/solid flow systems has advanced significantly during the last decade, including the flow of mixtures of particles. This book also addresses the advances in the kinetic theory for particle flow systems.

Chapters 1–3 of this book address multiphase flow basic equations; the advances in multiphase kinetic theory; phenomena associated with gas/solid and gas/liquid flows; and computational transport phenomena. Chapters 4–13 demonstrate the unique application of the multiphase approach to the design of several chemical, energy, biological, pharmaceutical, and environmental related processes. These topics are not merely "case studies." They demonstrate cohesively and constructively how the innovative application of multiphase transport equations produces a novel and unique approach for design and scale-up of these processes.

Even though many books on multiphase flow have been written, none of them clearly shows how the basic multiphase equations can be used in the design of multiphase processes. The authors represent a combination of nearly two centuries of experience and innovative application of multiphase transport representing hundreds of publications and several books. This book serves to encapsulate the essence of their wisdom and insight.

This book can be used as a senior-year elective or graduate-level textbook; by professionals in the design of processes that deal with a variety of multiphase processes in the chemical, energy, bioengineering, and pharmaceutical industries; by practitioners and experts in multiphase science in the area of CFD at US national laboratories, international universities, research laboratories and institutions, and in the chemical, pharmaceutical, and petroleum industries.

The following is a summary of the book contents:

Chapter 1 deals with the derivation of the basic governing equations for conservation of mass, momentum, and energy for multiphase systems. The Appendix complements Chap. 1 and is included to help understand the mathematics of conservation laws.

Chapter 2 provides continuity, momentum, and energy conservation equations, and constitutive equations for fluid/particle flow systems (see Tables 2.3–2.5). The conservation equations and constitutive relations are general and can be applied to all regimes of fluid/particle flow, from a very dilute particle volume fraction to the packed-bed regime. Kinetic theory of gases was extended to multiphase particulate flow where the interactions between particles are not conserved. The fundamentals of the kinetic theory approach for derivation of constitutive equations for the regimes when particle collision is dominant and the frictional behavior of particles based on soil mechanics principles for dense particle flow are discussed. Gas/particle flows are inherently oscillatory and they manifest in non-homogeneous structures; thus, the effects of the presence of particle clusters in fluid/particle flow modeling and numerical simulation are presented. In addition, the kinetic theory approach has been extended to multi-type particulate flows.

Chapter 3 provides a brief overview and basic properties of gas/solid and gas/liquid two-phase flows, with some discussion of gas/solid and gas/liquid flow regimes.

Chapter 4 investigates the feasibility of using the circulating fluidized bed (CFB) and rotating fluidized bed polymerization reactors for process intensification and the significant increase in polymer production rate using CFD.

Chapter 5 presents the use of circulating fluidized beds for catalytic reactors. CFD simulations for a CFB gasification process and a CFB sulfur dioxide capture process are also demonstrated.

Chapter 6 deals with synthetic gas conversion to liquid fuel using slurry bubble column reactors. Design of such reactors using CFD is demonstrated. This fundamental approach leads to lower cost due to the elimination of all cooling tubes in the reactor design.

Chapter 7 demonstrates an efficient and cost-effective CO_2 removal system using solid sorbents in CFBs with a regeneration system that may be performed at lower pressure. It is demonstrated that essentially different solid sorbents can be used to remove carbon dioxide from the power plants, with minimal energy loss. This is due to the low decomposition temperature of the solid carbonates. Design of such a system can be achieved using a multiphase CFD approach to simulate CO_2 sorption and regeneration in the entire CFB system utilizing different solid sorbents at different operating conditions.

Chapter 8 shows how the high cost of silicon needed for solar collectors can be greatly reduced by using fluidized bed reactors operating at higher velocities with no bubble formation. The CFD reactor simulation of such a system is presented.

Chapter 9 describes the development of a new paradigm for modeling blood flow (hemodynamics) using the multiphase flow approach with the aim of understanding the initiation of atherosclerosis. In addition, the application of multiphase kinetic theory to model flow in right coronary and carotid arteries and the depletion of red blood cells in the vicinity of vessel walls are discussed.

Chapter 10 deals with multiphase flow modeling of explosive volcanic eruptions. The development of multiphase flow models described in this chapter has illuminated several aspects of the dynamics of explosive eruptions including more quantitative future scenarios. In addition, the modeling and simulation of the continuous variability of particle size distribution in a rising volcanic plume are presented using method of moments.

Chapter 11 demonstrates the capability of multiphase CFD simulation to provide a reliable tool to design, scale-up, and analyze the performance of airfoils and wind turbines at different wind speed and direction, and to assess the impact of rain on the performance of the wind turbines and airfoils.

Chapter 12 shows the application of multiphase flow simulation in pharmaceutical processes, focusing on the drying process. The CFD approach was used for design, scale-up, and performance optimization of three different scale pharmaceutical bubbling fluidized bed dryers.

Chapter 13 deals with the hydrodynamics of fluidization with surface charge. In this chapter, the force generated due to the charge of particles was included in the multiphase model by solving the Poisson's equation for the voltage generated in the fluidized bed. This model is capable of predicting the sheeting phenomenon.

The authors greatly appreciate those individuals who have contributed significantly to the completion of this book. They would like to acknowledge Dr. Brian Valentine from the U.S. Department of Energy for contributing Chap. 3; Drs. Augusto Neri, Tomaso Esposti Ongaro, Mattia de' Michieli Vitturi, and Matteo Cerminara from the Istituto Nazionale di Geofisica e Vulcanologia in Pisa, Italy, for contributing Chap. 10; and Piyawan Tiyapiboonchaiya from the Thai Polyethylene Co. Ltd., Rayong, Thailand, and Somsak Damronglerd from the Department of Chemical Technology, Faculty of Science, Chulalongkorn University, Bangkok, Thailand, for their contributions to Chap. 13. The authors would like to acknowledge Dr. Wen Ho Lee and Miss Farnaz Esmaeili Rad for their assistance. The third author (RWL) of this book wishes to thank Professor Sanjeev G. Shroff, Chair, Department of Bioengineering, Swanson School of Engineering, University of Pittsburgh, for reviewing Chap. 9. The first author of this book (HA) would like to acknowledge Drs. Abbasian, Teymour, Benyahia, Iddir, Ahmadzadeh, Strumendo, Jang, Abbasi, Cohan, and Ghadirian, and Mr. Cai for their input and research contribution during their tenure at IIT that was used in this book.

The research conducted by the authors presented in this book is mainly funded by the U.S. Department of Energy, the National Science Foundation, Argonne National Laboratory, and the Wanger Institute for Sustainable Energy Research (WISER) at

Illinois Institute of Technology (IIT). The authors would like to thank and greatly appreciate Ms. Margaret M. Murphy, IIT WISER Assistant Director and Program Outreach Manager, for editing this book.

Chicago, IL   Hamid Arastoopour
Chicago, IL   Dimitri Gidaspow
Darien, IL   Robert W. Lyczkowski

Contents

Chapter 1
Introduction to Multiphase Flow Basic Equations

Hamid Arastoopour, Dimitri Gidaspow, and Robert W. Lyczkowski

1.1 Introduction

Conservation laws for mass, momentum, and energy for multiphase flow can be derived using what is referred to by Aris [1] as the Reynolds transport theorem. Reynolds [2] derived this result in 1903 by invoking what he called Axiom I, which states:

> Any change whatsoever in the quantity of any entity within a closed surface can only be effected in one or other of two distinct ways: (1) it may be effected by the production or destruction of the entity within the surface, or (2) by the passage of the entity across the surface.

For multiphase flow systems, the concept of the volume fraction for phase i, ε_i, appears. For a single phase system $\varepsilon_i = 1$ and these equations reduce to those found in standard transport phenomena books, such as Bird et al. [3]. Two concepts are needed: conservation laws and constitutive equations. This section shows how these conservation laws are obtained from the Reynolds transport theorem in reference to Fig. 1.1.

The point (x^0, y^0, z^0) represents the spatial coordinates of the mass at some fixed time, t^0. Then, the spatial coordinates of the mass at any time t are,

$$x = x(t, x^o, y^o, z^o) \quad y = y(t, x^o, y^o, z^o) \quad z = z(t, x^o, y^o, z^o)$$

Following Aris [1], let us define a property per unit volume $\Im(t, x)$, where t is time and x is the position vector such that,

$$F(t) = \iiint\limits_{V(t)} \Im(t, x) dV \tag{1.1}$$

H. Arastoopour et al., *Transport Phenomena in Multiphase Systems*, Mechanical Engineering Series, https://doi.org/10.1007/978-3-030-68578-2_1

Fig. 1.1 Motion for a
system of constant mass

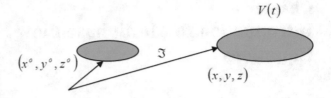

where $F(t)$ is the system variable quantity that can change with time. The balance made on $F(t)$ gives the Reynolds transport theorem,

$$\frac{d}{dt} \iiint\limits_{V(t)} \Im(t,x)dV = \iiint\limits_{V(t)} \left(\frac{\partial \Im}{\partial t} + \nabla \cdot \Im v_i \right) dV \tag{1.2}$$

For a more detailed discussion of the Reynolds transport theorem, the reader is referred to Appendix E in Gidaspow [4].

In multiphase flow, the volume occupied by phase i cannot be occupied by any other phase at the same position in space at the same time. This distinction introduces the concept of the volume fraction of phase i, ε_i. The volume of phase i, V_i, in a system of volume V, is [5],

$$V_i = \iiint\limits_{V(t)} \varepsilon_i dV \text{ where } \sum_{i=1}^{n} \varepsilon_i = 1 \tag{1.3}$$

Equation (1.3) defines into existence the volume fraction, ε_i, for phase i. Other authors have taken an alternative means of defining this quantity related to volume averaging. Based on the Crowe et al. [6] concept of defining the volume fraction for a dispersed phase flow at a point, one may define density at a point for a continuum. The volume fraction for the dispersed phase in some volume, δV, is defined,

$$\alpha_d = \lim_{\delta V \to \delta V^o} \delta V_d / \delta V$$

where δV_d is the volume of the dispersed phase in the volume, and δV^o is the limiting volume that ensures a stationary average, i.e., the volume should be large enough to contain a representative number of dispersed phase elements, even if it is decreased or increased slightly. Brennen [7] and Bowen [8] take a similar approach that shows how one would go about experimentally determining the dispersed phase volume fraction using, for example, particle image velocimetry (PIV). The fluid volume fraction in the porous media literature, called the void fraction or porosity, has been experimentally determined for decades and occurs in the Ergun equation [3]. This chapter is based upon and expands Chap. 1 by Arastoopour et al. [9].

1.2 Multiphase Conservations Laws

The Reynolds transport theorem discussed in Sect. 1.1 will be used to derive the multiphase balances of mass, momentum, and energy.

1.2.1 Mass Balances

The mass of phase i in a volume element V moving with phase i velocity is,

$$m_i = \iiint\limits_{V(t)} \varepsilon_i \rho_i dV \tag{1.4}$$

where ε_i is the volume fraction of phase i, and ρ_i is defined as the density of phase i. This volume fraction is the only new variable in multiphase flow. Traditionally, this volume fraction, ε_i, was derived by volume averaging as discussed above and by Jackson [10]. For multicomponent systems, $\varepsilon_i = 1$ because it is assumed that molecules of component i occupy the same space at the same time. Such an approximation cannot be made in multiphase flow. In multiphase flow, the volume fraction for phase i cannot be occupied by the remaining phases at the same position in space at the same time.

The balance on mass, m_i, moving with the velocity, v_i, is,

$$\frac{dm_i}{dt^i} = \frac{d}{dt^i} \iiint \varepsilon_i \rho_i dV = \iiint m_i' dV \tag{1.5}$$

where m_i' is the rate of production of phase i, and d/dt^i is the substantial derivative moving with phase i velocity.

The Reynolds transport theorem gives the continuity equation for phase i,

$$\frac{\partial(\varepsilon_i \rho_i)}{\partial t} + \nabla \cdot (\varepsilon_i \rho_i v_i) = m_i' \tag{1.6}$$

Conservation of mass requires,

$$\sum_{i=1}^{n} m_i' = 0 \tag{1.7}$$

In Eq. (1.3), the volume fraction of phase i was defined,

$$V_i = \iiint_{V(t)} \varepsilon_i dV \tag{1.8}$$

For incompressible fluid with no phase change, the continuity equation becomes

$$\frac{\partial \varepsilon_i}{\partial t} + \nabla \cdot (\varepsilon_i v_i) = 0 \tag{1.9}$$

where $i = 1, 2, \ldots n$ for the phases. An equation such as Eq. (1.9) does not appear in conventional transport phenomena theory books such as Bird et al. [3].

1.2.2 Momentum Balances

The rate of change of momentum of phase i moving with velocity v_i equals the forces acting on the system and can be mathematically expressed,

$$\frac{d}{dt} \iiint_{V(t)} \rho_i v_i \varepsilon_i dV = f_i \tag{1.10}$$

where the force f_i is,

$$f_i = \oiint_{S(t)} T_i da + \iiint_{V(t)} \rho_i F_i \varepsilon_i dV + \overline{p}_i dV + m'_i v_i dV \tag{1.11}$$

The first term represents the surface forces acting on the differential area, da; the second term represents the external forces; the third term represents the interaction forces between phases; and the last term represents the force due to the phase change. The stress tensor T_i for phase i has nine components and is written,

$$T_i = \begin{pmatrix} T_{ixx} & T_{ixy} & T_{ixz} \\ T_{iyx} & T_{iyy} & T_{iyz} \\ T_{izx} & T_{izy} & T_{izz} \end{pmatrix} \tag{1.12}$$

Figure 1.2 represents the stress tensor acting on the surfaces of volume of phase i.

The Reynolds transport theorem is applied to the left side of Eq. (1.10), and the divergence theorem ($\oiint_a T_i da = \iiint_V \nabla \cdot T_i dV$) is applied to the first term of Eq. (1.11). The result is the momentum balance for phase i,

Fig. 1.2 Stress tensor acting on the surfaces of volume of phase i

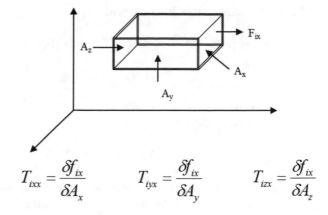

$$T_{ixx} = \frac{\delta f_{ix}}{\delta A_x} \qquad T_{iyx} = \frac{\delta f_{ix}}{\delta A_y} \qquad T_{izx} = \frac{\delta f_{ix}}{\delta A_z}$$

$$\frac{\partial(\rho_i \varepsilon_i v_i)}{\partial t} + \nabla \cdot (\rho_i \varepsilon_i v_i \otimes v_i) = \nabla \cdot T_i + \rho_i \varepsilon_i F_i + \bar{p}_i + m_i' v_i \tag{1.13}$$

A different form of the momentum equation has been used in the literature [11].

Using the continuity equation, Eq. (1.6), the momentum balance for phase i moving with phase velocity v_i becomes,

$$\rho_i \varepsilon_i \frac{dv_i}{dt^i} = \nabla \cdot T_i + \rho_i \varepsilon_i F_i + \bar{p}_i \tag{1.14}$$

where $\rho_i \varepsilon_i \frac{dv_i}{dt^i}$ is the acceleration of phase i, $\nabla \cdot T_i$ is the gradient of stress due to the surface forces, $\rho_i \varepsilon_i F_i$ is the body force, and $\bar{p}_i$ is the interaction forces between phases, called the momentum supply by Bowen [8]. Note that $\sum_{i=1}^{n\,phase} \bar{p}_i = 0$. The interaction terms between phases given by the expression,

$$\bar{p}_i = \sum_j \beta_j (v_j - v_i) \tag{1.15}$$

account for the drag force between the phases in terms of friction coefficients, β_j. Note that moving with the velocity, v_i, there is no force due to the phase change because the mass in the balance equation, Eq. (1.10), moves with velocity v_i. The simplest expression for the stress, T_i, is,

$$T_i = -P_i I \tag{1.16}$$

Each phase i has its own pressure, P_i, similar to Euler's equation for single phase flow.

1.2.2.1 Incompressible Viscous Flow

To meet the requirement of objectivity for each phase k,

$$T_k = T_k(\nabla^s v_k) \tag{1.17}$$

where $\nabla^s v_k$ is the symmetrical gradient of v_k.

Linearization of T_k gives,

$$T_k = A_k I + B_k(\nabla^s v_k) \tag{1.18}$$

For incompressible fluids, A_k is chosen to be the negative of the pressure of fluid phase k, $A_k = -P_k$, and the derivative of the traction T_k, with respect to the symmetric gradient with viscosity of fluid k, is,

$$B_k = 2\mu_k = \frac{\partial T_k}{\partial \nabla^s v_k} \tag{1.19}$$

where the traction for phase k is,

$$T_k = -P_k I + 2\mu_k \nabla^s v_k \tag{1.20}$$

and

$$\nabla^s v_k = \frac{1}{2}\left[\nabla v_k + (\nabla v_k)^T\right] \tag{1.21}$$

1.2.2.2 Incompressible Navier-Stokes Equation

Substitution of stress tensors for each phase k into the momentum balance, Eq. (1.13), gives the incompressible Navier-Stokes equation for each phase k,

$$\frac{\partial(\varepsilon_k \rho_k v_k)}{\partial t} + \nabla \cdot \varepsilon_k \rho_k v_k = \varepsilon_k \rho_k F_k + \sum_j \beta_j(v_j - v_k) + m'_k v_k - \nabla P_k$$
$$+ \mu_k \nabla^2 v_k \tag{1.22}$$

The friction coefficients, β_k, can be expressed in terms of standard drag coefficients [4]. In this model, the viscosity of each phase k, μ_k, is an input into the model. It can be either directly measured or obtained from kinetic theory [12].

1.2.2.3   Compressible Viscous Flow

For compressible viscous flow, there is another parameter, the bulk viscosity of phase k, λ_k,

$$A_k = -P_k + \lambda_k \nabla \cdot v_k \tag{1.23}$$

The stress for phase k is then expressed,

$$T_k = -P_k I + 2\mu_k \nabla^s v_k + \lambda_k I \nabla \cdot v_k \tag{1.24}$$

where $2\mu_k + 3\lambda_k = 0$, similar to that for single phase flow [1]. Hence, the compressible Navier-Stokes equation for each phase k in substantial derivative form becomes,

$$\varepsilon_k \rho_k \frac{dv_k}{dt^k} = \varepsilon_k \rho_k F_k + \sum_j \beta_j (v_j - v_k) - \nabla P_k$$

$$+ \nabla \left(2\mu_k \nabla^s v_k + \frac{2}{3} \mu_k I \nabla \cdot v_k \right) \tag{1.25}$$

1.2.3   Energy Balances

The energy balance for an open system with phase change for phase i can be written,

$$\frac{dU_i}{dt^i} = \frac{dQ_i}{dt} - P_i \frac{dV_i}{dt^i} + D_{iss} + \left(U_i^n + P_i^n / \rho_i^n \right) \frac{dm_i}{dt^i} \tag{1.26}$$

where U_i is the internal energy per unit mass, Q_i is the heat per unit mass flowing into the system, and the only work done by the system is the mechanical work due to the volume change, $P_i \, dV_i$ [4]. With phase change, there is inflow due to phase change consisting of the energy inflow, U_i^n, and pressure, P_i^n. The energy dissipation by the system is $D_{iss} = \int_{V(t)} u_i' dV \, D_{iss}$.

Heat transfer, Q_i, is related to the rate of heat transfer by conduction, q_i, by,

$$-\frac{dQ_i}{dt} = \oiint_{A(t)} (q_i \varepsilon_i da) / T_i q_i \varepsilon_i da = \iiint_{V(t)} (\nabla \cdot \varepsilon_i q_i) dV \tag{1.27}$$

In Eq. (1.27), the heat flux, q_i, need not be multiplied by the volume fraction of phase i.

The Reynolds transport theorem applied to the energy balance, Eq. (1.26), produces the energy equation for phase i in the form,

$$\frac{\partial(\varepsilon_i\rho_i U_i)}{\partial t} + \nabla \cdot (\varepsilon_i\rho_i v_i U_i) = -\nabla \cdot \varepsilon_i q_i - P_i\frac{\partial\varepsilon_i}{\partial t} - P_i\nabla \cdot \varepsilon_i v_i + h_i^n m_i' + u_i' \quad (1.28)$$

where u_i' is the internal energy per unit volume, h_i^n is the enthalpy flowing into the system, and q_i is the heat flux by conduction. There appears a term in Eq. (1.28), $P_i\frac{\partial\varepsilon_i}{\partial t}$, the work of expansion of the volume fraction of phase i, not found in single phase flow. See Gidaspow [4] for a discussion of this term.

1.2.3.1 Enthalpy Representation

The energy balance expressed in terms of enthalpy of phase i per unit of mass, h_i, in substantial derivative form becomes,

$$\varepsilon_i\rho_i\frac{dh_i}{dt^i} = -\nabla \cdot \varepsilon_i q_i + \varepsilon_i\frac{dP_i}{dt^i} + U_i + m_i'\left(h_i^n - h_i\right) \quad (1.29)$$

Equation (1.28) reduces to that found in standard transport phenomena textbooks for $\varepsilon_i = 1$ and $h_i^n = h_i$. The energy equation, Eq. (1.29), in enthalpy form does not have the work term found in Eq. (1.28).

1.2.3.2 Entropy Representation

The entropy balance for phase i moving with velocity v_i is,

$$\frac{d}{dt}\iiint\limits_{V(t)} \varepsilon_i\rho_i S_i dV \; + \; \oiint\limits_{A(t)} \frac{q_i\varepsilon_i da}{T_i} \; - \; \iiint m_i' S_i dV \; = \; \iiint \sigma_i dV \quad (1.30)$$

where $\frac{d}{dt}\iiint\limits_{V(t)}\varepsilon_i\rho_i S_i dV$ is the accumulation of entropy in the system, $\oiint\limits_{A(t)}\frac{q_i\varepsilon_i da}{T_i}$ is the rate of entropy outflow due to flow of heat across area A, $\iiint m_i' S_i dV$ is the rate of entropy at equilibrium due to the phase change, and $\iiint \sigma_i dV$ is the rate of entropy production.

The combination of entropy balance, Eq. (1.30), with the energy balance, Eq. (1.26), generates the following equation for entropy production in the system, which is zero for reversible processes and positive for irreversible processes,

$$\sigma_i = -\frac{1}{T_i} \nabla \cdot \varepsilon_i q_i + \nabla \cdot \frac{q_i \varepsilon_i}{T_i} + \frac{U_i'}{T_i} = -\frac{q_i \varepsilon_i \cdot \nabla T_i}{T_i^2} + \frac{U_i'}{T_i} \geq 0 \qquad (1.31)$$

1.2.4 Conservation of Species

To treat a fluid in which a chemical reaction occurs, such as a fluidized bed polymerization reactor described in Chap. 4, one must write mass balances for the individual components in a phase. The mathematical procedure outlined in Sect. 1.2.1 gives the balance [13],

$$\frac{\partial (\varepsilon_i Y_{ik} \rho_i)}{\partial t} + \nabla \cdot (\varepsilon_i Y_{ik} \rho_i v_{ik}) = \epsilon_i \dot{m}_{sik} \qquad (1.32)$$

where $i = 1, 2,\ldots n$ for the phases, $k = 1, 2,\ldots m$ for the components, and Y_{ik} is the weight fraction of component k in phase i. The right hand side of Eq. (1.32) is written in a more natural way in terms of a specific mass transfer rate $\dot{m}_{si}$. The term $\dot{m}_{sik}$ includes the transfer of component k from another phase and its creation by chemical reaction in phase i. The density, ρ_i, is a function of pressure and enthalpy of phase i and its composition.

The Appendix of this book provides a more detailed explanation of the mathematics of conservation laws used in the theories of multicomponent and multiphase flows.

1.3 Exercises

1.3.1 Ex. 1: Balances of Mass and Species in Multicomponent Systems

In Eq. (1.2.1), Bowen [8] writes the balance of mass, in the present notation, for constituent a (similar to that in Chap. 19 of Bird et al. [3]),

$$\frac{\partial}{\partial t} \int \rho_a dV + \oint \rho_a v_a ds = \int r_a dV$$

accumulation + net rate of outflow = rate of reaction

where dV is the element of volume, ds is the outward drawn vector element of area, and r_a is the rate of mass supplied per unit volume by the reaction.

1. Derive the conservation of species balance.
2. Obtain the conservation of mass equation.

For a one-dimensional example, see the combustion of hydrogen in a tube in Gidaspow Appendix D [4].

1.3.2 Ex. 2: Momentum Balances for a Multicomponent System

1. Compare the multiphase conservation of momentum equations derived in this chapter to those of Bowen [8], and to those in Appendix B of Bird et al. [3].
   In Eq. (1.4.1), Bowen [8] writes the linear momentum balance for constituent a in present notation,

$$\frac{\partial}{\partial t} \int \rho_a v_a dV + \oint \rho_a v_a (v_a \cdot d\mathbf{s}) = \oint \mathbf{T}_a d\mathbf{s} + \int (\rho_a \mathbf{b}_a + \bar{p}_a + r_a v_a) dV$$

accumulation + net outflow = surface + (body + momentum supply + reaction)

where $\mathbf{b}_a$ is the body force.

1.3.3 Ex. 3: Mixture Momentum Balance

1. Add the momentum balances in Exercise 2 and explain why the sum of the momentum supplies, denoted by $\bar{p}_a$, is zero. Also show that the reaction forces have to vanish.
   HINT: Write the momentum balance for the mixture moving with the component a using the chain rule of calculus,

$$\frac{d}{dt} v(t, x(t)) = \frac{\partial v}{\partial t} + \frac{\partial v}{\partial x} \frac{dx}{dt} = \frac{\partial v}{\partial t} + \frac{\partial v}{\partial x} v$$

This is often called the substantial derivative and is written as Dv/Dt. It can be generalized to n dimensions in Gidaspow Appendix E [4].

1.3.4 Ex. 4: Balance of Energy for a Multicomponent System

In Eq. (1.5.1), Bowen [8] writes the energy balance for constituent a as follows in present notation,

$$\frac{\partial}{\partial t} \int \rho_a (U_a + \frac{1}{2}v_a^2) dV + \oint \rho_a (U_a + \frac{1}{2}v_a^2) v_a ds = \oint \mathbf{T}_a^T v_a - Q_a) \cdot d\mathbf{s}$$

$$+ \int (\rho_a Q_{supply} + \rho_a v_a \mathbf{b}_a + U_a + v_a \overline{p}_a + r_a U_a + \frac{1}{2} r_a v_a^2) dV$$

where Q_{supply} is the external energy supply.

1. Obtain the partial differential equation for constituent a and compare it to the multiphase energy balances derived in this chapter and to the energy balances in Bird et al. [3].

Nomenclature

C_d	Drag coefficient
D_{iss}	Energy dissipation by means of friction
d_p	Particle diameter
F	Body force
h	Enthalpy per unit mass
$\mathbf{I}$	Identity matrix
m	Mass
m_i'	Rate of production of phase i
P	Pressure
$\overline{p}$	Interaction forces between phases or momentum supply
Q	Heat transfer
S	Entropy
T	Stress tensor
t	Time
V	System volume
$\vec{v}$	Velocity vector
U	Internal energy per unit mass
Y_{ik}	Weight fraction of component k in phase i

Greek Symbols

ε_i	Volume fraction of phase i
β	Interphase friction coefficient
μ	Viscosity
λ	Bulk viscosity
ρ	Density
$\Im$	General property per unit volume

References

1. Aris R (1989) Vectors, tensors, and the basic equations of fluid mechanics. Dover, New York
2. Reynolds O (1903) Papers on mechanical and physical subjects, vol III: the sub-mechanics of the universe. Cambridge University Press, Cambridge
3. Bird RB, Stewart WE, Lightfoot EN (2007) Transport phenomena: 2nd cdn. Wiley, New York

4. Gidaspow D (1994) Multiphase flow and fluidization: continuum and kinetic theory descriptions. Academic, San Diego
5. Gidaspow D (1977) Fluid-particle systems. In: Kakac S, Mayinger F (eds) Proceedings of NATO advanced study institute on two-phase flows and heat transfer, Istanbul, August 1976, vol 1. Hemisphere, New York, pp 115–128
6. Crowe C, Sommerfeld M, Tsuji Y (1998) Multiphase flows with droplets and particles. CRC Press LLC, Boca Raton
7. Brennen CE (2005) Fundamentals of multiphase flows. Cambridge University Press, Cambridge
8. Bowen RM (1976) Theory of mixtures. In: Eringen AC (ed) Continuum physics, vol III, part I. Academic, New York, pp 1–127
9. Arastoopour H, Gidaspow D, Abassi E (2017) Computational transport phenomena of fluid-particle systems. Springer, Cham, Switzerland
10. Jackson R (2000) The dynamics of fluidized particles. Cambridge University Press, Cambridge
11. Syamlal M, Musser J, Dietiker JF (2017) Two-fluid model in MFIX. In: Michaelides EE, Crowe CT, Schwarzkopf, JD (eds) Multiphase flow handbook, 2nd edn. CRC Press, Boca Raton, pp 242–274
12. Gidaspow D, Jiradilok V (2009) Computational techniques: the multiphase CFD approach to fluidization and green energy technologies. Nova, New York
13. Lyczkowski RW, Gidaspow D, Solbrig CW (1982) Multiphase flow models for nuclear, fossil and biomass energy production. In: Majumdar AS, Mashelkar RA (eds) Advances in transport processes, vol II. Wiley Eastern, New Delhi, pp 198–351

Chapter 2
Multiphase Flow Kinetic Theory, Constitutive Equations, and Experimental Validation

Hamid Arastoopour, Dimitri Gidaspow, and Robert W. Lyczkowski

2.1 Introduction

Originally, the kinetic theory was developed by Chapman and Cowling [1] for gases to predict the behavior of mass point molecules whose interaction energies are conserved. Nearly three decades ago, this theory was extended to particulate flow where the interactions between particles are not conserved. Savage and Jeffrey [2] were probably the first to apply the kinetic theory to rapidly deforming material in the form of smooth, hard spherical particles in order to develop the particle phase constitutive equation. In their derivation, to calculate the stress tensor arising from interparticle collisions, they assumed that the collisions between particles were purely elastic.

Although in some granular flows the restitution coefficient is restrained to values close to unity, its deviation from unity results in a significant variation in the properties of granular flow. This was shown first by Jenkins and Savage [3]. They extended the kinetic theory of an idealized granular mixture to predict the rapid deformation of granular material by including energy dissipation during collision for nearly inelastic particles. Later, Lun et al. [4] developed a theory that predicts the simple shear flow behavior for a wide range of restitution coefficients. Many models for granular flow were subsequently developed based on the kinetic theory approach, for example, Jenkins and Richman [5] and Gidaspow [6]. The kinetic theory has been extended to cohesive and multiproperty particle flow by Kim and Arastoopour [7] and Iddir and Arastoopour [8].

© Springer Nature Switzerland AG 2022
H. Arastoopour et al., *Transport Phenomena in Multiphase Systems*, Mechanical
Engineering Series, https://doi.org/10.1007/978-3-030-68578-2_2

2.2 Elementary Multiphase Kinetic Theory

The kinetic theory approach was reviewed more than 10 years ago in the 2002 Fluor-Daniel lecture by Gidaspow et al. [9], and the computational fluid dynamics (CFD) approach was described in Arastoopour's 1999 Fluor-Daniel lecture [10]. Since then, the kinetic theory approach has been extended to the flow of mixtures of particles of various sizes including rotation by Songprawat and Gidaspow [11] and Shuai et al. [12], to anisotropic flow using the method of moments by Strumendo et al. [13], and to continuous particle size distributions using population balances by Strumendo and Arastoopour [14].

2.2.1 Frequency Distributions

The frequency distribution of velocities of particles, f, is a function of position, $\mathbf{r}$, and the instantaneous velocity, $\mathbf{c}$, as well as time, t,

$$f = f\,(t, \mathbf{r}, \mathbf{c}) \tag{2.1}$$

The six coordinates consisting of the position, $\mathbf{r}$, and the velocity space, $\mathbf{c}$, are sufficient to determine the location of a particle since Newton's second law has six integration constants. The number of particles per unit volume, n, is given by the integral over the velocity space, $\mathbf{c}$,

$$n = \int f d\mathbf{c} \tag{2.2}$$

The mean value of a quantity ϕ, such as mass, momentum, energy and stress, is defined,

$$n\langle \phi \rangle = \int \phi f d\mathbf{c} \tag{2.3}$$

Hence the hydrodynamic velocity, $\mathbf{v}$, is the integral over all the velocity space,

$$\mathbf{v} = \frac{1}{n} \int \mathbf{c} f(\mathbf{c}) d\mathbf{c} \tag{2.4}$$

2.2.2   Peculiar Velocity and Transport

The transport of a quantity ϕ, such as energy, must be invariant under a change of frame. Hence it cannot be a function of the velocity, $\mathbf{c}$. Otherwise, it will have different values in different frames of reference. But $\mathbf{c}$-$\mathbf{v}$ is independent of the frame of reference. Hence the difference between the instantaneous and the hydrodynamic velocities, $\mathbf{C}$, is defined,

$$\mathbf{C} = \mathbf{c} - \mathbf{v} \tag{2.5}$$

In the kinetic theory of gases [1], this difference is called the peculiar velocity. Its mean is zero, since the mean of $\mathbf{c}$ is $\mathbf{v}$,

$$<\mathbf{C}> \ = \ <\mathbf{c} - \mathbf{v}> \ = \mathbf{v} - \mathbf{v} = 0 \tag{2.6}$$

This property is the same as that of the turbulent velocity, $\mathbf{v}$, defined as the instantaneous minus the time average velocity.

The flux vector of ϕ is defined, $n<\mathbf{C}\phi(\mathbf{C})>$. For example, if $\phi = E$, the internal energy, then the conduction flux, q, becomes $\mathbf{q} = n<E\mathbf{C}>$.

Since momentum is the mass, m, times the velocity, $\mathbf{C}$, the kinetic stress tensor, $\mathbf{P}_k$ is,

$$\mathbf{P}_k = n < \mathbf{C}m\mathbf{C} > = \rho < \mathbf{C}\mathbf{C} > \tag{2.7}$$

where the bulk density is $\rho = nm$. Table 2.1 shows the components of the stress tensor.

Table 2.1 Kinetic stress tensor

$$\mathbf{P}_k = \rho\langle\mathbf{CC}\rangle = \begin{pmatrix} \rho\langle C_x^2\rangle & \rho\langle C_x C_y\rangle & \rho\langle C_x C_z\rangle \\ \rho\langle C_y C_x\rangle & \rho\langle C_y^2\rangle & \rho\langle C_y C_z\rangle \\ \rho\langle C_z C_x\rangle & \rho\langle C_z C_y\rangle & \rho\langle C_z^2\rangle \end{pmatrix}$$

Since $n\langle\mathbf{CC}\rangle = \int \mathbf{CC}f d\mathbf{C}$,

$$\mathbf{P}_k = \begin{pmatrix} \rho\int C_x^2 f d\mathbf{C} & \rho\int C_x C_y f d\mathbf{C} & \rho\int C_x C_z f d\mathbf{C} \\ \rho\int C_y C_x f d\mathbf{C} & \rho\int C_y^2 f d\mathbf{C} & \rho\int C_y C_z f d\mathbf{C} \\ \rho\int C_z C_x f d\mathbf{C} & \rho\int C_z C_y f d\mathbf{C} & \rho\int C_z^2 f d\mathbf{C} \end{pmatrix}$$

where $\rho = $ BULK DENSITY $= \varepsilon_s \rho_s$ and where $f = f(\mathbf{C})$

NOTE that if $f(\mathbf{C}) = f_x(C_x) \cdot f_y(C_y) \cdot f_z(C_z)$ as in a Maxwellian distribution, then

$$P_{xy} = \rho\int C_x C_y f(\mathbf{C}) d\mathbf{C} = \rho \int\limits_{-\infty}^{\infty} f_z dC_z \int\limits_{-\infty}^{\infty} C_y f_y dC_y \int\limits_{-\infty}^{\infty} C_x f_x dC_x$$

But $\int\limits_{-\infty}^{\infty} C_x f_x dC_x = \langle C_x\rangle = \langle c_x - v_x\rangle = v_x - v_x = 0$

Hence, for a Maxwellian distribution, $P_{xy} = P_{zx} = P_{ij} = 0 \ (i \neq j)$.

For a Maxwellian velocity distribution, the kinetic viscosity is zero. The particle viscosity is non-zero due to collisions and due to a non-Maxwellian distribution.

The hydrostatic pressure, p, is the mean of the sum of the normal components of the stress tensor **p**,

$$\mathbf{p} = \frac{1}{3} \left(p_{xx} + p_{yy} + p_{zz} \right) \tag{2.8}$$

2.2.3 Granular Temperature and the Equation of State

In the kinetic theory of gases, the thermal temperature, T, is defined as the average of the random kinetic energy, with the conversion factor of the Boltzmann constant from Joules to degrees Kelvin,

$$k_B T = \frac{1}{3} m < C^2_x + C^2_y + C^2_z > \tag{2.9}$$

The Boltzmann constant has the value,

$$k_B = 1.3805 \times 10^{-23} \ \text{J/K} \tag{2.10}$$

The ideal gas law constant equals the very small Boltzmann constant, due to the small mass of the molecule, m, times the large value of the Avogadro's number, 6.023×10^{23}, the number of molecules per mole. Converting from Joules to calories gives the gas law constant of 1.987 cal/g mol K.

Elimination of the squares of the peculiar velocities in,

$$P = nm/3 < \mathbf{C}^2 > \tag{2.11}$$

and in the definition of temperature, Eq. (2.9), gives the ideal gas law equation of state,

$$P = n \, k_B T = (N/V) \, RT \tag{2.12}$$

where N is the number of moles, V the volume, and R is the gas constant.

The granular temperature is defined as the random kinetic energy of the particles without the conversion of Joules to degrees and can be defined either analogously to Eq. (2.9), or as kinetic energy per unit mass. Let θ be the granular temperature given by the random kinetic energy per unit mass,

$$\theta = \frac{1}{3}\langle C^2 \rangle = \frac{1}{3(n)} \int_{-\infty}^{\infty}\int_{-\infty}^{\infty}\int_{-\infty}^{\infty} \left(C_x^2 + C_y^2 + C_z^2 \right) dC_x dC_y dC_z \qquad (2.13)$$

In two dimensions, there are only two random velocities and $\langle C^2 \rangle$ would be divided by two. In one dimension, there is only one random velocity and the granular temperature is then simply the variance of the measured instantaneous velocities. However, its behavior is not the same as the three-dimensional granular temperature given by Eq. (2.13) [13]. The units of the granular temperature are $(m/s)^2$. These units are found to be convenient. The alternate definition with mass multiplying θ is not that convenient for a single particle size mixture. It may, however, be useful for a mixture of particles [6]. For a gaseous mixture of molecules, where there is no dissipation of energy, mass times velocity square of molecules is the same because there is only one temperature.

The equation of state for particles is obtained by eliminating $\langle C^2 \rangle$ between Eqs. (2.11) and (2.13),

$$p_s = nm\theta \qquad (2.14)$$

where the subscript, s, is added to emphasize that it is the solids pressure, and nm is the bulk density. In terms of the volume fraction of solids, ε_s, and the solids density, ρ_s, the ideal equation of state for particles becomes,

$$p_s = \varepsilon_s \rho_s \theta \qquad (2.15)$$

The more complete equation of state for particles containing the collisional contribution (see Sect. 2.4) has been verified experimentally by Gidaspow and Huilin [15].

2.2.4 FCC Equation of State

The experimental granular temperature for flow of fluidized catalytic cracking (FCC) particles is shown in Fig. 2.1. As the particle concentration increases, the granular temperature (turbulent kinetic energy of particles) increases, similar to the rise of thermal temperature upon compression of a gas. The decrease of granular temperature in the collisional regime is due to the decrease of the mean free path, which becomes zero in the packed state.

The particulate pressure measured with a specially designed transducer [15] is shown in Fig. 2.2. To construct a complete equation of state for FCC particles, IIT researchers used the statistical mechanics of liquid theory and their CCD camera system to determine radial distribution functions as a function of solids volume fraction.

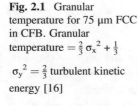

Fig. 2.1 Granular temperature for 75 μm FCC in CFB. Granular temperature $= \frac{2}{3}\sigma_x^2 + \frac{1}{3}$

$\sigma_y^2 = \frac{2}{3}$ turbulent kinetic energy [16]

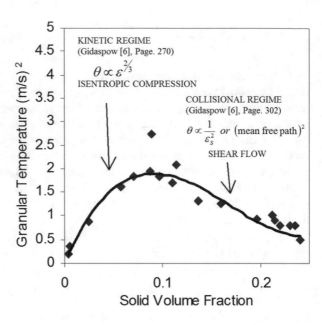

KINETIC REGIME
(Gidaspow [6], Page. 270)

$\theta \propto \varepsilon^{2/3}$

ISENTROPIC COMPRESSION

COLLISIONAL REGIME
(Gidaspow [6], Page. 302)

$\theta \propto \dfrac{1}{\varepsilon_s^2}$ or $(\text{mean free path})^2$

SHEAR FLOW

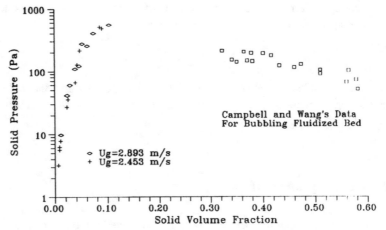

Fig. 2.2 Time average solid pressure for FCC particles [16]

Campbell and Wang's Data
For Bubbling Fluidized Bed

◇ Ug=2.893 m/s
+ Ug=2.453 m/s

Figure 2.3 presents the equation of state. It shows that the kinetic theory of granular flow is valid for flow of FCC particles below about 5%. For dilute flow, there is an analogy of the ideal gas law: The ratio of solids pressure to bulk density multiplied by the granular temperature is one, as a limit. For volume fractions above ~5%, the standard granular flow theory has to be corrected for a cohesive pressure, obtained from measurements of the radial distribution functions and granular temperature, using a modified Boltzmann relation [15].

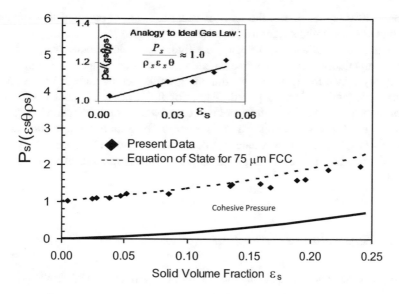

Fig. 2.3 Equation of state for 75 μm FCC particles determined in IIT CFB [16]. Dimensionless solid pressure: $\frac{P_s}{\rho_s \varepsilon_s \theta} = 1 + 2(1 + e)\varepsilon_s g_o \cdot (0.73\varepsilon_s + 8.957\varepsilon_s^2)$

2.2.5 Particle and Molecular Velocities

The average molecular velocity can be estimated from the definition of the thermal temperature given by Eq. (2.9). Multiplication of both side of Eq. (2.9) by Avagadro's number, A, and recalling that the gas law constant, $R = k_B A$,

$$Ak_B T = \frac{1}{3} mA < \mathbf{C}^2 >$$
(2.16)

Using the molecular weight, $M = m\,A = $ kg/molecule $\times$ molecules/mol in Eq. (2.16), gives the useful relation between molecular velocities and thermal temperature,

$$< \mathbf{C}^2 >= 3RT/M$$
(2.17)

Moving with the average velocity, in terms of meters/second the molecular velocity becomes,

$$< c^2 >= 158(T/M)^{1/2}$$
(2.18)

For carbon dioxide at 273 K, Eq. (2.18) gives the velocity as 294 m/s. For hydrogen, this velocity is multiplied by the ratio of the molecular weights to give 1845 m/s. The critical or sonic velocity is evaluated at a constant entropy and hence is about 20% higher due to the ratio of specific heats at constant temperature to constant volume in the square root relation in Eq. (2.18).

For particles, the average velocity, the hydrodynamic velocity, cannot be zero because an energy input is required to keep the particles in motion owing to their inelasticity. Thus, Eq. (2.13) shows that the granular temperature is of the order of the hydrodynamic velocity squared. Thus, for fluidization of small particles in a bubbling bed, the velocity is on the order of cm/s, while, for fluidization in risers, it is of the order of m/s. Due to dissipation of energy, the granular temperatures are not inversely proportional to the square root of masses, as shown for gases in Eq. (2.18), although they are smaller for large particles in a binary mixture of small and large particles [9].

2.2.6 Maxwellian Distribution

The Maxwellian distribution using the granular temperature can be shown [16] to be,

$$f(\mathbf{r}, \mathbf{c}) = \frac{n}{(2\pi\theta)^{3/2}} \exp\left(-\frac{(c-v)^2}{2\theta}\right) \tag{2.19}$$

This is the expression found in the kinetic theory of gases [1], with $k_B/m = 1$ and the thermal temperature replaced with the granular temperature, expressed in the units of velocity square.

2.2.7 Restitution Coefficients

Figure 2.4 shows typical restitution coefficients. A more complete treatment is found in [17]. During the impact of particles, the work of deformation can be expressed in terms of the elastic pressure, P_e, and the plastic pressure, P_p, and the deformation volume by means of the usual relation,

Fig. 2.4 Restitution coefficients for various materials [16]

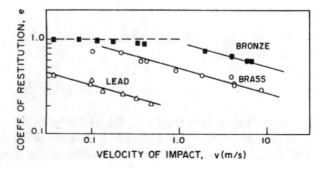

$$\text{work of deformation} = \int_0^V \left(P_e + P_p\right) dV \tag{2.20}$$

This work equals the relative velocity of impact square times half the mass. However, the relative kinetic energy equals only the integral of the elastic pressure. Hence, the restitution coefficient, e, can be expressed,

$$e^2 = \frac{v'^2}{v^2} = \frac{\int_0^V P_e dV}{\int_0^V \left(P_e + P_p\right) dV} \tag{2.21}$$

where v is the relative velocity before impact and v' is the rebound velocity. Equation (2.21) suggests that, for low velocities, where the plastic deformation is small, the restitution coefficient will be ~1. It also clearly shows that the restitution coefficient is a function of the material properties, as well as the dynamic properties associated with plastic flow. Indeed, data summarized by Johnson [17] show that, for hard materials, the restitution coefficients are nearly unity for impact velocities of 0.1 m/s and less (Fig. 2.4).

2.2.8 Frequency of Binary Collisions

The objective of this section is to derive the classical binary frequency of collisions corrected for the dense packing effect, as done with the factor χ in Chapman and Cowling's Chap. 16 [1]. Analogously to the single frequency distribution given by Eq. (2.1), a collisional pair distribution function, $f^{(2)}$, is introduced,

$$f^{(2)} = f^{(2)}(\mathbf{c}_1, \mathbf{r}_1, \mathbf{c}_2, \mathbf{r}_2) \tag{2.22}$$

It is defined such that,

$$f^{(2)} d\mathbf{c}_1 d\mathbf{c}_2 d\mathbf{r}_1 d\mathbf{r}_2 \tag{2.23}$$

is the probability of finding a pair of particles in the volume $d\mathbf{r}_1 d\mathbf{r}_2$ centered on points $\mathbf{r}_1$ and $\mathbf{r}_2$ with velocities within the ranges $\mathbf{c}_1$ and $\mathbf{c}_1 + d\mathbf{c}_1$ and $\mathbf{c}_2$ and $\mathbf{c}_2 + d\mathbf{c}_2$. Figure 2.5 shows the geometry of collisions in the spherical coordinates. It is that given by Savage and Jeffrey [2] with a generalization to two rigid spheres of unequal diameters.

Collecting all terms, the collision frequency becomes,

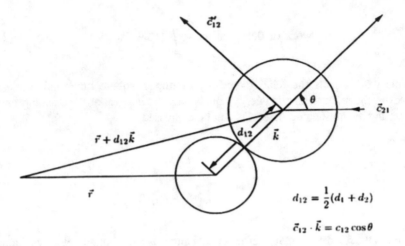

$$d_{12} = \frac{1}{2}(d_1 + d_2)$$

$$\vec{c}_{12} \cdot \vec{k} = c_{12}\cos\theta$$

Fig. 2.5 Geometry of a collision of two spheres, d_1 and d_2 [16]

$$N_{12} = 4n_1 n_2 d_{12}^2 g_0 \sqrt{\pi\theta} \tag{2.24}$$

This agrees with Chapman and Cowling's [1] classical kinetic theory of gases result,

$$N_{12} = 2n_1 n_2 d_{12}^2 \left(\frac{2\pi k_B T m_0}{m_1 m_2}\right)^{1/2} \tag{2.25}$$

for dilute gases,

$$m_0/m_1 = 2 \tag{2.26}$$

and,

$$k_B/m = 1 \tag{2.27}$$

in the definition of the granular temperature. Equations (2.24) and (2.25) show the difference between the granular and the thermal temperature. In the latter, the Boltzmann constant k_B serves to assign the scale of the temperature. In Eq. (2.24), g_0 becomes infinite at maximum packing, but at this point the granular temperature is expected to approach zero, leaving N_{12} undefined.

2.2.9   Mean Free Path

In the dilute theory of gases, the intuitive concept of the mean free path plays a very important role. Through the use of this concept, transport coefficients, such as viscosity, are obtained that are surprisingly close to those obtained from the exact theory.

In Sect. 2.2.8, it was assumed that the two particles undergoing collision were of different diameters, as shown in Fig. 2.5. To derive the mean free path for a single particle, IIT researchers restricted themselves to the case where the two particles have the same diameter, d_p. In this case, the mean time between successive collisions, called the collision time, τ, is obtained as shown in Eq. (2.28) for dilute flow where g_0 is equal to one in Eq. (2.24) and with $N_{22} = N_{11}$,

$$\tau = n_1 / N_{11} \tag{2.28}$$

Substituting Eq. (2.24) with $n_2 = n_1$ into Eq. (2.28) results in,

$$\tau = 1 / \left(4 n_1 d_{12}^2 \sqrt{\pi \theta} \right) \tag{2.29}$$

The mean free path, l, is the product of the average velocity, $\langle c \rangle$, and the collision time as,

$$l = \langle c \rangle \tau \tag{2.30}$$

where

$$\langle c \rangle = \left(\frac{8 \theta}{\pi} \right)^{1/2} \tag{2.31}$$

Using Eq. (2.29) with $d_{12} = d_p$ and $n_1 = n$, the mean free path becomes independent of θ,

$$l = \frac{1}{\pi \sqrt{2} n_1 d_p^2} = \frac{0.707}{\pi n d_p^2}. \tag{2.32}$$

Using the relation,

$$\varepsilon_s = \frac{1}{6} \pi d_p^3 n \tag{2.33}$$

one obtains a formula for the mean free path that gives the intuitively correct dependence of the mean free path on the diameter. Solving for n in Eq. (2.33) and substitution into Eq. (2.32), the man free path becomes,

$$l = \frac{1}{6\sqrt{2}} \frac{d_p}{\varepsilon_s} \qquad (2.34)$$

2.2.10 Elementary Treatment of Transport Coefficients

2.2.10.1 Diffusion Coefficients

It is well known in kinetic theory of gases that a simple, non-rigorous treatment of transport phenomena produces surprisingly accurate values of transport coefficients [1]. Such a treatment is presented below.

A collisional interpretation was used. Hence,

$$\Delta Q = l\langle v_A \rangle \frac{d\rho_A}{dx} \qquad (2.35)$$

Fick's law of diffusion in mass units is,

$$\rho_A(v_A - v) = -D\frac{d\rho_A}{dx} \qquad (2.36)$$

where D is the diffusion coefficient and v, the mass average velocity, is zero in this case. It is clear that,

$$D = l\langle v_A \rangle \qquad (2.37)$$

$$D = mean\ free\ path \times fluctuating\ velocity$$

Using Eqs. (2.31) and (2.34),

$$D = \frac{l}{3\sqrt{\pi}} \frac{\sqrt{\theta}d_p}{\varepsilon_s} \qquad (2.38)$$

The mean free path is,

$$l = \frac{1}{6\sqrt{2}} \frac{d_p}{\varepsilon_s} \qquad (2.38a)$$

The average of the oscillating velocity is related to the average velocity by granular temperature,

$$\langle v_A \rangle = \langle c \rangle = \left(\frac{8\theta}{\pi} \right)^{1/2} \tag{2.39}$$

For molecules, the fluctuating velocity can be estimated from,

$$\langle c^2 \rangle^{\frac{1}{2}} = \sqrt{\frac{3RT}{M}} \tag{2.40}$$

Hence,

$$D = l \sqrt{\frac{3RT}{M}} \tag{2.41}$$

At 273 K, the diameter of a CO_2 molecule is 3×10^{-10} m, the mean free path is 2×10^{-7} m, and, with a fluctuating velocity of 394 m/s, the diffusion coefficient can be estimated using Eq. (2.41) 7.88×10^{-5} m²/s. This compares with the experimental value of 1.4×10^{-5} m²/s. The equation gives only the order of magnitude for D. It does, however, give the correct dependence on the molecular weight and density through the correct dependence of the mean free path on density. Hence for liquids, the diffusivity is three orders of magnitude smaller. For particles, the diffusivity can be calculated by direct measurement of particle velocity and computation using the autocorrelation method [18].

In a multiphase CFD approach, dispersion coefficients are not an input into the codes as they are in the convection-dispersion model, but IIT researchers computed them for fluidization in agreement with the data [19].

2.2.10.2   Viscosity

To obtain the viscosity of the particulate phase, let the momentum flux be,

$$Q = \rho v \langle v \rangle \tag{2.42}$$

where Q = momentum/volume × average of oscillating velocity.
For a constant density, ρ, the change in momentum flux is,

$$\Delta Q = l \rho \langle v \rangle \frac{dv}{dx} \tag{2.43}$$

The viscosity for fully developed incompressible flow is defined,

$$\text{shear} = \mu \frac{dv}{dx} \tag{2.44}$$

The momentum transport, ΔQ, equals the force per unit area,

$$\Delta Q = \text{shear} \tag{2.45}$$

Therefore, the viscosity assumes the form,

$$\mu = l\rho\langle v\rangle \tag{2.46}$$

From Eq. (2.37),

$$\mu = \rho D \tag{2.47}$$

as in the kinetic theory of gases. Therefore, using Eq. (2.38), a simple formula for the collisional viscosity is,

$$\mu = \left(\frac{1}{3\sqrt{\pi}}\right)\rho_p d_p \sqrt{\theta} \tag{2.48}$$

viscosity = mean free path × fluctuating velocity

For smooth rigid spherical molecules of diameter, d_p,

$$\mu_l = \frac{5}{16d_p^2}\left(\frac{k_B m T}{\pi}\right)^{1/2} \tag{2.49}$$

where $m = \rho_p \frac{\pi}{6} d_p^3$.

To convert T to granular temperature, θ, let $\frac{k_B}{m} = 1$. Thus,

$$\text{DILUTE } \mu_s = \frac{5\sqrt{\pi}}{96}\rho_p d_p \theta^{1/2} \tag{2.50}$$

Multiplying Eq. (2.50) by $(\rho_p \varepsilon_p)$ and dividing by ε_p,

$$\mu_s = \frac{5\sqrt{\pi}}{96} \cdot (\rho_p \varepsilon_s) \cdot \left(\frac{d_p}{\varepsilon_s}\right) \cdot \theta^{1/2} \tag{2.51}$$

viscosity = constant × bulk density × mean free path × oscillation velocity

For critical flow experiments [6], it was found that $\theta \cong 1$ m/s for 100 μm diameter particles having a density of 1500 kg/m³. The viscosity, μ_s, estimated by Eq. (2.51) becomes $\approx 1.5 \times 10^{-3}$ kg/(m s) (15 cP). This value is 15 times that of water and is in

Fig. 2.6 FCC powder viscosity measured in three different ways. Circles: CCD camera method; triangles: pressure drop; squares: Brookfield viscometer [16]

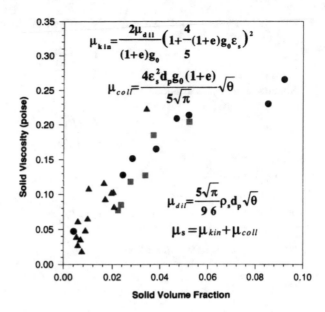

close agreement with the measurements [9]. The collisional viscosity includes the effect of particle concentration.

In Fig. 2.6, the circles represent the viscosity measurements as a function of solids volume obtained for FCC particles using the charged couple diode (CCD) camera method. The particle viscosity expressions for kinetic viscosity, μ_{kin}, and collisional viscosity, μ_{coll}, in this figure were obtained from kinetic theory of dense gases modified to apply to the granular flow in terms of granular temperature. The viscosity was obtained by substituting the measured granular temperature and the solids volume fractions into Eq. (2.51). The squares represent data obtained by Miller and Gidaspow [20] using a Brookfield viscometer, and the triangles represent the data from the shear drop method. Figure 2.6 shows a reasonable agreement between the three methods.

2.2.10.3   Thermal Conductivity

Here,

$$\Delta Q = l \frac{dQ}{dx} \tag{2.52}$$

Heat flux is,

$$Q = \rho U \langle v \rangle \tag{2.53}$$

$$Q = \text{density} \times \text{internal energy} \times \text{average of fluctuating velocity}$$

$$\Delta Q = l\langle v\rangle \rho \frac{dU}{dx} \tag{2.54}$$

Specific heat, c_v, is defined,

$$c_v = \left(\frac{\partial U}{\partial T}\right)_v \tag{2.55}$$

By the chain rule,

$$\Delta Q = l\langle v\rangle \rho c_v \frac{dT}{dx} \tag{2.56}$$

Fourier's law of heat conduction is,

$$q = -k\frac{dT}{dx} \tag{2.57}$$

Thus, Eqs. (2.56) and (2.57) show that,

$$conductivity = mean\ free\ path \times oscillation\ velocity \times density$$
$$\times specific\ heat \tag{2.58}$$

If enthalpy is used in place of U,

$$k = \mu c_p \tag{2.59}$$

The thermal diffusivity is,

$$\alpha = \left(\frac{k}{\rho c_p}\right) = \rho\langle v\rangle \tag{2.60}$$

There is a rough agreement between this simplified theory and the measurements [9].

2.2.11 Boundary Conditions

At the inlet and outlet, all properties should be defined based on the specific physics and assumption of the problem. For the gas phase, no-slip and non-penetrating wall conditions may be considered. For the solid phase, the slip boundary condition is the recommended boundary condition [21],

$$\bar{\tau}_s = -\frac{\pi}{6}\sqrt{3\phi}\frac{\varepsilon_s}{\varepsilon_{s,\,max}}\rho_s g_0\sqrt{\theta_s}\,\vec{v}_{s,para} \tag{2.61}$$

where $\vec{v}_{s,para}$ is the particle slip velocity parallel to the wall. ϕ is the specularity coefficient between the particle and the wall, which is defined as the average fraction of relative tangential momentum transferred between the particle and the wall during a collision. The specularity coefficient varies from zero (smooth walls) to one (rough walls). A proper value based on the particles and wall properties should be assumed. For a specularity coefficient tending toward zero, a free slip boundary condition for the solids tangential velocity is imposed at a smooth wall boundary [22].

Johnson and Jackson [21] proposed the following wall boundary condition for the total granular heat flux,

$$q_s = \frac{\pi}{6}\sqrt{3\phi}\frac{\varepsilon_s}{\varepsilon_{s,\,max}}\rho_s g_0\sqrt{\theta_s}\,\vec{v}_{s,para}\cdot\vec{v}_{s,para} - \frac{\pi}{4}\sqrt{3}\frac{\varepsilon_s}{\varepsilon_{s,\,max}}$$
$$\times\left(1 - e_{sw}^2\right)\rho_s g_0\theta_s^{3/2} \tag{2.62}$$

The dissipation of solids turbulent kinetic energy by collisions with the wall is specified by the particle-wall restitution coefficient, e_{sw}. A high value of the specularity coefficient implies high production at the wall, and a value of e_{sw} close to unity implies low dissipation of granular energy at the wall. It is expected that the specularity coefficient and the particle-wall restitution coefficient need to be calibrated for a given gas/particle flow system because the specularity coefficient cannot be measured and e_{sw} can be measured only with some difficulty [22]. Equations (2.61) and (2.62) could be written,

$$v_{s,w} = -\frac{6\mu_s\varepsilon_{s,\,max}}{\sqrt{3}\pi\phi\rho_s\varepsilon_s g_0\sqrt{\theta}}\frac{\partial v_{s,w}}{\partial x} \tag{2.63}$$

$$\theta_w = -\frac{\kappa\theta}{\gamma_w}\frac{\partial\theta_w}{\partial x} + \frac{\sqrt{3}\pi\phi\rho_s\varepsilon_s v_{s,slip}^2 g_0\theta^{3/2}}{6\varepsilon_{s,\,max}\gamma_w} \tag{2.64}$$

where

$$\gamma_w = \frac{\sqrt{3}\pi\left(1 - e_{sw}^2\right)\varepsilon_s\rho_s g_0\theta^{3/2}}{4\varepsilon_{s,\,max}} \tag{2.65}$$

These boundary conditions do not take into account electrostatic forces. In reality, particles tend to stick to the wall.

2.3   Drag Expressions

Syamlal [23] derived expressions for the particle-particle friction term for dense multiphase systems consisting of a binary mixture of particles using concepts from a very simplified kinetic theory. This approach has been partially verified experimentally but requires additional research to further confirm its validity.

2.4   Multiphase Flow Experimental Verification

2.4.1   Experimental

To test the validity and the accuracy of the kinetic theory model, a two-story riser was built at IIT with a splash plate on top of the riser to obtain symmetry [16]. Figure 2.7 shows the IIT circulating fluidized bed with splash plate and measurement equipment. The γ-ray source was used to measure the particle concentration. The particle velocities were measured using kinetic theory-based PIV.

2.4.2   Kinetic Theory-Based PIV

Figure 2.8 illustrates IIT researchers' improved PIV method of obtaining instantaneous velocities for a binary mixture of glass beads. This technique was recently described fully for flow of 530 μm glass beads by Tartan and Gidaspow [24]. In the CFB regime, a curtain of solids at the pipe wall restricts the use of LDV for obtaining velocities in the core of the riser, a vertical pipe. Hence a probe, shown in Fig. 2.8 was used. Figure 2.9 shows the typical streak images captured by a CCD camera. The velocity is the length of the streak divided by exposure time. The order of the colors on the rotating transparency establishes the direction. The study in [24] was generalized to a mixture of two sizes of particles. Large particles form thicker streaks than the small particles. To obtain radial profiles, a probe was inserted into the riser. The size of the probe was varied to establish an optimum balance between its hydrodynamic interference and a sufficient number of streaks (1030) in a picture to obtain meaningful statistics of velocity averages and their variances. Figure 2.10a, b shows typical instantaneous axial and radial velocities measured by a particle image velocity meter in the riser.

The hydrodynamic velocity, v, was calculated from measurement of the instantaneous velocity c by,

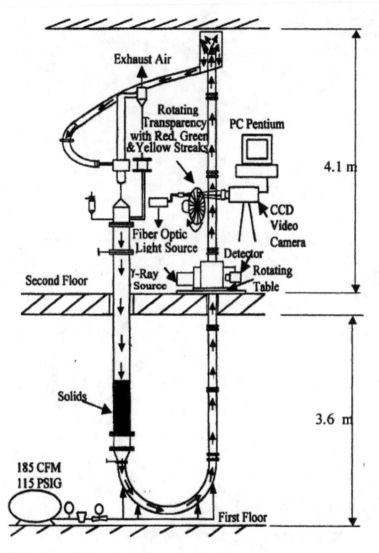

Fig. 2.7 IIT circulating fluidized bed with splash-plate [16]

$$v_i(r,t) = \frac{1}{n} \sum_{k=1}^{n} c_{ik}(r,t) \tag{2.66}$$

The kinetic stresses were calculated by,

Fig. 2.8 The PIV system
with probe [16]

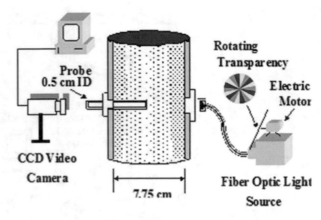

Fig. 2.9 Typical streak
images captured by CCD
camera. Exposure
time = 0.001 s, typical
length = 2 mm [16]

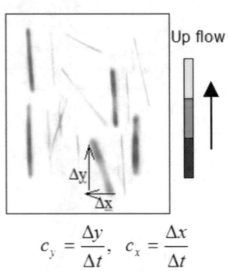

$$c_y = \frac{\Delta y}{\Delta t}, \quad c_x = \frac{\Delta x}{\Delta t}$$

$$< C_i C_i > (r,t) = \frac{1}{n} \sum_{k=1}^{n} (c_{ik}(r,t) - v_i(r,t))(c_{ik}(r,t) - v_i(r,t)) \qquad (2.67)$$

where n is the total number of streaks in each frame and $C_i = c_i - v_i$. The particle
stress, $\langle C_z C_z \rangle$ in the direction of flow is much larger than the tangential and the
radial stresses, similar to turbulent flow of gas in a pipe [25, 26], but are an order of
magnitude larger. The orders of magnitude of larger particle stresses for fluidization
are at the expense of an order of magnitude larger pressure drop.

The particle Reynolds stresses are calculated from the hydrodynamic velocity, v,
by,

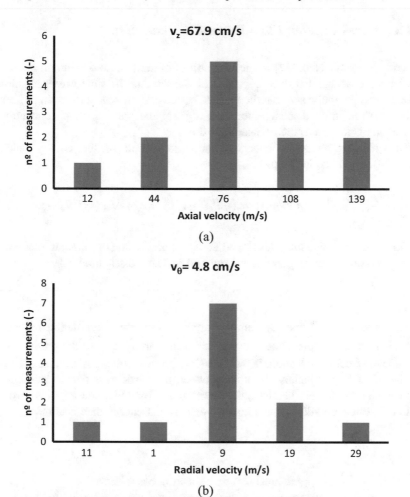

Fig. 2.10 Typical axial (**a**) and radial (**b**) velocities [16]

$$\overline{v_i'v_i'}(r) = \frac{1}{m} \sum_{k=1}^{m} \left(v_{ik}(r,t) - \overline{v}_i(r)\right)\left(v_{ik}(r,t) - \overline{v}_i(r)\right) \tag{2.68}$$

The laminar granular temperature is the average of normal stresses,

$$\theta(r,t) = \frac{1}{3} < C_z C_z > + \frac{1}{3} < C_r C_r > + \frac{1}{3} < C_\theta C_\theta > \tag{2.69}$$

2.4.3   Core-Annular Flow Regime Explanation

Recently Benyahia et al. [27] studied the ability of multiphase continuum models to predict the core-annulus flow. Figure 2.11 shows that the time-averaged particle velocity was parabolic and that the particle concentration was uniform in the center of the 7.6 cm inner diameter tube and high at the wall. Such a concentration distribution is known as the core-annular flow.

The granular temperate energy balance is similar to the energy balance, Eq. (1.24), discussed in Chap. 1,

$$\frac{3}{2}\left[\frac{\partial(\rho_s\varepsilon_s\theta)}{\partial t} + \nabla \cdot (\rho_s\varepsilon_s\theta v_s)\right] = (-P_s\mathbf{I} + \tau_s) : \nabla v_s + \nabla \cdot (k_s\nabla\theta) - \gamma \quad (2.70)$$

For steady state and fully developed flow, the production of granular temperature equals the conduction of granular energy and inelastic dissipation,

$$(-P_s\mathbf{I} + \tau_s) : \nabla v_s + \nabla \cdot (\kappa_s\nabla\theta) - \gamma = 0 \quad (2.71)$$

The production of granular temperature due to particle oscillation reduces to $\mu\left(\frac{\partial v}{\partial r}\right)^2$. In transport phenomena texts, this term corresponds to the production of heat due to viscous dissipation. In a suspension, particle collisions can be assumed to be elastic. During collisions the fluid between the particles has to be pushed out, requiring a large force. Therefore, the dissipation of granular energy, γ, is zero.

In cylindrical coordinates, the balance of granular energy then becomes,

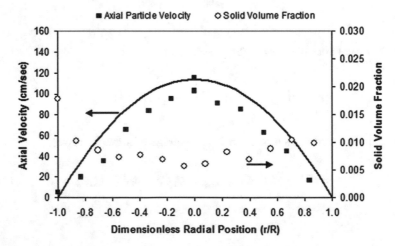

Fig. 2.11 Developed time-averaged particulate-phase axial velocity for 530 μm diameter glass beads in the IIT two-story riser [16]. $U_g = 4.9$ m/s, $W_s = 14.2$ kg/m^2 s

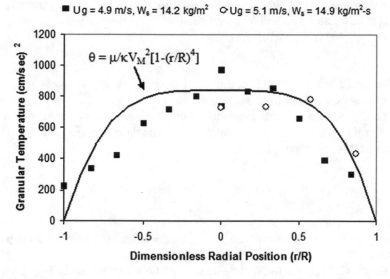

Fig. 2.12 A comparison of measured granular temperature in the IIT two-story riser to the analytical solution [16]

$$\kappa\frac{d}{dr}\left(\frac{rd\theta}{dr}\right) = \mu\left(\frac{\partial v}{\partial r}\right)^2 \tag{2.72}$$

The measured velocity distribution in our riser as shown in Fig. 2.11 is,

$$v = 2\bar{v}\left[1 - \left(\frac{r}{R}\right)^2\right] \tag{2.73}$$

Similar parabolic distributions have been reported by Berruti et al. [28].

Substituting Eq. (2.73) into Eq. (2.72), the balance of granular temperature can be expressed,

$$\kappa\frac{d}{dr}\left(\frac{rd\theta}{dr}\right) = -16\mu\bar{v}^2\frac{r^2}{R^4} \tag{2.74}$$

Integrating Eq. (2.74) with zero wall granular temperature, the solution of the above equation is,

$$\theta = \bar{v}\left(\frac{\mu}{\kappa}\right)\left[1 - \left(\frac{r}{R}\right)^4\right] \tag{2.75}$$

Figure 2.12 shows that measured granular temperatures can be approximated by the fourth order parabolic equation obtained by solving the granular temperature derived above.

In a reasonably dilute flow, the solids pressure shown in Fig. 2.3 can be approximated to be a linear function of the granular temperature,

$$p_s = \varepsilon_s \rho_s \theta \tag{2.76}$$

Equation (2.76) corresponds to the use of the ideal gas law for molecules, where the particles are far apart from each other.

In developed flow, the radial variation of this pressure is approximately zero. This approximation allows one to obtain the very simple expression for the particle volume fraction distribution,

$$\varepsilon_s = \frac{P_s \kappa}{\rho_s \mu_s \bar{v}^2} \frac{1}{\left[1 - \left(\frac{r}{R}\right)^4\right]} \tag{2.77}$$

where, μ_s is the particle viscosity, κ is the granular conductivity, and $\bar{v}$ is the particle velocity.

Equation (2.77) explains the core-annular particle distribution in Fig. 2.11. This expression is not valid at the tube wall. The kinetic theory shows that the core annular regime is independent of the pipe radius, not believed to be so by the oil industry until measurements were finally made.

With a few simplifications [29], this equation can give the ratio of the number of particles per unit volume, n, to its inlet into the system, n_{inlet}, by,

$$\frac{n}{n_{inlet}} = \frac{1}{\left[1 - \left(\frac{r}{R}\right)^4\right]} = \frac{1}{\left[1 - \left(\frac{r}{R}\right)^4\right]} \tag{2.78}$$

where $n = \frac{6\varepsilon_s}{\pi d_p^3}$. These simplifications result in the following:

- From kinetic theory, for dilute flow, $\frac{\mu_s}{\kappa_s} \cong \frac{4}{15}$.
- The inlet granular temperature can be approximated by

$$\theta_{inlet} = \frac{\mu_s}{\kappa_s} \bar{v}^2 \text{ since } \frac{\mu_s}{\kappa_s} \cong 1$$

Equation (2.78) also describes the platelet distribution for blood flow in a fully developed flow regime [29].

2.4.4   Turbulent Granular Temperature

There are two granular temperatures as shown in Table 2.2. The laminar granular temperature per unit bulk density that is computed using the granular temperature

Table 2.2 A comparison of laminar and turbulent granular temperatures (m^2/s^2) in dilute and dense regimes of rises for flow 5.4 μm FCC $(v_g = 3.5$ m/s) and 1094 μm diameter ceramic alumina particles $(v_g = 19$ m/s)

Section	Height (m)	Solids volume fraction (−)	Laminar granular temperature (m^2/s^2)	Turbulent granular temperature (m^2/s^2)
Geldart D particles				
Bottom	2	0.32	0.1	1.20
Top	6	0.02	1.1	1.25
Geldart A particles				
Bottom	2	0.202	0.001	0.558
Top	6	0.048	0.142	1.675

equation in CFD codes, such as ANSYS Fluent [30], the DOE MFIX code [31], and the IIT [16]; and turbulent granular temperature computed from the normal Reynolds stresses. For dilute riser flow, the granular kinetic theory agrees well with CFD experiments [24].

Unfortunately for commercially useful dense flow, the turbulent granular temperatures exceed the laminar granular temperatures (Table 2.2). In this table, the values of solids volume fractions and granular temperatures for flow of Geldart D particles are from the investigation of Kashyap et al. [32] who performed computations done for UOP for high solids flux in the IIT riser. The flow was in a solids slugging regime. The computations were done using the standard drag model [6]. The values for Geldart A particles are for the riser of Wei et al. [33], with computations done using a correction for the drag, derived using the energy minimization principle [34].

Figure 2.13 shows that, in the dense bubbling bed, the dimensionless turbulent granular temperature, called "bubble-like" [35], is represented by the solid circles and is almost an order of magnitude larger than the laminar or particle granular temperature. For the dilute riser, the turbulent granular temperatures for both 156 and 530 μm diameter particles were smaller than the laminar granular temperatures, which agrees with the theoretical analytical solution for the granular temperature equation for elastic particles (Fig. 2.12).

2.5 Flow Regime Computation

An excellent review of flow regimes before wide use of computational fluid dynamics was given by Berruti et al. [28]. One-dimensional two-phase models of more than three decades ago required a specification of measured flow regimes [36]. The two- and three-dimensional models used today can successfully compute these flow regimes. Tables 2.3, 2.4 and 2.5 summarize the governing equations for gas/solids flow based on the kinetic theory model for constitutive relations. Furthermore,

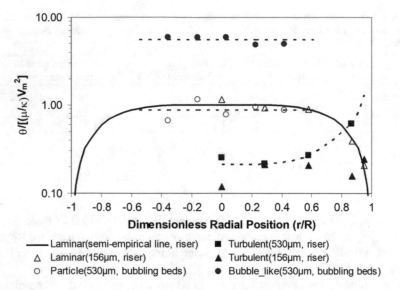

Fig. 2.13 Measured dimensionless laminar and turbulent granular temperatures in the IIT two-story riser in dilute flow and in rectangular bubbling beds [16]

Arastoopour et al. [37] provided additional discussion on drag treatment for nonhomogeneous gas/particles flow and the EMMS approach.

Figure 2.14 shows batch fluidization flows regimes for nanoparticles and Geldart A (aerated), B (bubbling), and C (cohesive) particles. The batch-fluidization flow of 10 nm Tullanox nanoparticles occurs without bubbles due to the formation of clusters [38]. Geldart C particles are small cohesive particles, which fluidize with the formation of small bubbles. Group A particles form aerated beds with small bubbles, whereas Group B particles fluidize with the formation of large bubbles. Details can be found in [39]. As the gas velocity is increased, the particles are blown out of the bed.

To obtain continuous particle flow, the particles are fed into the bed through one or more jets. Figure 2.15 summarizes our computed fluidization flow regimes. Turbulent fluidization is the name given to the flow regime in which there exists a dense phase at the bottom of the bed and a dilute particle phase on the top. The volume fractions of the particles in the bottom and top sections of the bed were estimated by Matsen [40] using the drift-flux model, knowing the gas and particles flow rates. IIT researchers had seen a sharp interphase between the dilute and dense portions of the bed. Our code describes the axial experimental measurements of the solid volume fractions of Wei et al. [33]. In the fast fluidization regime, large clusters are formed which descend near the wall. Dense suspension flow is formed at high gas velocities and high solid fluxes. In this regime, there is a core-annular flow. For large particles (d_p = 1093 µm) and very high velocities, slugging fluidization was observed and computed. In a reactor, high velocities are needed to obtain high production rates [41, 42]. If the velocity is too high, the pneumatic transport flow regime will occur with a low catalyst concentration.

Table 2.3 The kinetic theory CFD model

Continuity equations with no phase change

Gas phase

$$\frac{\partial}{\partial t}\left(\varepsilon_g\rho_g\right) + \nabla \cdot \left(\varepsilon_g\rho_g\vec{v}_g\right) = 0$$

Solid phases ($k = 1, 2$)

$$\frac{\partial}{\partial t}\left(\varepsilon_k\rho_k\right) + \nabla \cdot \left(\varepsilon_k\rho_k\vec{v}_k\right) = 0$$

Momentum equations without bulk viscosity

Gas momentum

$$\frac{\partial}{\partial t}\left(\varepsilon_g\rho_g\vec{v}_g\right) + \nabla \cdot \left(\varepsilon_g\rho_g\vec{v}_g\vec{v}_g\right) = -\nabla P + \sum_{k=1}^{N}\beta_{gk}\left(\vec{v}_k - \vec{v}_g\right) + \nabla \cdot 2\varepsilon_g\mu_g\nabla^s\vec{v}_g + \varepsilon_g\rho_g\vec{g}$$

Particulate phases, $k = (1, \ldots, N)$

$$\frac{\partial}{\partial t}\left(\varepsilon_k\rho_k\vec{v}_k\right) + \nabla \cdot \left(\varepsilon_k\rho_k\vec{v}_k\vec{v}_k\right) = \beta_{gk}\left(\vec{v}_g - \vec{v}_k\right) + \sum_{l=1}^{N}\beta_{kl}\left(\vec{v}_l - \vec{v}_k\right) - P_s + \nabla \cdot 2\varepsilon_k\mu_k\nabla^s\vec{v}_k + \varepsilon_k\rho_k\vec{g}$$

Energy equations

Gas phase

$$\frac{\partial}{\partial t}\left(\varepsilon_g\rho_gH_g\right) + \nabla \cdot \left(\varepsilon_g\rho_gH_g\vec{v}_g\right) = \left(\frac{\partial P}{\partial t} + \vec{v}_g \cdot \nabla P\right) + \sum_{k=1}^{N}h_{vk}\left(T_k - T_g\right) + \nabla \cdot \left(K_g\varepsilon_g\nabla T_g\right)$$

Solid phases

$$\frac{\partial}{\partial t}\left(\varepsilon_k\rho_kH_k\right) + \nabla \cdot \left(\varepsilon_k\rho_kH_k\vec{v}_k\right) = h_{vk}\left(T_g - T_k\right) + \nabla \cdot \left(K_k\varepsilon_k\nabla T_k\right)$$

Granular temperature

$$\frac{3}{2}\left[\frac{\partial(\rho_s\varepsilon_s\theta)}{\partial t} + \nabla \cdot (\rho_s\varepsilon_s\theta v_s)\right] = (-P_s\mathbf{I} + \tau_s) : \nabla v_s + \nabla \cdot (k_s\nabla\theta) - \gamma$$

Table 2.4 Constitutive equations

Particle pressure (for frictional case of granular flow, see Sect. 2.7)

$$P_s = \varepsilon_s\rho_s\theta + 2\rho_s(1 + e)\varepsilon_s^2 g_0\theta$$

Shear particle viscosities (for frictional case of granular flow, see Sect. 2.7)

$$\mu_s = \frac{4}{5}\varepsilon_s\rho_s d_p g_0(1 + e)\left(\frac{\theta}{\pi}\right)^{1/2} + \frac{10\rho_s d_p\sqrt{\theta\pi}}{96\varepsilon_s(1+e)g_0}\left[1 + \frac{4}{5}g_0\varepsilon_s(1 + e)\right]^2$$

Bulk particle viscosity (for frictional case of granular flow, see Sect. 2.7)

$$\zeta_s = \frac{4}{3}\varepsilon_s\rho_s d_p g_0(1 + e)\left(\frac{\theta}{\pi}\right)^{1/2}$$

Radial distribution function

$$g_0 = \left[1 - \left(\frac{\varepsilon_s}{\varepsilon_{s,\max}}\right)^{1/3}\right]^{-1}$$

Granular conductivity

$$k_s = \frac{150\rho_s d_p\sqrt{\theta\pi}}{384(1+e)g_0}\left[1 + \frac{6}{5}\varepsilon_s g_0(1 + e)\right]^2 + 2\rho_s\varepsilon_s^2 d_p(1 + e)g_0\sqrt{\frac{\theta}{\pi}}$$

Particle stress

$$\tau_s = \varepsilon_s\mu_s\left(\nabla v_s + \nabla v_s^T\right) + \varepsilon_s\left(\xi_s - \frac{2}{3}\mu_s\right)\nabla \cdot v_s\mathbf{I}$$

Dissipation

$$\gamma = \frac{12\left(1-e^2\right)g_0}{d_p\sqrt{\pi}}\rho_s\varepsilon_s^2\theta^{3/2}$$

Table 2.5 Drag and heat transfer coefficients

Gas-solid drag coefficients $k = (1, \ldots, N)$

$\beta = 150 \frac{\varepsilon_s^2 \mu_g}{\varepsilon_g^2 d_p^2} + 1.75 \frac{\rho_g \varepsilon_s}{\varepsilon_g d_p} |\vec{v}_g - \vec{v}_s|$ when $\varepsilon < 0.74$

$\beta = \frac{3}{4} C_d \frac{\rho_g \varepsilon_s |\vec{v}_g - \vec{v}_s|}{d_p} Hd$ when $\varepsilon \geq 0.74$

$C_d = \frac{24}{Re}\left(1 + 0.15\,Re^{0.687}\right)$ $Re < 1000$

$C_d = 0.44$ $Re \geq 1000$

where Re is the Reynolds number, $Re = \frac{\varepsilon_g \rho_g d_p |\vec{v}_g - \vec{v}_s|}{\mu_g}$

For fast fluidization H_d may be expressed in terms of ε

$$Hd = \begin{cases} -0.5760 + \dfrac{0.0214}{4(\varepsilon - 0.7463)^2 + 0.0044} & (0.74 < \varepsilon \leq 0.82) \\[2mm] -0.0101 + \dfrac{0.0038}{4(\varepsilon - 0.7789)^2 + 0.0040} & (0.82 < \varepsilon \leq 0.97) \\[2mm] -31.8295 + 32.8295\varepsilon & (\varepsilon > 0.97) \end{cases}$$

Particle-particle drag coefficients (for alternative equation, see Sect. 2.9)

$\beta_{\substack{kl \\ k, l \neq f}} = \frac{3}{2}\alpha(1 + e)\frac{\rho_k \rho_l \varepsilon_k \varepsilon_l (d_k + d_l)^2}{\rho_k d_k^3 + \rho_l d_l^3}|\vec{v}_k - \vec{v}_l|$

Gas-particle heat transfer (for equation, see Sect. 2.10)

For $\varepsilon_g \leq 0.8$

$Nu_s = (2 + 1.1Re^{0.6}Pr^{1/3})S_s$ $Re \leq 200$

$\quad = 0.123\left(\frac{4Re}{d_s}\right)^{0.183} S_s^{0.17}$ $200 < Re \leq 2000$

$\quad = 0.61Re^{0.67}Ss$ $Re > 2000$

For $\varepsilon_g > 0.8$

$Nu_s = (2 + 0.16Re^{0.67})S_s$ $Re \leq 200$

$\quad = 8.2Re^{0.6}S_s$ $200 < Re \leq 2000$

$\quad = 1.06Re^{0.457}S_s$ $Re > 2000$

where,

$Re = \frac{\varepsilon_g \rho_g |\vec{v}_g - \vec{v}_s| d_s}{\mu_g}$

$S_s = \varepsilon_s \frac{6}{d_s}, Nu = \frac{h_{vs} d_s}{k_g}$

2.6 Wave Propagation

The speeds of compression waves through a fluidized bed of 75 μm FCC particles between the minimum bubbling and the minimum fluidization velocities were determined by measuring the times of arrival of compression zones using a gamma ray densitometer. The theory of characteristics shows that this speed, of the order of 1.4 m/s, represents the maximum velocity of discharge of non-fluidized FCC particles. The velocity of these compression waves was used to calculate the solids stress modulus of the fluidized bed via the one-dimensional mass and momentum balances for granular flow. Additionally, measurements of the pressure wave propagation produced a sonic velocity that is an order of magnitude higher than the speeds of the solids compression wave. The homogeneous sonic velocity, of the order of 20 m/s, gives the maximum achievable velocity for the circulation of FCC particles in a standpipe of an oil refinery.

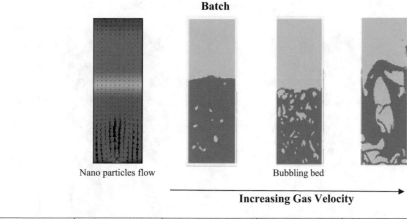

Batch

Nano particles flow   Bubbling bed

Increasing Gas Velocity →

	Nano	**C**	**A**	**B**
v (cm/s)	2	0.5	3	35
d_p (μm)	0.01	20	71	200
ρ (g/cm³)	2.22	2	1.42	2.10

Fig. 2.14 A summary of basic batch fluidization flow regimes computed at IIT [16]

Using the complete equation of state for FCC particles and the relationship between the compression wave velocity and the solids stress modulus, the computed granular temperature is two orders of magnitude lower than the square of the wave propagation velocity and agrees with literature values [43].

2.6.1 Compression Wave Theory

The experiment is based on the granular flow theory presented by Gidaspow [6]. The derivation is initiated by stating the one-dimensional mass and momentum balances for the granular flow of solids.

Mass balance for the solids phases,

$$\frac{\partial(\rho_b)}{\partial t} + \frac{\partial}{\partial y}(\rho_b v_s) = 0 \tag{2.79}$$

Momentum balance for the solids phases,

Continuous Particle Flow

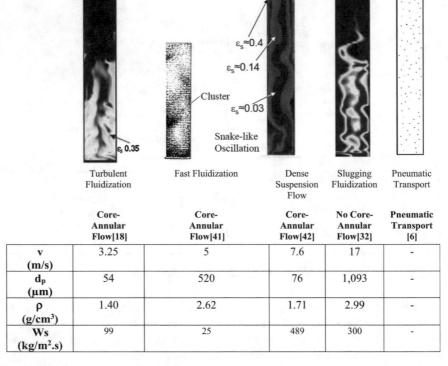

Fig. 2.15 A summary of continuous basic fluidization flow regimes computed at IIT [16]

$$\frac{\partial(\rho_b v_s)}{\partial t} + \frac{\partial}{\partial y}(\rho_b v_s v_s) = -\frac{\partial \sigma_s}{\partial y} + g(\rho_s - \rho_g)\varepsilon_s - \beta_B(v_s - v_g) \qquad (2.80)$$

Rearranging and substituting into matrix form yields,

$$\begin{pmatrix} \dfrac{\partial \rho_b}{\partial t} \\[2mm] \dfrac{\partial v_s}{\partial t} \end{pmatrix} + \begin{pmatrix} v_s & \rho_b \\[2mm] \dfrac{G'}{\rho_b} & v_s \end{pmatrix} \begin{pmatrix} \dfrac{\partial \rho_b}{\partial y} \\[2mm] \dfrac{\partial v_s}{\partial y} \end{pmatrix} = \left(\dfrac{\varepsilon_s(\rho_s - \rho_f)g}{\rho_b} - \dfrac{\beta_B(v_s - v_g)}{\rho_b} \right) \qquad (2.81)$$

The characteristic determinant for these equations is,

$$\begin{vmatrix} v_s - \lambda & \rho_b \\[2mm] \dfrac{G'}{\rho_b} & v_s - \lambda \end{vmatrix} = 0 \qquad (2.82)$$

The characteristic direction for a fluidized system of particles is,

$$\lambda = \frac{dx}{dt} = v_s \pm \sqrt{\frac{G}{\rho_s}} = v_s \pm \sqrt{G'} \tag{2.83}$$

Based on this characteristic equation for one-dimensional flow, the solids stress modulus, G, may be back-calculated from the solids velocity and the characteristic velocity. Similar to compressible gas flow, the maximum flow occurs when this characteristic is zero [6]. Then, the maximum discharge velocity from a pipe equals the square root of the measured solids stress modulus. The measured values of the propagation velocities are close to the reported maximum velocities of discharge of particles. Accurate values of discharge velocities are not available in the literature due to the sensitivity to the void fraction at discharge. The void fractions can be measured with gamma ray, x-ray, or other densitometers, but, in view of experimentally observed difficulties, such as the noise due to intermittent emissions, this has not been the practice in fluidization.

For dilute systems, such as fluidized nanoparticles [43, 44], the ideal equation of state for the particles ($\sigma_s = \rho_s \varepsilon_s \theta$) can be substituted into $G(\varepsilon_s) = \left(\frac{\partial \sigma_s}{\partial \varepsilon_s}\right)_{SorT}$ as an approximation, (assuming zero net solids velocity) yielding,

$$G' = \theta \tag{2.84}$$

Therefore, for dilute systems, the granular temperature can be determined by substituting this result into Eq. (2.83),

$$\theta = \left(\frac{dx}{dt}\right)^2 = v_w^2 \tag{2.85}$$

Now, for dense systems, such as fluidized beds of FCC particles that will be discussed, the full equation of state for the particles [15] could be utilized. The full equation of state for the particles is,

$$\sigma_s = \rho_s \varepsilon_s \theta + 2\rho_s \theta(1 + e)g_o \varepsilon_s^2 - 0.73\rho_s \theta \varepsilon_s^2 - 8.957\rho_s \theta \varepsilon_s^3 \tag{2.86}$$

Inserting this relationship into $G(\varepsilon_s) = \left(\frac{\partial \sigma_s}{\partial \varepsilon_s}\right)_{SorT}$ yields,

$$G(\varepsilon_s) = \rho_s \theta \left[1 + 2(1 + e)g_o \varepsilon_s \left(2 + 0.33g_o \left(\frac{\varepsilon_s}{\varepsilon_{sm}}\right)^{0.333} \right) - 1.46\varepsilon_s - 26.871\varepsilon_s^2 \right] \tag{2.87}$$

Substituting the relationship for the wave velocity into this equation and solving for granular temperature yields,

$$\theta = \frac{v_w^2}{\left[1 + 2(1 + e)g_o\varepsilon_s\left(2 + 0.33g_o\left(\frac{\varepsilon_s}{\varepsilon_{sm}}\right)^{0.333}\right) - 1.46\varepsilon_s - 26.871\varepsilon_s^2\right]} \quad (2.88)$$

Therefore, if the wave velocity and solids volume fraction are known, the granular temperature may be calculated.

2.6.2 Experimental Equipment

The experimental setup is shown in Fig. 2.16. The experiment was conducted in a 7.62 cm inner diameter riser approximately 6 m tall. The riser is an acrylic pipe to allow visual observations of the fluidization phenomena by the experimenter. The pipe extended through two stories of the building with a collector box and a splash plate at the riser top to give symmetric flow. Air was injected into the column at two points below the distributor plate. Fluidization air was injected into the pipe below the distributor to fluidize the initial bed. Air was also used to induce the compression wave that propagates through the riser. The air that induced the compression wave was controlled with a 7.62 cm solenoid valve located below the particle distributor. The particles were added to the riser and supported by a 316 stainless steel wire mesh.

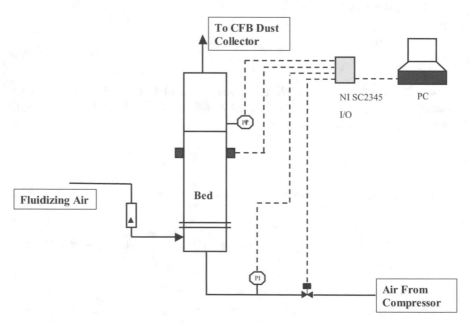

Fig. 2.16 Schematic diagram for riser experiment [16]

The experiment was initiated by adding a specific amount of solid particles to the riser. The height of the non-fluidized bed and the weight of material added were recorded. A constant flow rate of air was then injected into the riser to fluidize the particle bed. The new fluidized bed height for the bed was recorded. The passage of the compression wave was detected using a gamma ray densitometer for the FCC particles. The voltage generated by the detector was inversely proportional to the solids volume fraction of the fluidized bed. The gas pressures were measured using 0–689 kPa WIKA pressure sensors. The electrical signals were collected using the National Instruments data collection system LabVIEW. The voltage signal was sent to an NI SCC-AI05 2 channel 100 mV analog input module. This analog input module resided in a NI SC-2345 portable shielded module carrier. The NI SC-2345 was connected to a NI PCI-6221 M Series Multifunction DAQ card in a PC. Data were collected every millisecond and saved into an Excel spreadsheet utilizing LabVIEW software to direct and store the data. Next, a quick acting valve was actuated. The valve was controlled by a personal computer. By recording the time the valve was opened, the time required for the solids wave to travel to the sensor, the distance to the sensor, and the solids volume fraction of the bed, a correlation of G, or G′ versus volume fraction could be developed.

2.6.3   Pressure Wave Theory

The theoretical derivation for the velocity of a pressure wave through a fluidized bed is presented in [6, 45]. Gidaspow's derivation [6] starts with stating the mixture momentum balance for homogeneous flow in a gas/solids system.

$$\rho_m \frac{dv}{dt} = -\frac{dP}{dx} - \rho_m g \tag{2.89}$$

For the continuous phase, say the gas phase, the equation of state is,

$$\rho_g = \rho_g(T, P) \tag{2.90}$$

The isothermal sound speed, C_g, is,

$$C_g^2 = \left(\frac{\partial P}{\partial \rho_g}\right)_T \tag{2.91}$$

A similar speed of sound can be defined for constant mixture entropy. The continuity equation for the gas phase with this speed of sound and no phase change is,

$$\frac{d\varepsilon}{dt} + \frac{\varepsilon}{\rho_g C_g^2} \frac{dP}{dt} + \varepsilon \frac{\partial v}{\partial x} = 0 \tag{2.92}$$

Likewise, the continuity equation for the incompressible particulate phase is,

$$-\frac{d\varepsilon}{dt} + \varepsilon_s \frac{\partial v}{\partial x} = 0 \tag{2.93}$$

Summation of these last two equations yields,

$$\frac{\varepsilon}{\rho_g C_g^2} \frac{dP}{dt} + \frac{\partial v}{\partial x} = 0 \tag{2.94}$$

Differentiation of this equation with respect to time yields,

$$\frac{d}{dt} \frac{\partial v}{\partial x} = -\frac{d}{dt} \left(\frac{\varepsilon}{\rho_g C_g^2} \frac{dP}{dt} \right) \tag{2.95}$$

Differentiating the momentum balance with respect to x assuming no body forces yields,

$$\frac{d}{dt} \frac{\partial v}{\partial x} = -\frac{d}{dx} \left(\frac{1}{\rho_m} \frac{dP}{dx} \right) \tag{2.96}$$

Equating the previous two equations yields,

$$\frac{d}{dt} \left(\frac{\varepsilon}{\rho_g C_g^2} \frac{dP}{dt} \right) = \frac{d}{dx} \left(\frac{1}{\rho_m} \frac{dP}{dx} \right) \tag{2.97}$$

If the velocity is small, the convective derivative becomes partial with respect to time, and this equation reduces to the wave equation for the fluid pressure. Therefore, this equation shows that the pressure propagates with a mixture velocity, C_m, equal to,

$$C_m^{-2} = \frac{\rho_m \varepsilon}{\rho_g C_g^2} \tag{2.98}$$

where

$$\rho_m = \varepsilon \rho_g + \varepsilon_s \rho_s \tag{2.99}$$

Assuming non-dilute solids loading ($\varepsilon_s \rho_s > \varepsilon \rho_g$), Eq. (2.98) reduces to,

$$C_m = \frac{C_g}{\sqrt{(\varepsilon \varepsilon_s)}} \sqrt{\frac{\rho_g}{\rho_s}} \tag{2.100}$$

Therefore, there exists a minimum mixture velocity that depends on the void fraction of the mixture. This equation also shows that the sonic velocity of the mixture is much lower than the sonic velocity for a pure gas. The theory of characteristics [6] shows that this mixture velocity is the maximum discharge velocity of fluidized particles out of fluidized standpipes in refineries. This places a severe limit on the capacity of oil refineries that are being operated near their maximum capacity in the United States. The assumption in this theory is that the gas and the particles flow at the same velocity, not counter flow as is sometimes the case in standpipes. The second important assumption is that there is no settling in standpipes. Settling should decrease only the discharge velocity since the critical velocity will drop from about 20 m/s and the pressure wave propagation velocity to ~1 m/s when the particles settle. Blockage of the standpipe by bubbles may also limit the flow in a poorly designed standpipe. However, CFD simulations show that bubbles are swept out of standpipes in a good design. It may be possible to obtain supersonic flow in a converging-diverging standpipe and thus overcome the sonic limitation.

2.6.4 Pressure Wave Experimental Results

Likewise, the pressure wave of the gas through the bed was also measured in another experiment. The pressure as a function of time [43] shows that the pressure rises much quicker than the solids volume fraction in the riser. The calculated speed of the gas through the riser for this experiment is,

$$C_g = \frac{\Delta x}{\Delta t} = \frac{(3.66m)}{(3.321 - 3.152s)} = 21.6 m/s \tag{2.101}$$

This result is compared to the work of Roy et al. [45] in Fig. 2.17. This value is higher than the value presented for the sonic velocity by Roy et al.

Figure 2.18 presents the experimental gas pressure velocity as a function of the fluidized bed void fraction. The measurement lies reasonably close to the theoretical line presented in this graph.

Next, the granular temperature was calculated using Eq. (2.88) for a dense fluidized bed. The results of these calculations are presented in Fig. 2.19, along with past experimentally determined granular temperatures for other research. The granular temperature for these experiments in a dense fluidized bed lies right in line with the granular temperature measured by Polashenski and Chen [46] and reviewed by Gidaspow and Mostofi [47].

Fig. 2.17 Sonic velocity
through a fluidized bed [43]

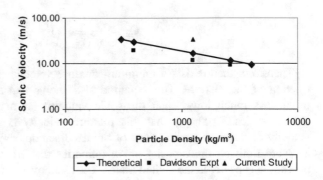

Fig. 2.18 Critical velocity
through a fluidized bed [16]

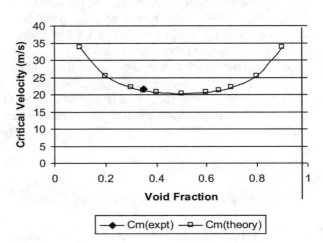

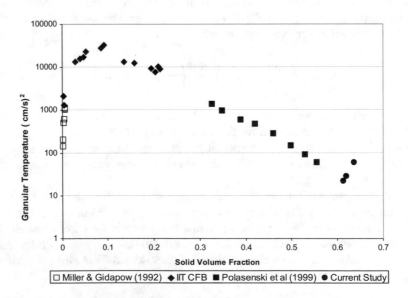

Fig. 2.19 Granular temperature for FCC particles [16]

2.7   Frictional Behavior of Granular Matters

Many industrially important gas/solid systems often include slow and dense solid flows, which is dominated by sustained frictional contacts between the particles. For simulations of these dense flows, in addition to the kinetic and collisional stresses (i.e., kinetic theory), the model should account for the frictional stresses, which could be calculated based on soil mechanics principles.

Makkawi and Ocone [48] have reviewed the modeling approaches to include the frictional effects. The most common approach to consider the effect of frictional stresses is the kinetic-frictional model based on the addition of stress from the two limiting regimes at a critical solid volume fraction (ε_{cr}) [21, 31, 48–50].

$$\tau_s = \tau_{kinetics-Collision} + \tau_{friction} \qquad \text{where } \tau_{friction} = 0 \quad \text{for } \varepsilon < \varepsilon_{cr} \quad (2.102)$$

$$\mu_s = \mu_{kinetics-Collision} + \mu_{friction} \qquad \text{where } \mu_{friction} = 0 \quad \text{for } \varepsilon$$
$$< \varepsilon_{cr} \qquad (2.103)$$

In general, $\mu_s = \mu_{kin} + \mu_{col} + \mu_{fric}$

This approach is based on the assumptions of Savage [51] that consider that the solid stress comes from the kinetic, collisional, and frictional contributions in an additive manner, where the frictional contributions appear only at higher solid volume fractions (i.e., greater than 0.5).

Most of the reported frictional stress models in the literature [48] and CFD codes such as Ansys and MFIX are based on the critical state theory of soil mechanics, where the shear stress τ_{fr} is described in terms of a frictional viscosity based on the work of Schaeffer [52]. Under a normal stress, a well-compacted granular material will shear only when the shear stress attains a critical magnitude. This is described by the Mohr-Coulomb law based on the laws of sliding friction. However, the Mohr-Coulomb law does not provide any information on how the granular material deforms and flows; rather, it describes the onset of yielding [53].

Schaeffer [52] derived the following expression for the frictional stress by assuming the system to be perfectly rigid plastic, incompressible, non-cohesive, Coulomb powder with a yield surface of von Mises type, and parallel eigenvectors of the strain rate and stress tensors as,

$$\tau_{fr} = \frac{\sqrt{2}p_c \sin \phi}{|S|} S \qquad (2.104)$$

or equivalently,

$$\mu_{fr} = \frac{p_c \sin \varphi}{2\varepsilon_s \sqrt{II_{2D}}} \qquad (2.105)$$

where ϕ is the angle of internal friction, II_{2D} is the second invariant of the deviatoric stress tensor, S, and p_c is the critical state pressure. According to Srivastava and Sundaresan [53], p_c increases monotonically with ε and is expected to become very large (i.e., diverge) as ε approaches random close packing ε_{max}. Various expressions have been proposed for the functional dependence of p_c on ε in the literature [21, 31, 51, 52, 54, 55, 56, 57].

Johnson and Jackson [21] proposed a critical state solid frictional pressure that allows for a slight compressibility with very limited particle concentration change [48]. The Johnson and Jackson correlation for frictional pressure can be written,

$$p_f = p_c$$
$$= \begin{cases} 0 & \text{elsewhere} \\ Fr.\dfrac{(\varepsilon_s - \varepsilon_{min})^q}{(\varepsilon_{max} - \varepsilon_s)^p} & \varepsilon_{min} = 0.5 \leq \varepsilon_s \leq \varepsilon_{max} = 0.63 \quad , q = 2, p = 3 \end{cases}$$

(2.106)

where Fr is a coefficient with different values reported in the literature from 0.05 to 5 [21, 58]. The coefficient Fr was modified by others [59, 60] assuming to be a function of the volume fraction as $Fr = 0.1\varepsilon_s$ while limiting the solid volume fraction to values less than 0.629 to prevent divergence.

Note that p_c is the critical state frictional pressure and many studies assumed the critical state frictional pressure is equal to solid frictional pressure p_f [41, 61, 62], although clearly it may not be an accurate assumption.

Srivastava and Sundaresan [54] modified the Schaeffer expression for the frictional stress and also the Johnson and Jackson [21] expression for the frictional pressure (see Eq. 2.108) to approximately account for strain rate fluctuations and slow relaxation of the assembly to the yield surface following Savage's [51] argument of existence of fluctuations in the strain rate, even in purely quasi-static flow. The standard deviation σ of the fluctuations is related to the granular temperature of the powder θ (taken from the rapid granular flow regime) and the particle diameter d_p, as $\sigma = b\frac{\theta^{\frac{1}{2}}}{d_p}$, where b is a constant of order unity.

Laux [50] suggested a correlation of the following form,

$$\mu_{fr} = \left[\frac{3\left| \xi_s \nabla . \vec{u}_s - \frac{p_f}{\varepsilon_s} \right|}{2\sqrt{3.II_{2D}}} \right] \left(\frac{6 \sin \phi}{9 - \sin^2 \phi} \right)$$

(2.107)

where p_f is calculated based on the following equation [54],

$$\frac{p_f}{p_c} = \left(1 - \frac{1}{n\sqrt{2}\sin\phi}\frac{\nabla.\vec{u}_s}{\sqrt{S:S+\left(\theta/d_p^2\right)}}\right)^{\frac{-1}{n-1}}$$ (2.108)

The angle of repose and, in turn, the frictional forces for solid particles are significantly affected by the value of the compactness factor (n). The value of compactness factor (n) may be determined by comparing the experimental angle of repose with the calculated values for solid packing [63]. Based on the above equation, if the granular material dilates as it deforms, $\nabla.u_s > 0$, then $p_f < p_c$ if it compacts as it deforms, $\nabla.u_s < 0$, then $p_f > p_c$ and when it deforms at constant volume, $\nabla.u_s = 0$, which is the critical state, $p_f = p_c$. This behavior is in line with the experimental measurements [64]. The value of n (the compactness factor) is different in the dilation and compaction parts of the system. Srivastava and Sundaresan [54] suggested a value of $\frac{\sqrt{3}}{2}\sin\varphi$ for the dilation branch to ensure that the granular assembly is not required to sustain tensile stress on the yield surface. They also pointed out that n for the compaction branch can be any value marginally greater than one. They suggested that a value of 1.03 be used when no additional information is available. This value is measured for glass beads by Jyotsna and Rao [65]. It is worth mentioning that decreasing the value of n in the compactness branch will cause more deviation from the critical state frictional pressure. In other words, n may be an indicator for non-linearity of the τ-σ relation.

According to Dartevelle [53], the plastic potential theory combined with the critical state approach can successfully describe the phenomenon of dilatancy, consolidation, and independence between the rate-of-strain-tensor and the stress tensor. Using this approach, assuming a slightly compressible, dry, non-cohesive, and perfectly rigid plastic system, the author derived the flowing expression for the frictional viscosity,

$$\mu_{fr} = \frac{p_s\sin^2\phi}{\varepsilon_s\sqrt{4\sin^2\phi.II_{2D} + \left(\nabla.\vec{u}_s\right)^2}}$$ (2.109)

and solid phase bulk viscosity,

$$\xi_s = \frac{p_s}{\varepsilon_s\sqrt{4\sin^2\phi.II_{2D} + \left(\nabla.\vec{u}_s\right)^2}}$$ (2.110)

Equation (2.109) reduces to Eq. (2.105) if, $\nabla.\vec{u}_s = 0$ which corresponds to the critical state of soil mechanics and linear τ-σ relation. For a detailed discussion and derivation of the models see Ghadirian and Arastoopour [63].

Nikolopolous et al. [61] have pointed out that the numerical results indicate that the values calculated by the Laux (Eq. 2.107) and Dartevelle (Eq. 2.109) expressions are of the same order of magnitude for values of solid volume fractions lower than 0.5. However, the Laux expression predicts higher solid frictional viscosity compared to the Dartevelle model for solid volume fractions higher than 0.55. Nikolopolous et al. [61] also showed that the results of simulations using the Dartevelle model are less accurate compared to the results of simulations using the Laux model in calculating the angle of repose.

2.8   Drag Force for Homogeneous and Non-homogeneous Flow of the Particle Phase

Gas-particle flows are inherently oscillatory and they manifest in non-homogeneous structures. Thus, if one sets out to solve the microscopic two-fluid model equations for gas/particle flows, grid sizes of less than 10-particle diameter become essential [66, 67]. For most devices of practical (commercial) interest, it is nearly impossible to resolve all heterogeneous flow structures in large-scale industrial risers using a computational grid size of the order of a few particle diameters. In addition, such extremely fine spatial grids and small time steps are unaffordable and require significant computational time and use of significant computational facilities. Thus, the effect of the large-scale structures using coarse grids must be accounted for through appropriate modifications of the closures (i.e., drag model).

One of the major sources of numerical inaccuracy for the two-fluid model (TFM) originates from the models used for the calculation of drag force [62, 68, 69]. The homogeneous drag models [6, 70–72] assume a homogeneous structure inside the control volumes, which is not valid due to the formation of clusters (dense phase) in the concentrated particulate phase (e.g. $\epsilon_s > 2\%$). The effective drag coefficient in the coarse grid simulations will be lower than that in the homogeneous TFM to reflect the tendency of the gas to flow more easily around the clusters (bypass the clusters) than through a homogeneous distribution of the particles [66, 73–75].

Qualitatively, this is equivalent to an effectively larger apparent size for the particles. Therefore, any coarse grid continuum simulation of gas/solid flows should include subgrid corrections to the homogeneous drag force acting on the particles. As a matter of fact, the overprediction of drag force by the homogeneous models is significant and can lead to overprediction of solid circulation rates and underestimation of pressure drops in circulating fluidized beds (CFB), and overprediction of bed expansion in bubbling fluidized beds (BFB) [37].

Arastoopour and Gidaspow [76] were the first to include the effect of clusters in the drag force between phases in gas/solid systems by assuming an effective particle diameter larger than the actual particle diameter and therefore reducing the drag force between phases. Recently, several approaches have been proposed to account for the effect of the small, unresolved scales on the interphase momentum exchange when using the TFMs on coarse computational grids. Among them, two approaches have gained significant attention in the literature: filtered (or subgrid) and energy minimization multi-scale (EMMS).

Igci et al. [65] and Milioli et al. [77] derive residual correlations from filtering fully resolved simulations on a two-dimensional (2D) periodic domain with several average particle volume fractions. They showed that the filtered TFM approach has shown promise to be a tool to simulate gas/particle flows of fluid catalytic cracking (FCC) particles in the industrial-scale riser of a CFB. Benyahia and Sundaresan [78] also showed that the subgrid models for coarse grid simulations of continuum models may also be used for coarse grid simulations of discrete particle models.

The EMMS approach [75, 79–81] on the other hand, is based on the assumption that heterogeneous structures (i.e., clusters) with different sizes form and contribute to the drag reduction between the gas and particulate phases. The resulting underdetermined set of equations is then solved by minimizing a function, called the stability condition. Physically, the stability criterion is the net energy exchange between phases to suspend and transport the solids.

2.8.1   Filtered or Subgrid Model

The TFM equations are coarse grained through a filtering operation that amounts to spatial averaging over some chosen filter length scale. In these filtered (coarse grained) equations, the consequences of the flow structures occurring on a scale smaller than a chosen filter size appear through residual correlations for which one must derive or postulate constitutive models [66]. In principle, the filtered equations should produce a solution with the same macroscopic features as the finely resolved kinetic theory model solution; however, obtaining this solution should come at less required computational time.

According to Igci et al. [66], if $\varepsilon_s(y,t)$ denotes the particle volume fraction at location, y, and time, t, is obtained by solving the microscopic TFM, the filtered particle volume fraction $\overline{\varepsilon_s}(x, t)$ can be defined as,

$$\overline{\varepsilon_s}(x,t) = \int_V G(x,y)\varepsilon_s(y,t)dy \qquad (2.111)$$

where $G(x,y)$ is a weight function that depends on x and y, and V denotes the region over which the gas/particle flow occurs. The weight function satisfies $\int G(x,y) dy = 1$. By choosing how rapidly $G(x,y)$ decays with distance measured from x, one can change the filter size.

For example, filtered gas/particle interaction force includes a filtered gas/particle drag force and a term representing correlated fluctuations in particle volume fraction and the (microscopic TFM) gas phase stress gradient,

$$\beta_{fil} = \frac{32\,Fr_f^{-2} + 63.02Fr_f^{-1} + 129}{Fr_f^{-3} + 133.6Fr_f^{-2} + 66.61Fr_f^{-1} + 129}\beta_{micro} \qquad (2.112)$$

where β_{micro} is the drag coefficient in the microscopic TFM, and Fr_f is the Froude number based on the filter size and is defined,

$$Fr_f = \frac{v_t^2}{g\Delta_f} \qquad (2.113)$$

where v_t is the terminal settling velocity and Δ_f is the filter size.

For further discussions on dependence of the residual correlations on the filter size, filtered particle volume fraction, and filtered slip velocity, all of which serve as a marker for the extent of subfilter-scale non-homogeneity, see Igci et al. [66] and Milioli et al. [77].

2.8.2 Energy Minimization Multi-scale (EMMS) Approach

The energy minimization multi-scale (EMMS) approach was first proposed by Li and Kwauk [79] based on the coexistence of both dense and dilute regions in a CFB reactor. The model parameters are found by minimization of the mass specific energy consumption for suspending and transporting the particles as the stability criteria for flow structure inside the reactor [81].

The EMMS model is able to account for heterogeneous solid structures and cluster formation in the system. Benyahia [81] concluded that use of the EMMS-based drag model is accurate and necessary for the prediction of the averaged solid mass and pressure profile along the fully developed flow region of the riser. The EMMS model calculates the heterogeneity factor, H_d. Then the drag expression can be expressed,

$$\beta_{gs} = \frac{3}{4}\frac{(1 - \varepsilon_g)\varepsilon_g}{d_p}\rho_g|v_g - v_s|C_{D0}.H_d \qquad (2.114)$$

Ghadirian and Arastoopour [75] calculated the heterogeneity factor H_d (the ratio of drag force for non-homogeneous solid phase flow using the EMMS approach to drag force calculated using the Wen and Yu drag expression for a homogeneous solid phase flow system) as a function of voidage at different specific slip velocities for flow of gas and particles with 185 μm dimeter and 2500 kg/m³ density. Figure 2.20 shows the calculated heterogeneity factor H_d at slip velocities of 0.5, 1, and 2 m/s. As Fig. 2.20 shows, in very dilute regions of the system, the solid flow pattern approaches toward homogeneous flow. At regions with a solid volume fraction of less than 0.1, H_d initially decreases sharply and then, for a wide range of solid volume fraction greater than 0.15, it levels off at a value of about 0.02 for slip

Fig. 2.20 Heterogeneity factor (H_d) as a function of voidage at different slip velocities. (This figure was originally published in [75] and has been reused with permission)

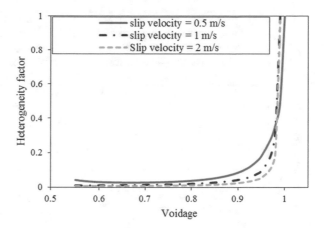

velocities between 0.5 and 2 m/s. This makes the EMMS calculated drag force for the non-homogeneous solid phase significantly lower than the prediction of any homogeneous drag model. The sudden decrease in H_d is because of the presence of clusters that allow the gas to bypass the solids and therefore results in a considerable decrease in the drag force between phases. Figure 2.20 also suggests that variations in H_d with respect to the solid volume fraction are more significant than H_d variations with respect to the slip velocity. Therefore, the effect of slip velocity variation may be neglected in most of the gas/solid flow systems.

Ghadirian and Arastoopour [75] simulated bed expansion using 2D TFM CFD equations for both homogeneous and non-homogeneous particle phases. They concluded that using a non-homogeneous drag expression, such as EMMS, predicts the bed expansion with noticeably higher accuracy (20% or less), while homogeneous models used in their study continued to over predict the bed expansion by up to about 70% in comparison with the correlation developed based on the experimental data of Krishna [82].

Figure 2.21 shows the bed expansion factor (final bed height/initial bed height) for the EMMS (developed for 185 μm and 2500 kg/m³ density particles) and two homogeneous models as well as the experimentally based correlation of Krishna [82]. To demonstrate the effect of particle type (particle size and density) on the heterogeneity of the system, the results of another set of simulations using the EMMS approach derived for FCC particles [83] are also shown Fig. 2.21. In the latter case, the EMMS approach was derived by Lu et al. [83] for FCC particles (75 μm diameter and 1500 kg/m³), but the resulting heterogeneity factor was used to simulate 185 μm bed expansion and 2500 kg/m³ particle density.

Figure 2.21 shows that the homogeneous models predict a very high value for bed expansion with about 70% deviation from the experimental correlation. Using the EMMS model, the bed expansion factor shows only less than 10% deviation from the experimental values that could be within the experimental error. This graph also shows that the bed expansion calculated based on the EMMS approach derived for

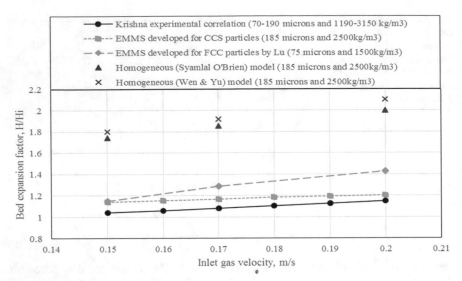

Fig. 2.21 Comparison of bed expansion factor as a function of inlet gas velocity using different drag models with experimental data. (This figure was originally published in [75] and has been reused with permission)

FCC particles improves the bed expansion predictions compared to the homogeneous model. It predicts experimental values within a 20% deviation.

2.9 Modeling of Multi-type Particle Flow Using the Kinetic Theory Approach

Fluid/particle systems are composed of particles of different properties, in which transfer of momentum and segregation by size or density occur during the flow [84–86]. The researchers developed a hydrodynamic model for a mixture of gas and a multi-size solid phase. They applied it to simulate one-dimensional flow in a vertical pneumatic conveying line. They showed that the particle size has a great effect on the pressure drop and choking velocity and that particles segregate along the vertical transport line.

The experiment of Savage and Sayed [87] showed the stresses in a shear cell for a mono-size mixture of polystyrenes beads were about five times higher than those for a binary mixture. Jenkins and Mancini [88] extended the kinetic theory of dense gases to a binary mixture of idealized granular material for the low dissipation case. Jenkins and Mancini [89] presented an extension of the kinetic theory for a binary mixture of smooth, nearly elastic spheres. Alam et al. [90] generalized the model of Willits and Arnarson [91] for a mixture of particles having different mass and size. However, the model proposed by them was limited to energy non-equipartition. Zamankhan [92] concluded that energy non-equipartition must be included in mixtures with different particle properties. Wildman and Parker [93] and Feitosa and Menon [94] experimentally confirmed the coexistence of two granular temperatures when the binary mixture was exposed to external vibrations. Huilin et al. [95, 96] developed a model for two-size particles with different granular temperatures; however, they used an approach that takes the arithmetic average of the particle properties in the collisional operator and the momentum source vanished. Garzó and Dufty [97] solved the kinetic equation for systems away from equilibrium. This approach could capture not only the energy non-equipartition, but also the flow behavior for a wide range of restitution coefficients. Such a model is restricted to dilute systems where the radial distribution function is close to unity. Iddir and Arastoopour [8] and Iddir et al. [98] extended the kinetic theory to a multi-type (size and/or density) mixture, assuming a non-Maxwellian velocity distribution and energy non-equipartition. Each particle type is represented by a phase, with an average velocity and a fluctuating energy or granular temperature. This means that the interaction between the different type particle phases is at the interface. They assumed that the deviation from the Maxwellian velocity distribution is in each individual particulate phase; however, they assumed Maxwellian velocity distribution at the interface. Then, they solved the Boltzmann's equation for each particulate phase using the Chapman-Enskog procedure by adding a perturbation to the Maxwellian velocity distribution function. In a similar analysis, Willits and Arnarson [91] solved the Boltzmann's equation based on the Maxwellian reference state and the revised Enskog equation. Although the range of applicability of Iddir and Arastoopour's work and Willits and Arnarson's work is the same, the major differences between the two studies are non-equipartition and unequal particle size properties considered in the Iddir and Arastoopour model. Willits and Arnarson assumed a multi-component mixture where all the particles fluctuate about the same mass average velocity and have the same granular temperature. Iddir and Arastoopour [8] considered a multiphase granular flow where each phase is represented by particles having different properties, velocities, and granular temperatures. The following is the model developed by Iddir and Arastoopour [8] that has been incorporated in the MFIX computer code [99].

2.9.1   Multi-type Particle Flow Equations

2.9.1.1   Continuity Equation

The continuity equation for the solid phase i can be written,

$$\frac{\partial \varepsilon_i \rho_i}{\partial t} + \nabla \cdot \left(\varepsilon_i \rho_i \vec{v}_i \right) = 0 \tag{2.115}$$

$\varepsilon_i \rho_i = n_i m_i$ is the mass of phase I per unit volume of mixture, ε_i is its solid volume fraction, and ρ_i is density of phase i.

Here $\vec{v}_i = <\vec{c}_i>$ is the mean velocity of the particle i.

The instantaneous velocity $\vec{c}_i$ is defined as the sum of the average velocity, $\vec{v}_i$, and peculiar velocity, $\vec{C}_i$,

$$\vec{c}_i = \vec{v}_i + \vec{C}_i \text{ with } <\vec{C}_i> = 0$$

2.9.1.2 Momentum Equation

The momentum equation for phase i may be expressed,

$$\varepsilon_i \rho_i \frac{D}{Dt}(\vec{v}_i) + \nabla \cdot \left(\sum_{p=1}^{N} \overline{P}_{cip} + \overline{P}_{ki} \right) - \frac{\varepsilon_i \rho_i}{m_i} \vec{F}_{iext} = \sum_{p=1}^{N} \vec{F}_{Dip} \qquad (2.116)$$

where

$$\frac{D}{Dt} \text{ is the material derivative,} \qquad (2.117)$$

$$\overline{P}_{ki} = \rho_i \varepsilon_i <\vec{C}_i \vec{C}_i> \text{ is the kinetic pressure tensor,} \qquad (2.118)$$

$$\overline{P}_{cip} = \chi_{cip}\left(m_i \vec{c}_i\right) \text{ is the collisional pressure tensor, and} \qquad (2.119)$$

$$\vec{F}_{Dip} = \gamma_{cip}\left(m_i \vec{c}_i\right) \text{is the collisional momentum source.} \qquad (2.120)$$

2.9.1.3 Fluctuating Energy Equation

The fluctuating energy equation for solid phase, i, can be expressed,

$$\frac{3}{2} \frac{\varepsilon_i \rho_i}{m_i} \frac{D\theta_i}{Dt} + \nabla \cdot \left(\vec{q}_{ki} + \sum_{p=1}^{N} \vec{q}_{cip} \right) - \left(\overline{P}_{ki} + \sum_{p=1}^{N} \overline{P}_{cip} \right) : \nabla \vec{v}_i$$

$$= \sum_{p=1}^{N} \left(N_{ip} - \vec{v}_i \cdot \vec{F}_{Dip} \right) \qquad (2.121)$$

$$\theta_i = \frac{1}{3} m_i <\vec{C}_i \cdot \vec{C}_i> \qquad (2.122)$$

θ_i is the granular temperature or the fluctuating granular energy of the solid phase, i,

$$\vec{q}_{ki} = \rho_i \varepsilon_i < \vec{C}_i C_i^2 > \text{ is the kinetic heat flux,} \qquad (2.123)$$

$$\vec{q}_{cip} = \chi_{cip}(m_i c_i^2) \text{ is the collisional energy flux, and} \qquad (2.124)$$

$$N_{ip} = \gamma_{cip}(m_i c_i^2) \text{ is the collisional energy dissipation flux.} \qquad (2.125)$$

In the above governing equations, the relevant variables describing the flow field are the average velocities, the solid volume fractions, and the granular temperatures evaluated at location, $\vec{r}$, of the center of the particle at time, t.

2.9.1.4 Kinetic Equation

The kinetic equations that characterize the flow of a multi-phase system are,

$$\left(\frac{\partial}{\partial t} + \vec{c} \cdot \nabla + \frac{\vec{F}_{iext}}{m_i} \cdot \nabla_{\vec{c}}\right) f_i^1 =$$

$$\sum_{p=1}^{N} \frac{d_{ip}^2}{4} \iint \left[g_{ip}\left(\vec{r}, \vec{r} + \vec{k}d_{ip}/2\right) f_i^1\left(\vec{c}_{i1}, \vec{r}, t\right) f_i^1\left(\vec{c}_{i2}, \vec{r} + \vec{k}d_{ip}/2, t\right)\right.$$

$$\left. -g_{ip}\left(\vec{r}, \vec{r} - \vec{k}d_{ip}/2\right) f_p^1\left(\vec{c}_{p1}, \vec{r}, t\right) f_p^1\left(\vec{c}_{p2}, \vec{r} + \vec{k}d_{ip}/2, t\right)\right]\left(\vec{c}_{ip} \cdot \vec{k}\right) d\vec{c}_p d\vec{k}$$

$$(2.126)$$

where $g_{ip}\left(\vec{r}\right)$ is the spatial-pair radial distribution function when the particles, i, and, p, are in contact. A solution of Eq. (2.126) near the equilibrium was obtained using the Chapman-Enskog method [1, 100],

$$f_i^1 = f_i^0(1 + \phi_i) \qquad (2.127)$$

where, f_i^0, is the Maxwellian velocity distribution,

$$f_i^0 = n_i \left(\frac{m_i}{2\pi\theta_i}\right)^{3/2} \exp\left[-\frac{m_i C_i^2}{2\theta_i}\right] \qquad (2.128)$$

and ϕ_i is a perturbation to the Maxwellian velocity distribution. It is a linear function of the first derivative of n_i, θ_i, and $\vec{v}_i$. Note that ϕ_i is function of the phase mean velocity, $\vec{v}_i$, and not the total flow velocity, because, as mentioned in the introduction, each kind of particle is treated as a separate phase and the interaction is at the interface. The radial distribution function $g_{ip}(\varepsilon_i, \varepsilon_p)$ describes a multi-size mixture of hard spheres at contact. Iddir and Arastoopour [8] modified the Lebowitz [101] radial distribution function. This approach is in agreement with the results of the molecular dynamics (MD) simulation obtained by Alder and Wainwright [102] at both lower and higher solid volume fractions. This equation can be written,

$$g_{ip}(\varepsilon_i, \varepsilon_p) = \frac{\left[d_p g_{ii}(\varepsilon_i, \varepsilon_p) + d_i g_{pp}(\varepsilon_i, \varepsilon_p) \right]}{2d_{ip}} \tag{2.129}$$

where

$$g_{ii}(\varepsilon_i, \varepsilon_p) = \frac{1}{\left(1 - (\varepsilon_i + \varepsilon_p)/\varepsilon_{\max} \right)} + \frac{3d_i}{2} \sum_{p=1}^{N} \frac{\varepsilon_p}{d_p} \tag{2.130}$$

The expression of $g_{pp}(\varepsilon_i, \varepsilon_p)$ is obtained by simply interchanging the indices, i, and, p.

For a more detailed explanation for constitutive relation expressions for all solid phases and application of multi-type particle model to gas/particle flow systems [8, 98].

2.10   Heat Transfer

2.10.1   Fluid/Particles Heat Transfer in Fluid/Particles Flow Systems

In order to improve the design and control of gas/solid processes such as: fluid/particle flow systems, packed and fluidized beds, a detailed understanding of heat transfer as well as mass and momentum transfers are needed. In the literature, there are several experimental and theoretical studies to obtain an expression for the fluid/particle Nusselt number and heat transfer coefficient and at different flow regimes [16, 103–106]. Kothari [103], based on all available data, obtained an expression for the Nusselt number in bubbling fluidized bed systems as a function of only the particle Reynolds number.

There are also expressions for the Nusselt number as a function of both the Reynolds and Prandtl numbers for a single particle [104] and a packed bed of particles [105], and for regimes between single particle and packed beds such as a fluidized bed [16, 106].

The following expressions for the Nusselt number was presented by Gidaspow and Jiradilok [16],

For $\varepsilon_g \leq 0.8$,

$$Nu_s = \left(2 + 1.1 \, \text{Re}^{0.6} \text{Pr}^{1/3} \right) S_s \quad \text{Re} \leq 200 \tag{2.131}$$

$$= 0.123 \left(\frac{4\,\mathrm{Re}}{d_s} \right)^{0.183} S_s^{0.17} \quad 200 < \mathrm{Re} \leq 2000 \qquad (2.132)$$

$$= 0.61\,\mathrm{Re}^{0.67} Ss \quad \mathrm{Re} > 2000 \qquad (2.133)$$

for $\varepsilon_g > 0.8$,

$$Nu_s = \left(2 + 0.16\,\mathrm{Re}^{0.67} \right) S_s \quad \mathrm{Re} \leq 200 \qquad (2.134)$$

$$= 8.2\,\mathrm{Re}^{0.6} S_s \quad 200 < \mathrm{Re} \leq 2000 \qquad (2.135)$$

$$= 1.06\,\mathrm{Re}^{0.457} S_s \quad \mathrm{Re} > 2000 \qquad (2.136)$$

where,

$$\mathrm{Re} = \frac{\varepsilon_g \rho_g \left| \vec{v}_g - \vec{v}_s \right| d_s}{\mu_g} \qquad (2.137)$$

$$S_s = \varepsilon_s \frac{6}{d_s} \qquad (2.138)$$

$$Nu = \frac{h_{vs} d_s}{k_g} \qquad (2.139)$$

Gunn [106] developed an expression for the Nusselt number to describe mass and heat transfer within the porosity range of 0.35–1.0 and applicable up to a Reynolds number of 10^5 and Prandtl number range of $(0.6 < \mathrm{Pr} < 380)$. The Gunn expression is,

$$Nu_s = \left(7 - 10\varepsilon_g + 5\varepsilon_g^2 \right) \left(1 + 0.7\,Re_s^{0.2} Pr^{1/3} \right)$$
$$+ \left(1.33 - 2.4\varepsilon_g + 1.2\varepsilon_g^2 \right) Re_s^{0.7} Pr^{1/3} \qquad (2.140)$$

Recently Buist et al. [107] showed that their experimental and numerical results based on Direct Numerical Simulation (DNS) for different solid volume fractions were a remarkable comparison to the Gunn [106] expression.

2.10.2   Wall Heat Transfer in Fluid/Particles Flow Systems

To design better fluidized bed reactors for new processes, such as for CO_2 capture from flue gases or for the production of pure silicon for solar collectors, wall-to-bed heat transfer must be improved. The values of the wall-to-bed heat transfer coefficients limit the size of the fluidized bed reactors. Since the temperatures inside such reactors are nearly constant, the wall-to-bed heat transfer coefficients are simply the thermal conductivity divided by the small boundary layer thickness. To obtain a high thermal conductivity, the turbulent granular temperature and solids volume fraction must be as high as possible. IIT researchers' kinetic theory-based CFD code [16, 108] computes the thermal temperature and its standard deviations. The standard deviation is the highest near the walls, in the thermal boundary layer, in agreement with measurements done in the IIT two-story circulating fluidized bed. The average standard deviations of the temperature and velocities are 0.75 K and 0.15 m/s, respectively. The heat transfer coefficient obtained by this method and from an overall energy balance is 0.39 kW/m^2 K. If the standard deviation of velocity is increased from 0.15 to 1 m/s, the heat transfer coefficient becomes 2.6 kW/m^2 K, still well below boiling heat transfer coefficients, but much larger than the usual CFB heat transfer coefficients. This example demonstrates the potential of CFD to improve fluidized bed reactors.

2.11   Mass Transfer

2.11.1   Mass Transfer Coefficients

The conventional additive resistance concept permits us to compute the mass transfer coefficient. At steady state, the external mass transfer in terms of global rate is equated to the mass transfer from the bulk gas to catalyst surface [109–112] by,

$$k_{mass\ transfer} a_v \left(C_{O_3} - C_{O_3,surface} \right) = k_{reaction} C_{O_3,surface} \tag{2.141}$$

where $k_{mass\ transfer}$ is the mass transfer coefficient, a_v is the external surface per volume of catalyst, and $C_{O_3,surface}$ is the surface molar concentration of ozone. Eliminating $C_{O_3,surface}$ from Eq. (2.141) and expressing the global reaction rate per unit mass of catalyst, r_p, in terms of C_{O_3} yields,

$$r_p \rho_s \varepsilon_s = \frac{1}{\frac{1}{k_{mass\ transfer} a_v} + \frac{1}{k_{reaction}}} C_{O_3} \varepsilon_s = K C_{O_3} \varepsilon_s \tag{2.142}$$

The result from Eq. (2.142) give the overall mass transfer coefficient, $K, \frac{1}{K} = \frac{1}{k_{mass\ transfer} a_v} + \frac{1}{k_{reaction}}$. The Sherwood number, Sh, is then,

$$Sh = \frac{k_{mass\ transfer} d_p}{D} \tag{2.143}$$

where d_p is the diameter of the catalyst particle and D is molecular diffusivity. Bolland and Nicolai [111] have analyzed their ozone decomposition data in a fluidized bed using this method.

Experimental data or equivalent CFD computation allow us to obtain the effective rate constant, K, by approximate integration of the species balance,

$$\ln C_{O_3} = \ln C_{O_3,0} - \frac{K\varepsilon_s}{v_y} Y \tag{2.144}$$

where the subscript "0" is the initial molar concentration of ozone.

2.11.2 Low Sherwood Number But Good Mass Transfer in Fluidized Beds

An example of the Sherwood number and mass transfer coefficient calculation is illustrated for the reaction rate constant of 39.6 s^{-1} at a height of 3.5 m for the ozone decomposition simulation found in [9]. The computed time-averaged and natural logarithm of computed time-averaged ozone molar concentrations are displayed in Figs. 8.3 and 8.4, respectively. The slope for this case is -0.7584 and the intercept is 13.8290. From the simulation, the computed volume fraction of the solid phase is 0.1289, the computed volume fraction of the gas phase is 0.871, and the velocity of the gas phase in the axial direction was 6.0115 m/s. Substitution of all these values into Eq. (2.144),

$$\frac{K(0.1289)}{(6.0115)(0.8711)} = 0.7584 \tag{2.145}$$

produces the overall mass transfer coefficient, $K = 30.81$ s^{-1}.

The mass transfer coefficient is calculated from Eq. (2.135)

$$\text{with } d_p = 76 \times 10^{-6} \text{ m}$$
$$a_v = \left(3 \times 4\pi \left(\text{particle radius}^2\right)\right) / \left(4\pi \left(\text{particle radius}^3\right)\right)$$
$$= 3/\text{particle radius}$$
$$= 3/\left(d_p/2\right)$$
$$= 3/\left((76 \times 10^{-6})/2\right) = 78947.37 \text{ m}^{-1}$$
$$\text{to give } k_{reaction} = 39.60 \text{ s}^{-1}$$

Note that the overall resistance, K, and the reaction resistance, $k_{reaction}$, are close to each other. This implies that the mass transfer resistance is small.

Therefore, $k_{mass\ transfer}a_v = 138.71$ s^{-1}, and $k_{mass\ transfer} = 0.0018$ m/s.

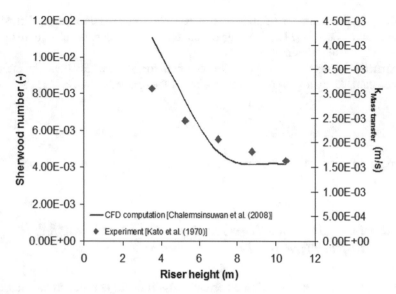

Fig. 2.22 Sherwood number as a function of riser height [16]

The Sherwood number is calculated from Eq. (2.143) with $D = 2.88 \times 10^{-5}$ m^2/s, Sh = 0.0046.

Figure 2.22 shows a comparison between the Sherwood number calculated using the CFD approach with the experimental data of Kato et al. [115]. In Fig. 2.22, the behavior of the Sherwood number is similar to that of diffusion in a channel in fully developed laminar flow, called the Graetz problem, except for the much lower Sherwood number. There, the Sherwood number is large in the diffusion boundary layer and reaches a constant value of about two for large heights. Our explanation [19, 113, 114] is the formation of clusters shown for the fast fluidization regime in Fig. 2.15.

If the Sherwood number is defined based on the cluster size rather than particle diameter, the low Sherwood number in Fig. 2.22 becomes close to the theoretical value of 2, since the ratio of cluster size to particle diameter can be as much as 1000. This theory suggests that the cluster size should be reduced to eliminate potential mass transfer limitation that may require much taller reactors than those computed from the reaction rate data.

2.12 Exercises

2.12.1 Ex. 1: Alternate Definition of Granular Temperature

An alternate definition of granular temperature is one closer to that used in the kinetic theory of gases involving kinetic energy rather than the variance of velocity of the particles used here. The Boltzmann constant then is unity in the gas theory for the particles used in multiphase flow. The definition is as follows for particles of phase i,

$$\theta_i = 1/3 \, m_i < \mathbf{C}_i^2 >$$

The Maxwellian frequency distribution is,

$$f_i = n_i \, [m_i/2\pi\theta_i]^{1/2} \exp\left[-m_i/2\theta_i(c_i - v_i)^2\right]$$

1. Find the expressions for the two mean values of the velocity in terms of the above new definitions of the granular temperature:

   (a) $<c_i>$
   (b) $<c_i^2>^{1/2}$

2.12.2   Ex. 2: Diffusion Coefficients and Viscosities

1. Develop the expressions for a binary mixture of mass m_1 and mass m_2 for:

   (a) The diffusion coefficient, D
   (b) The viscosity, μ

   in terms of the new granular temperature definition in Exercise 1.

2.12.3   Ex. 3: Collision Theory for Reactions and Burning Rate

According to the collision theory, the rate of reactions is proportional to the collision frequency given in this chapter and the Arrhenius activation energy, $\exp(-E/RT)$, where E is the activation energy, R is the gas constant, and T is the absolute temperature. Smith [116] shows that this simple model gives the correct order of magnitude of the rate of reaction for the reaction, $2HI = H_2 + I_2$.

1. Check the numerical calculations presented by Smith.
2. Williams [117] shows that the burning rate is proportional to the square root of the rate of reaction of fuels. The burning rate of gasoline is 16.1 m/s and that of hydrogen is 48.3 m/s. Show that the simple collision theory demonstrates that such a rate for combustion of hydrogen in an engine is a reasonable value.

2.12.4 Ex. 4: Apollo 13 Oxygen Tank Explosion

The Apollo 13 moon mission life support system was powered by fuel cells using liquid hydrogen and oxygen. The relief valve on the oxygen tank was designed based on the maximum oxygen flow rate in the gas phase which is of the order of 300 m/s. However, due to the near zero gravity condition, oxygen in the gas and liquid phases flowed out of the tank when the tank was overheated. The mixture velocity of this two-phase mixture is of the order of magnitude of 30 m/s, as shown in this chapter. Therefore, the pressure in the tank rose and the tank exploded.

1. Formulate the one-dimensional mixture mass and momentum equations for the pressure and velocities.
2. Find the characteristic directions.
3. Show that the mixture velocity, C_m, is as follows,

$$C_m{}^2 = (\rho_g/\rho_l)[(1 - \varepsilon)/\varepsilon]C_g{}^2$$

where ρ_g, and ρ_l are the densities of gas the liquid, respectively, C_g is the gas phase velocity, and ε is the gas phase volume fraction since ρ_g is much smaller than ρ_l, the mixture density is much smaller than the gas critical velocity [36]. Hence, conclude that this was the cause of the failure of Apollo 13 to go to the moon.

2.12.5 Ex. 5: One-Dimensional Gas/Solid Flow

1. Express one-dimensional steady state two phase flow of gas and particles equations using the pressure drop in the gas phase model in the following form,

$$\frac{d\vec{y}}{d\vec{x}} = f\left(\vec{y}\right)$$

2. The above form of equations enables us to make a quantitative analysis of certain special characteristics of the physical process [e.g., if the right side of the above equation shows $\frac{d\vec{y}}{d\vec{x}}$ is unbounded]. Obtain such critical conditions.

2.12.6 Ex. 6: Three-Dimensional Gas/Solid Flow

Consider gas and group B particles of average diameter d_p entering from the bottom into a rectangular riser section of a circulation fluidized bed (with cross-sectional area dimensions of a and b and height of L). The riser is operating at a non-homogeneous particle phase regime. Gas and particles flow vertically (consider z as vertical direction and x and y as directions perpendicular to the z direction). Inlet

($z = 0$) gas and solid velocities, pressure, and solid volume fraction are V_{g1}, V_{s1}, P_1, and ϵ_{s1}, respectively. Initially, there is no particle in the riser with stationary gas at pressure P_0.

1. Develop transient, three-dimensional (x, y, and z) isothermal governing equations (continuity and momentum equations for each phase and the granular temperature equation for the particle phase). Assume laminar Newtonian flow for the gas phase and the kinetic theory for the particle phase flows.
2. Write down all constitutive equations and all needed boundary and initial conditions for both gas and particles phases. Clearly mention any assumption that you are making and why. Write down the proper expression for drag forces and particle phase viscosity. Assume the particle phase is incompressible and the gas phase obeys ideal gas equation of state.

2.12.7 Ex. 7: Two-Dimensional Gas/Solid Flow

A mixture of gas and particles of size D_p, initial volumetric concentration and pressure of ε_{s0} and P_0 flows through a vertical tube with decreasing diameter as shown in the following figure. The inlet gas and particle velocities are V_{g0} and V_{s0}, respectively.

1. Develop the transient two-dimensional governing equations (continuity and momentum) and boundary conditions for this flow system.
2. Develop governing equations for the one-dimensional steady state case.

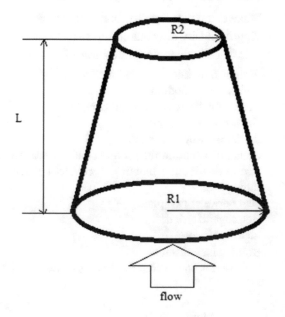

flow

2.12.8   Ex. 8: Two-Dimensional Transient Gas/Solid Flow

1. Write down two-dimensional transient continuity and momentum equations
   along with constitutive relations (closures) for three phase flow of liquid, gas,
   and spherical particles of diameter d_p in a vertical cylindrical channel of diameter
   D and length L. Initially, there is no gas and solid in the channel and liquid is
   stationary at pressure P_0. Assume the inlet mass fluxes and pressure are w_{g1}, w_{s1},
   w_{L1}, and P_1, respectively. Inlet solid and gas volume fractions are 1% and 2%;
   thus, you may consider that gas flows in the form of bubbles. What drag
   expressions are you proposing and why? Clearly express all independent variables and boundary and initial conditions.

Nomenclature

D	Diffusivity
A	Avogadro's number
c	Velocity space
C_D	Drag coefficient
$\vec{c}_s$	Instantaneous particle velocity in i direction
$\vec{C}_s$	Peculiar particle velocity
$d_{ip} = (d_i + d_p)/2$	Average diameter of particles i and p
d_s and, d_p	Diameter of particle
E	Internal energy
E	Restitution coefficient
e_{sw}	Particle–wall restitution coefficient
F	Frequency distribution of velocities of particles
$\vec{F}_{Dip}$	Momentum source (drag between solid phases)
f_s^0	Maxwellian velocity distribution function
f_s^1	Single velocity distribution function
f_{ip}^2	Pair velocity distribution function
$G(x,y)$	Weight function in filtered model
G	Solids stress modulus
G'	Ratio of solid stress to solid density
g	Gravity acceleration
g_0	Radial distribution function
g_{ip}	Radial distribution function at contact between particles i and p
g_{ss}	Radial distribution function at contact between particles of the same phase
H	Gap between two plates
H_d	Heterogeneity index
$\bar{I}$	Identity tensor
k_B	Boltzmann constant
$\vec{k} = \vec{k}_{ip}$	Unit vector connecting the centers of two particles

K_f	Turbulent kinetic energy of fluid
k_s	Granular conductivity
M	Molecular weight
m_s	Mass of particle of phase s
$m_0 = m_i + m_p$	Total mass of two colliding particles
N	Number of moles
n	Compact factor
n_s	Number density of phase s
N_{ip}	Energy dissipation
$\overline{P}_c$	Collisional pressure tensor
$\overline{P}_k$	Kinetic pressure tensor
p_c	Critical state pressure
p_f	Frictional pressure
Q	Momentum flux
$\vec{q}_c$	Collisional flux
$\vec{q}_k$	Kinetic flux
R_j	Rate of heterogeneous reaction between phases
Re_p	Particle Reynolds number
$\langle S \rangle$	Average strain rate
$\boldsymbol{r}$	Position
S	Deviatoric stress tensor
T	Average of random kinetic energy
$\mathbf{v}$	Hydrodynamic velocity
$\mathbf{v_t}$	Terminal velocity
$\vec{v}$	Center of mass mean velocity
$\vec{v}_s$	Mean velocity of phase s
$\vec{V}_0$	Velocity of plates
$\vec{v}_{s,para}$	Particle slip velocity parallel to wall
v_{ts}	Solid particle terminal velocity
y_i	Species mass fraction

Subscripts

1	Large
2	Small
cr	Critical
fr	Frictional
g	Gas phase
col	Collisional
kin	Kinetic
max	Maximum
T	Total
s	Solid phase
w	Wall

Greek Symbols

χ_{ip}	Collisional flux
ε_s	Solid volume fraction of phase s
ε_T	Total solid volume fraction
ϕ_s	Perturbation to Maxwellian distribution function
γ_{ip}	Collisional source of granular temperature
μ	Viscosity
μ_{ss}	Granular viscosity of phase s
μ_{ip}	Mixture granular viscosity
ρ_s	Solid density of phase s
θ_s	Fluctuating granular energy
θ_m	Mixture fluctuating granular energy
τ_N	Mixture normal stress
τ_s	Solid phase shear stress
τ_t	Mixture shear stress
ψ_s	Property of particle
θ	Granular temperature (random kinetic energy per unit mass)
ϕ	Mean values of a quantity
ϕ	Specularity coefficient
ϕ	Angle of internal friction
ϕ_{gs}	Granular energy exchange between phases
ρ	Bulk density
εg	Volume fraction of gas
ε_s	Volume fraction of solids
ρ_s	Solids density
II_{2D}	Second invariant of the deviatoric stress tensor
γ	Shear rate
β	Interface momentum exchange coefficient
β_{fill}	Filtered drag coefficient
β_{micron}	Drag coefficient n microscopic two-fluid model
γ	Collisional energy dissipation
ξ	Solid phase bulk viscosity
Δ_f	Filter size

References

1. Chapman S, Cowling TG (1960) The mathematical theory of non-uniform gases. Cambridge at the University Press, Cambridge, UK
2. Savage SB, Jeffrey DJ (1981) The stress tensor in a granular flow at high shear rates. J Fluid Mech 110:255–272
3. Jenkins J, Savage SB (1983) Theory for the rapid flow of identical, smooth, nearly elastic, spherical particles. J Fluid Mech 130(1):187–202
4. Lun CKK, Savage SB, Jeffrey DJ, Chepurniy N (1984) Kinetic theories for granular flow: inelastic particles in Couette flow and slightly inelastic particles in a general flow field. J Fluid Mech 140:223–256

5. Jenkins JT, Richman MW (1985) Kinetic theory for plane flows of a dense gas of identical, rough, inelastic, circular disks. Phys Fluids 28(12):3485–494
6. Gidaspow D (1994) Multiphase flow and fluidization: continuum and kinetic theory description. Academic Press, San Diego, California, USA
7. Kim H, Arastoopour H (2002) Extension of kinetic theory to cohesive particle flow. Powder Technol 122(1):83–94
8. Iddir H, Arastoopour H (2005) Modeling of multitype particle flow using the kinetic theory approach. AIChE J 51(6):1620–1632
9. Gidaspow D, Jung J, Singh RK (2004) Hydodaynamics of fluidization using kinetic theory: an emerging paradigm: 2002 Fluor-Daniel plenary lecture. Powder Technol 148:123–141
10. Arastoopour H (2001) Numerical simulation and experimental analysis of gas/solid flow systems: 1999 Fluor-Daniel plenary lecture. Powder Technol 119(2–3):59–67
11. Songprawat S, Gidaspow D (2010) Multiphase flow with unequal granular temperatures. Chem Eng Sci 65:1134–1143
12. Shuai W, Zhenhua H, Huilin L, Guodong L, Jiaxing W, Pengfei X (2012) A bubbling fluidization model using kinetic theory of rough spheres. AIChE J 58:440–455
13. Strumendo M, Gidaspow D, Canu P (2005) Method of moments for gas-solid flows: application to the riser. In: Cen K (ed) Circulating fluidized bed technology VIII. New York: Intern. Acad. Pub., New York, pp 936–942
14. Strumendo M, Arastoopour H (2010) Solution of population balance equations by the finite size domain complete set of trial functions method of moments (FCMOM) for inhomogeneous systems. Ind Eng Chem Res 49:5222–5230
15. Gidaspow D, Huilin L (1998) Equation of state and radial distribution functions of FCC particles in a CFB. AIChE J 44(2):279–291
16. Gidaspow D, Jiradilok V (2009). Computational techniques: the multiphase CFD approach to fluidization and green energy technologies. Nova Science Publishers, New York USA
17. Johnson KL (1985) Contact mechanics. Cambridge University Press, Cambridge, UK
18. Jiradilok V, Gidaspow D, Breault RW (2007) Computation of gas and solids dispersion coefficients in turbulent risers and bubbling beds. Chem Eng Sci 62(13):3397–3409
19. Kashyap M, Gidaspow D (2012). Dispersion and mass transfer coefficients in fluidized beds. Lap Lambert Academic Publishing, Saarbruchen, Germany
20. Miller A, Gidaspow D (1992) Dense vertical gas-solid flow in a pipe. AIChE J 38:1801–1815
21. Johnson PC, Jackson R (1987) Frictional-collisional constitutive relations for granular materials, with application to plane shearing. J Fluid Mech 176:67–93
22. Benyahia S, Syamlal M, O'Brien TJ (2005) Evaluation of boundary conditions used to model dilute, turbulent gas/solids flows in a pipe. Powder Technol 156(2):62–72
23. Syamlal M (1985) Multiphase hydrodynamics of gas-solids flow. PhD thesis, Illinois Institute of Technology, Chicago, IL
24. Tartan M, Gidaspow D (2004) Measurement of granular temperature and stresses in risers. AIChE J 50:1760–1775
25. Schlichting HT (1960) Boundary-layer theory, 4th ed. McGraw Hill, New York, USA
26. Kim J, Moin P, Moser R (1987) Turbulence statistics in fully developed channel flow at low Reynolds number. J Fluid Mech 177:133–166
27. Benyahia S, Syamlal M, O'Brein TJ (2007) Study of ability of multiphase continuum models to predict core-annulus flow. AIChE J 53(10):2549–256
28. Berruti F, Chaouki J, Godfroy L, Pugsley TS, Patience GS (1995) Hydrodynamics of circulating fluidized bed risers: a review. Canadian J Chem Eng 73(5):579–602
29. Gidaspow D, Chandra V (2014) Unequal granular temperature model for motion of platelets to the wall and red blood cells to the center. Chem Eng Sci 117:107–113
30. ANSYS, Inc., FLUENT user's guide. Canonsburg, PA
31. Syamlal, M, Rogers W, O'Brien TJ (1993) MFIX documentation theory guide: technical note. DOE/METC-94/1004, NTIS/DE94000087, U.S. Department of Energy, Office of Fossil

Energy, Morgantown Energy Technology Center Morgantown, WV, National Technical Information Service, Springfield, VA

32. Kashyap M, Gidaspow D, Koves WJ (2011) Circulation of Geldart D type particles: part I-high flux solids measurements and computation under solids slugging conditions. Chem Eng Sci 66:183–206

33. Wei F, Lin H, Chang Y, Wang Z, Jin Y (1998) Profiles of particle velocity and solids faction in a high density riser. Powder Technol 100:183–189

34. Li J, Kwauk M (1994) Particle-fluid two-phase flow. Metallurgical Industry Press, Beijing, China

35. Jung J, Gidaspow D, Gamwo IK (2005) Measurement of two kinds of granular temperatures, stresses and dispersion in bubbling beds. Ind Eng Chem Res 44(5):1329–1341

36. Lyczkowski RW, Gidaspow D, Solbrig W (1982) Multiphase flow-models for nuclear, fossil and biomass energy conversion. In: Mujumdar AS, Mashelkar RA (eds) Advances in transport processes (1982), Wiley-Eastern Publisher, New York, pp 198–351

37. Arastoopour H, Gidaspow D, Abbasi E (2017) Computational transport phenomena of fluid-particle systems. Springer, Cham, Switzerland

38. Jung J, Gidaspow D (2002) Fluidization of nano-size particles. J Nanoparticle Research 4:483–497

39. Gelderbloom S, Gidaspow D, Lyczkowski RW (2003) CFD simulation of bubbling and collapsing fluidized bed experiments for three Geldart groups. AIChE 49:844-858

40. Matsen JM (2000) Drift flux representation of gas-particle flow. Powder Technol 111 (1-2):25–33

41. Tsuo YP, Gidaspow D (1990) Computation of flow patterns in circulating fluidized beds. AIChE J 36:885–896

42. Sun B, Gidaspow D (1999) Computation of circulating fluidized-bed riser flow for the fluidization VIII benchmark test. Ind Eng Chem Res 38:787–792

43. Driscoll MC (2007). A Study of the fluidization of FCC and nanoparticles in a rectangular bed and a riser. PhD thesis, Illinois Institute of Technology, Chicago

44. Gidaspow, D, Driscoll M (2007) Wave propagation and granular temperature in fluidized beds of nanoparticles. AIChE J 53(7):1718–1726

45. Roy R, Davidson JF, Tuponogov VG (1990) The velocity of sound in fluidised beds. Chem Eng Sci 45(11):3233–3245

46. Polashenski W, Chen JC (1999) Measurement of particle stresses in fast fluidized beds. Ind Eng Chem Res 38:705–713

47. Gidaspow D, Mostofi R (2003) Maximum carrying capacity and granular temperature of A, B, C particles. AIChE J 49(4):831–843

48. Makkawi Y, Ocone R (2005) Modelling the particle stress at the dilute-intermediate-dense flow regimes: a review. KONA Powder and Particle J 23(0):49–63

49. Ocone R, Sundaresan S, Jackson R (1993) Gas-particle flow in a duct of arbitrary inclination with particle-particle interactions. AIChE J 39(8):1261–1271

50. Laux H (1998) Modeling of dilute and dense dispersed fluid-particle flow. Trondheim, Norway, Norwegian University of Science and Technology

51. Savage SB (1998) Analyses of slow high - concentrations flows of granular materials. J Fluid Mech 377:1–26

52. Schaeffer DG (1987) Instability in the evolution equations describing incompressible granular flow. J Diff Eq:19–50

53. Dartevelle S (2003) Numerical and granulometric approaches to geophysical granular flows. Houghton, MI, Michigan Technological University

54. Srivastava A, Sundaresan S (2003) Analysis of a frictional-kinetic model for gas-particle flow. Powder Technol:72–85

55. Atkinson, JH, Bransby PL (1978) The mechanics of soils: an introduction to critical state soil mechanics. McGraw-Hill Book Company, London, New York

56. Tardos GI (1997) A fluid mechanistic approach to slow, frictional flow of powders. Powder Technol 92(1):61–74
57. Prakash JR, Rao KK (1988) Steady compressible flow of granular materials through a wedgeshaped hopper: the smooth wall, radial gravity problem. Chem Eng Sci 43(3)
58. O'Brien TJ, Mahalatkar K, Kuhlman J (2010) Multiphase CFD simulations of chemical looping reactors for CO_2 capture. 7th International Conference on Multiphase Flow, ICMF 2010, Tampa, FL USA
59. Nikolopoulos A, Nikolopoulos N, Charitos A, Grammelis P, Kakaras E, Bidwe AR, Varela G (2013) High-resolution 3-D full-loop simulation of a CFB carbonator cold model. Chem Eng Sci 90:137–150
60. Abbasi E, Abbasian J, Arastoopour H (2015) CFD–PBE numerical simulation of CO_2 capture using MgO-based sorbent. Powder Technol 286:616–628
61. Nikolopoulos A, Nikolopoulos N, Varveris N, Karellas S, Grammelis P, Kakaras E (2012) Investigation of proper modeling of very dense granular flows in the recirculation system of CFBs. Particuology 10(6):699–709
62. Abbasi E, Arastoopour H (2011) CFD simulation of CO_2 sorption in a circulating fluidized bed using deactivation kinetic model. In: Knowlton TM (ed) Proceeding of the Tenth International Conference on Circulating Fluidized Beds and Fluidization technology, CFB-10, ECI, New York, pp 736–743
63. Ghadirian E, Arastoopour H (2017) Numerical analysis of frictional behavior of dense gas–solid systems. Particuology, 32:178–190
64. Das BM (1997) Advanced soil mechanics (2nd ed). Taylor & Francis, Washington DC
65. Jyotsna R, Rao KK (1997) A frictional-kinetic model for the flow of granular materials through a wedge-shaped hopper. Journal Fluid Mech 346:239–270
66. Igci Y, Pannala S, Benyahia S, Sundaresan S (2008) Validation studies on filtered model equations for gas-particle flows in risers. Ind Eng Chem Res 51(4):2094–2103
67. Benyahia S (2012) Fine-grid simulations of gas-solids flow in a circulating fluidized bed. AIChE J 58(11):3589–3592
68. Nikolopoulos A, Atsonios K, Nikolopoulos N, Grammelis P, Kakaras E (2010) An advanced EMMS scheme for the prediction of drag coefficient under a 1.2 MW CFBC isothermal flow, part II: numerical implementation. Chem Eng Sci 65(13):4089–4099
69. Jang J, Rosa C, Arastoopour H (2010) CFD simulation of pharmaceutical particle drying in a bubbling fluidized bed reactor. In: Kim S.D. et al. (ed) Fluidization XIII, ECI, New York, pp 853–860
70. Arastoopour H, Pakdel P, Adewumi M (1990) Hydrodynamic analysis of dilute gas-solids flow in a vertical pipe. Powder Technol 62:163–170
71. Syamlal M, O'Brien TJ (2003). Fluid dynamic simulation of O_3 decomposition in a bubbling fluidized bed. AIChE Journal 49(11):2793–2801
72. Wen CY, Yu YH (1966) Mechanics of fluidization. Chem Eng Prog Symp Series 62: 100–111
73. Benyahia S (2009) On the effect of subgrid drag closures. Ind Eng Chem Res 49(11): 5122-5131
74. Sarkar A, Xin S, Sundaresan S (2014) Verification of sub-grid filtered drag models for gas-particle fluidized beds with immersed cylinder arrays. Chem Eng Sci 114:144–154
75. Ghadirian E, Arastoopour H (2016) CFD simulation of a fluidized bed using the EMMS approach for the gas-solid drag force. Powder Technol 288:35-44
76. Arastoopour H, Gidaspow D (1979) Analysis of IGT pneumatic conveying data and fast fluidization using a thermohydrodynamic model. Powder Technol 22(1):77–87
77. Milioli CC, Milioli FE, Holloway W, Agrawal K, Sundaresan S (2013) Filtered two-fluid models of fluidized gas-particle flows: new constitutive relations. AIChE J 59(9):3265–3275
78. Benyahia S, Sundaresan S (2012) Do we need sub-grid scale corrections for both continuum and discrete gas-particle flow models? Powder Technol 220:2–6
79. Li J, Kwauk M (1994) Particle-fluid two-phase flow: the energy-minimization multi-scale method. Metallurgy, Industry Press, Beijing

80. Wang W, Li J (2007) Simulation of gas-solid two-phase flow by a multi-scale CFD approach: extension of the EMMS model to the sub-grid level. Chem Eng Sci 62(1):208–231
81. Benyahia S (2012) Analysis of model parameters affecting the pressure profile in a circulating fluidized bed. AIChE J 58(2):427–439
82. Krishna BSVSR (2013) Predicting the bed height in expanded bed adsorption column using RZ correlation. Bonfring Int J Ind Eng and Mgmt Sci 3(4):107
83. Lu B, Wang W, Li J (2009) Searching for a mesh-independent sub-grid model for CFD simulation of gas-solid riser flows. Chem Eng Sci 64(15):3437–3447
84. Arastoopour H, Lin SC, Weil, SA (1982) Analysis of vertical pneumatic conveying of solids using multiphase flow models. AIChE J 28(3):467–473
85. Cutchin, J, Arastoopour H (1985) Measurement and analysis of particle interaction in a concurrent pneumatic conveying system. Chem Eng Sci 40(7):1134–1143
86. Arastoopour H, Wang CH, Weil SA (1982) Particle-particle interaction force in a dilute gas-solid system. Chem Eng Sci 37(9):1379–1386
87. Savage SB, Sayed M (1984) Stresses developed by dry cohesionless granular materials sheared in an annular shear cell. J Fluid Mech 142:391–430
88. Jenkins JT, Mancini F (1987) Balance laws and constitutive relations for plane flows of a dense, binary mixture of smooth, nearly elastic, circular disks. J App Mech 54(1):2734
89. Jenkins JT, Mancini F (1989) Kinetic theory for binary mixtures of smooth, nearly elastic spheres. Phys Fluids A: Fluid Dynamics (1989-1993) 1(12):2050–2057
90. Alam M, Willits JT, Arnarson BÖ, Luding S (2002) Kinetic theory of a binary mixture of nearly elastic disks with size and mass disparity. Phys Fluids (1994-present) 14 (11):4085–4087
91. Willits JT, Arnarson BÖ (1999) Kinetic theory of a binary mixture of nearly elastic disks. Phys Fluids (1994-present) 11(10):3116–3122
92. Zamankhan P (1995) Kinetic theory of multicomponent dense mixtures of slightly inelastic spherical particles. Phys Rev E 52(5):4877
93. Wildman RD, Parker DJ (2002) Coexistence of two granular temperatures in binary vibrofluidized beds. Phys Rev Lett 88(6):064301
94. Feitosa K, Menon N (2002) Breakdown of energy equipartition in a 2D binary vibrated granular gas. Phys Rev Lett 88(19):198301
95. Huilin L, Gidaspow D, Manger E (2001). Kinetic theory of fluidized binary granular mixtures. Phys Rev E 64(6):061301
96. Huilin L, Wenti L, Rushan B, Lidan Y, Gidaspow D (2000) Kinetic theory of fluidized binary granular mixtures with unequal granular temperature. Physica A: Statistical Mechanics and its Applications 284(1):265–276
97. Garzó V, Dufty JW (2002) Hydrodynamics for a granular binary mixture at low density. Phys Fluids (1994-present), 14(4):1476–1490
98. Iddir H, Arastoopour H, Hrenya CM (2005) Analysis of binary and ternary granular mixtures behavior using the kinetic theory approach. Powder Technol 151(1):117–125
99. Benyahia S (2008) Verification and validation study of some polydisperse kinetic theories. Chem Eng Sci 63(23):5672–5680
100. Ferziger JH, Kaper HG (1972) Mathematical theory of transport processes in gases. North Holland Publishing Company
101. Lebowitz JL (1964) Exact solution of generalized Percus-Yevick equation for a mixture of hard spheres. Phys Rev 133(4A):A895
102. Alder BJ, Wainwright TE (1967) Velocity autocorrelations for hard spheres. Phys Rev Let 18 (23):988
103. Khotari KA (1967) M.S. Chem. Engineering, Illinois Institute of Technology, Chicago
104. Ranz WE, Marshall WR (1952) Evaporation from drops. Chem Eng Prog 48(3):141–146
105. Ranz WE (1952) Friction and transfer coefficients for single particles and packed beds. Chem Eng Prog 48(5):247–253

106. Gunn, DJ (1978) Transfer of heat or mass to particles in fixed and fluidised beds. Int J Heat Mass Transf 21(4):467–476
107. Buist, KA, Backx BJGH, Deen NG, Kuipers JAM (2017) A combined experimental and simulation study of fluid-particle heat transfer in dense arrays of stationary particles. Chem Eng Sci 169:310–320
108. Chaiwang P, Gidaspow D, Chalermsinsuwan B, Piumsomboon P (2014) CFD design of a sorber for CO_2 capture with 75 and 375 micron particles. Chem Eng Sci 105:32–45
109. Fogler HS (1999) Elements of chemical reaction engineering. Prentice Hall, New Jersey
110. Levenspiel O (1999) Chemical reaction engineering. John Wiley and Sons, New York
111. Bolland O, Nicolai R (2001) Describing mass transfer in circulating fluidized beds by ozone decomposition. Chem Eng Comm 187:1–21
112. Welty, JR, Wicks CE, Wilson RE, Rorrer G (2001) Fundamentals of momentum, heat and mass transfer. John Wiley and Sons, New York
113. Chalermsinsuwan B, Piumsomboon P, Gidaspow D (2008a) Kinetic theory based computation of PSRI riser – Part 1. Estimate of mass transfer coefficient. Chem Eng Sci 64:1195–1211
114. Chalermsinsuwan B, Piumsomboon P, Gidaspow D (2008b) Kinetic theory based computation of PSRI riser – Part II. Computation of mass transfer coefficient with chemical reaction. Chem Eng Sci 64:1212–1222
115. Kato K, Kubota H, Wen CY (1970) Mass transfer in fixed and fluidized beds. Chem Eng Prog Symp Series 105:87–99
116. Smith, JM (1970) Chemical engineering kinetics. McGraw-Hill Book Company, New York
117. Williams, FA (2018) Combustion theory, 2nd edn. CRC Press: Taylor & Francis Group, Boca Raton, Florida

Chapter 3
Multiphase Flow Phenomena (Gas/Solid and Gas/Liquid Systems)

Brian G. Valentine

3.1 Introduction

In this chapter, the basic properties of gas/solid and gas/liquid two-phase flows are examined; the serious student of the subject will have to study more comprehensive texts and monographs related to the subject, some of which are discussed in the references at the end of this chapter.

Two-phase flow systems are considerably more complex than single-phase fluid flows, and two-phase flows exhibit phenomena that are not observed in single-phase fluid systems. This makes the prediction of two-phase flow momentum and heat transfer difficult, and intuition is frequently not a useful guide, even in the case of dilute two-phase mixtures. Many computational fluid dynamics (CFD) models and computer programs are available for the analysis of two-phase flows. In many cases, although CFD analysis is the most scientific and comprehensive approach, using different CFD packages for two-phase flow predictions can result in different predictions of the fluid flow and heat transfer behavior due to the different constitutive relations used in these packages. It is, therefore, left to the discernment of the user to evaluate the validity of the results. These computational tools are not discussed in this chapter.

Two-phase flow analyses for gas/solid and gas/liquid systems have relied heavily on correlations, and these correlations, in turn, are predominately based on experiments. In some cases, theory guides the chosen form of the correlations. In other cases, the form of the correlation chosen to represent multiphase flow phenomena is a matter of convenience, but the most important aspect of the use of multiphase fluid flow and heat transfer correlations is the understanding of the limitations and ranges of applicability of the correlations. This is especially important in the use of two-phase flow correlations for the design of process systems that rely on two-phase flows. In some cases, complete ranges of validity are given with the correlations and, in some cases, limitations are incomplete. Correlation validity is also influenced by factors that are sometimes not reported with the correlation, such

© Springer Nature Switzerland AG 2022
H. Arastoopour et al., *Transport Phenomena in Multiphase Systems*, Mechanical Engineering Series, https://doi.org/10.1007/978-3-030-68578-2_3

as pipe diameter limitations and pipe roughness. Discernment and comparison of correlation predictions is left to the design engineer.

3.2   Gas/Solid Flows

3.2.1   Introduction to Gas/Solid Flows

Gas/solids flows are applied extensively in industry for pneumatic transport of solids; combustion or gasification of solid fuels, waste or biomass; carbon capture; pharmaceutical processes; and gas phase catalytic chemical reactions. The most important flow regime is fluidization of solid particles in a column bed of particles by upward flow of gas through the bed. Only particles of a certain range of sizes and densities can be fluidized and, if particles are too small, it is difficult to achieve stable fluidization (Group C particles). If particles are too big, they cannot be fluidized at all (Group D particles).

When convenient, small and low-density particles (powders) in bulk are best transported by gas flows in pipes. This eliminates the problems caused by spreading of dust when trying to move the solid by other means. Static electricity is a problem with combustible materials, however, and is the first consideration in the design of air transport of powders in industry.

Larger particles (like peas) cannot be fluidized, and upward flow of gas through a column of large particles will sometimes circulate the particles downward toward the edge of the pipe. This is called spouting and is occasionally applied in industry to dry or coat agricultural products and other materials. Mass and heat transfer are not optimized with spouting, however, and the process is not as important as fluidization.

3.2.2   Fluidization Concepts and Flow Regimes

By far, the most important industrial process involving flow of a gas through a column of particles is fluidization. Although there are different regimes of fluidization [1], the "fluidized bed" generally behaves, in a sense, like a liquid phase. Figure 3.1 shows schematic diagram of a typical fluidized bed. As gas is passed upward through the column of particles, the bed expands and the height of the bed increases, and, at certain gas speed (minimum fluidization velocity, U_{mf}), the forces of drag, buoyancy, and gravity are in equilibrium. When the bed is fluidized, the pressure drop across the fluidized bed remains almost constant and independent of the gas speed above the minimum fluidization velocity for non-cohesive particles. A fluidized bed thus will achieve hydrostatic equilibrium with another fluidized bed connected to it. This property finds industrial application in fluidized combustion and catalytic processes.

Fig. 3.1 Schematic diagram of a typical fluidized bed

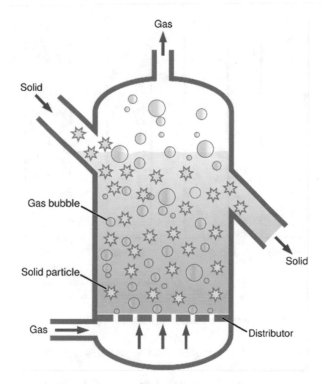

A remarkable property of the fluidized bed is an approximately uniform temperature profile when heat is generated in the bed by a reaction such as combustion. The reason is the fluidized bed exhibits very high thermal conductivity. Thus, some exothermic reactions can be carried out on an industrial scale in a fluidized bed that would be impossible on a large scale by other means of gas/solid contacting. The heat generated by such reactions can be integrated for use in endothermic reactions and to make process steam.

Figure 3.2 shows different gas/solid flow regimes. It can be seen that gas will begin to expand the packed bed of particles in the bed in Case (a) until it reaches to minimum fluidization; the weight of the particles will be balanced by total drag exerted to the particles in the bed ($U = U_{mf}$), where U is superficial fluid velocity. Case (b) is called bubbling fluidized bed, in which the bubbles are formed at relatively low gas velocities ($U > U_{mf}$). Case (c) is called fluidized at slugging regime, in which the bubbles' size is becoming equal to the diameter of the pipe, their rise velocity is controlled by the size of the bed diameter, and the bubbles become slugs of gas ($U > U_{mf}$). Case (d) is called turbulent regime of the fluidized bed, in which the flow is chaotic and is referred to as the transition from bubbling or slugging to pneumatic transport ($U \gg U_{mf}$). Case (e) is called fast fluidization regime in which the gas velocity is even higher than U_t, the terminal velocity of the individual particles ($U > Ut$), and particles form clusters. Case (f) is called pneumatic transport in which the gas velocity is much higher than the terminal

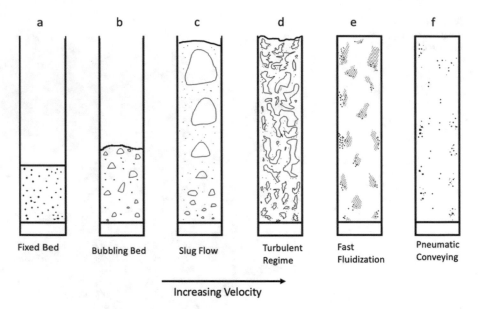

Fig. 3.2 Different fluidization regimes

velocity of the particles and the particles mainly move in a dispersed suspended manner ($U > U_t$).

On the other hand, liquid will smoothly fluidize the particles, and this smooth fluidization is quite independent of the liquid upward velocity through the bed due to the smaller difference between solid and liquid densities. Momentum and heat transfer of the bubbles can be analyzed, however, and the analysis becomes quite complex. The ratio,

$$(\rho_s - \rho_f)/\rho_f \tag{3.1}$$

where ρ_f and ρ_s are the density of the fluid and solid, respectively, can be used as an approximation to help discern the two types of fluidization that can appear. When the ratio is <10, fluidization is generally particulate and when the ratio is >10, fluidization is generally prone to bubbling. The nature of the bubbling regimes depends on the density and the size of the particles and the velocity of the fluid upward through the bed, as well as the column diameter, and has been studied extensively.

3.2.3 Geldart Particle Classification

Solid particles that can be fluidized have been classified by Geldart [2] according to (equivalent) particle diameter and density. Figure 3.3 shows the Geldart particle classification diagram. Geldart Group A particles (aeratable) have typical densities

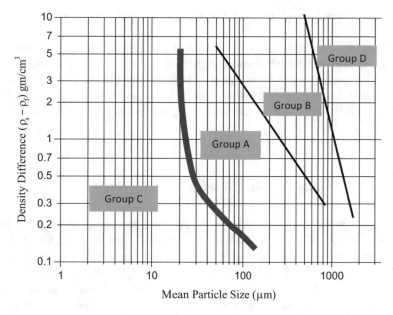

Fig. 3.3 Geldart particle classification diagram. (This figure was published in [2] and has been redrawn and reused here)

of 1.5–3 g/cm^3 and are most easily stably fluidized. The Group A particle diameters and densities do not promote large air bubbles in the fluidized bed, the particles can be transported pneumatically easily, and of solid materials that can be fluidized, they are the most widely prepared particle sizes for fluidized industrial purposes such as catalysts. The Group B or "sand like" particles tend to be larger in size than the A particles, and beds of these size particles will form gas bubbles easily—such that the minimum fluidization velocity is close to the minimum velocity needed for bubble formation. Slugging (Fig. 3.2) is likely with Group B particles at higher gas velocities, especially in smaller diameter beds, when wall friction can stabilize a gas bubble. Fluidized bed processes for combustion and gasification of biomass, waste, or coal usually involve Geldart Group B size particles, although biomass particles such as ground rice hulls are less dense than coal and can be fluidized for combustion or gasification as Group A particles. Group C particles are cohesive, insofar as the particles are so small with high surface area that electrostatic and van der Waals interactions cause the particles to agglomerate, and actual fluidization occurs with clusters of particles of larger size. Such fine particles are not easily fluidized, and particle drifting due to the high buoyancy is likely even from natural convection. Channeling of the fluid through a bed of very fine particles is likely and special designs of distributors are needed. Recent interest in "nanoparticles" has led to research in fluidization regimes of Geldart Group C particles for a variety of applications. Geldart Group D particles are too large, generally, to be fluidized—insofar as friction and drag cannot be overcome by buoyancy forces to fluidize them. Beds of such large particles will generally operate in spouting regime—meaning that fluid will channel through the center of the bed and particles move down near the walls.

Geldart assigned the classes A, B, C, and D as "letter grades" of particles of various densities and diameters according to the ease of achieving true fluidization of particles in a packed bed. It is also noted that other factors, such as surface properties, physical and chemical properties of particles, and particle shapes can influence fluidization behavior, in addition to average "particle diameter" and particle density shown in Fig. 3.3.

3.2.4   Standpipes, Non-mechanical Valves, and Cyclones

A standpipe connects two fluidized beds, one bed positioned above the other, to move solid particles by gravity from the bed at higher elevation and lower pressure to the fluidized bed below at higher pressure and lower elevation. The force of gravity overcomes the difference in pressure between the two fluidized beds. If the gas from the lower bed is to be prevented from entering the elevated bed, then the solid particles through the standpipe must move through the standpipe faster than the gas moves.

The standpipe may be operated with the particles fluidized in the standpipe or with the particles in a packed bed state. To achieve fluidization, the mean fluidization velocity of the particles in the bed must be smaller than the difference between the downward particle speed and the upward gas speed. If the upward gas speed exceeds the downward particle speed, then the minimum fluidization velocity must be smaller than the difference between the gas and particle speed for the particles to remain fluidized. In the case in which the gas and particle motion are both downward through the standpipe, the minimum fluidization velocity must remain greater than the difference in gas speed and particle speed. In such a case, there will be a pressure seal to the upper bed to prevent gas transfer from the lower to the upper bed. To maintain a fluidized state of the particles in the standpipe, gas is often injected into the standpipe. Careful consideration must be given to the direction and magnitude of gas flow into the standpipe, as well as the classification of particles according to Geldart. Particle bubbling is also possible in a standpipe, which inhibits the particle flow through the standpipe. Sometimes conditions for bubbling in the standpipe are unavoidable, but the entrance of the particles to the standpipe should avoid bubbles that are in a fluidized bed.

Particles such as catalysts and pulverized coal are frequently circulated between beds. For example, a fluid catalytic cracker used for cracking residual petroleum in a gas phase to lighter gasoline fractions of petroleum using zeolite catalysts regenerates the catalyst by burning off the coke that accumulates on the catalyst in a second fluidized bed that uses the heat generated by the cracking process. Control of the catalyst particle flow between beds is achieved with valves, and the valves can be mechanical or non-mechanical. A non-mechanical valve operates by controlling the gas pressure difference between fluidized beds.

Fig. 3.4 Schematic
diagram of stand pipe and L
valve for solid transfer from
low pressure to high
pressure fluidized beds

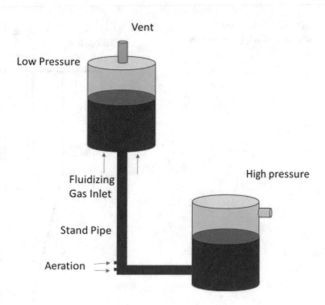

The "L" shaped portion of the standpipe in Fig. 3.4 can function as a non-mechanical valve to control solid particle flow between the beds [3, 4]. When used as a non-mechanical valve, additional gas can be injected into it. The valve serves to start or stop particle flow from one bed to the other.

Since Geldart Group A particles are easily fluidized, the flow of particles of this size range is not easily controlled by the non-mechanical valve because fluidized particles flow just like a fluid and the valve cannot control the particles flowing through it to stop them, no matter how much additional gas is injected into the valve. Thus, the non-mechanical valve operates best with larger sized particles when control of the particles is needed.

Gas exiting a fluidized bed usually entrains particles, and a cyclone is typically used to separate the particles from the bed. The most common type of cyclone used in conjunction with a fluidized bed is a reverse-flow cyclone. Figure 3.5 shows a schematic diagram of a reverse flow cyclone.

Air with entrained particles enters the cyclone, which undergoes tangential motion through the conical section of the cyclone and separates solid particles by means of the added centrifugal force of the motion [5]. The flow of particles and gas entering the cyclone is downward. Larger particles are returned to the fluidized bed, smaller particles (dust) may be collected in filter bags. The flow of gas through the cyclone is reversed, however, because the upward momentum of the gas exceeds the downward momentum carried by the particles, hence, the term "reverse flow." One cyclone separator is typically not enough for a large fluidized bed, and a train of three cyclones is typically employed. Very small particles are typically not captured by the cyclones, however, and other methods must be employed to capture this dust. In addition to the cost of the lost solid material, air pollution from particulate dust must also be considered. There are stringent controls of dust released to the atmosphere in

Fig. 3.5 Schematic diagram of a typical reverse flow cyclone

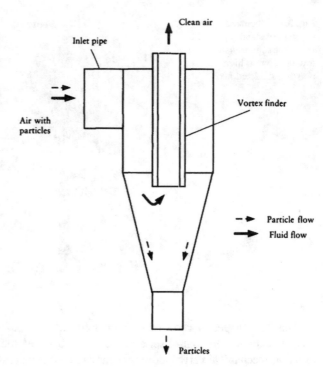

an industrial operation and regulations are based on particle sizes. Thus, in many cases, electrostatic particle separators and other means are employed.

3.3 Gas/Liquid Flows

3.3.1 Overview and Fundamental Relations of Gas/Liquid Two-Phase Flows

Gas/liquid flows are a very important class of two-phase flow because they are present in virtually all chemical and petroleum processes, as well as heating, ventilation, and air conditioning (HVAC) systems that involve steam and water, vapor compression cycle refrigeration, humidification, and other processes. The basic consideration of conveyance of a two-phase mixture is often difficult to analyze. Because gases or vapors cannot be pumped, the transport of a two-phase mixture from one point to another in pipes is usually possible only by vapor compression, system pressures generated in boilers, or by gravity. Calculating the amount of gas and liquid material of a two-phase mixture transported from one point to another in a pipe requires knowledge of the state of the mixture at each point of the pipe, which, in turn, requires knowledge of the pressure drop in the pipe and any heat transfer that may be present. A complete theory to accomplish this is not available at

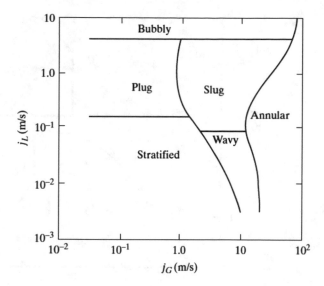

Fig. 3.6 Mandhane-Gregory-Aziz's map for horizontal co-current gas liquid flow [7], redrawn in SI units [8]

this time, and, in any analysis of simultaneous gas/liquid flows, various assumptions are made to make analysis at all possible. A fundamental approach and multiphase flow equations are carefully considered in Govier and Aziz [6].

Flow regimes are different for horizontal and vertical two-phase flow, and these are different for flows of the phases in different directions; that is, co-current, counter-current, upward, or downward. Flow "regimes" refer to the state of the gas/liquid mixture at particular speeds of the phases in the pipes. For example, gas and water flowing co-currently in a long horizontal pipe can have a stratified configuration at low gas and liquid speeds, where the liquid flows co-currently along the bottom of the pipe, and the gas moves co-currently on top of the liquid. At higher gas velocities, waves that form at the interface can have a strong effect on the pressure drop. At high liquid rates and low gas rates, long bubbles form at the top of the pipe. At high liquid rates and high gas rates, a pattern is exhibited whereby slugs of highly aerated liquid move down the pipeline at the gas speed. This flow regime is usually avoided in practice because it results in vibrations that can sometimes burst pipes. At very high gas speeds, an annular pattern is observed whereby part of the liquid flows along the pipe as a film and part as droplets entrained in the gas flow. In the annular flow regime, there is an exchange of liquid between the wall layer and the drops in the gas core. At very high gas speeds, a homogenous bubbly or foam pattern exists. At such large gas speeds, the two-phase flow may be analyzed as a single-phase flow (sometimes referred to as "high Reynolds number flow"). Flow in the other regimes is said to be "dispersed."

The different flow regimes of two-phase flows in pipes are typically summarized in "flow maps" with coordinates usually related to gas and liquid speeds. Two such flow maps are presented in Figs. 3.6 and 3.7.

The coordinate variables on these maps are related to fluid speeds and are defined in the next section. Before providing a simple analysis of gas/liquid flow regimes of

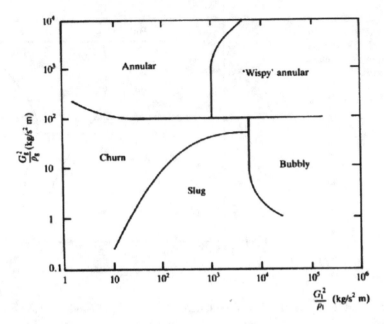

Fig. 3.7 Hewitt and Roberts's map for vertical co-current up-flow [9], redrawn in SI units [10]

dispersed flows, some differences between flows of single and two phases together are discussed.

3.3.2 Some Differences Between Gas/Liquid Flows and Single-Phase Fluid Flows

Single-phase fluid flow is considered incompressible for flow speeds much below the Mach number of the flow (ratio of fluid speed to speed of sound in the fluid). As the flow of a single phase in a pipe approaches the speed of sound, the fluid density is not constant (independent of position or time in the case of a transient flow). When a fluid of a single phase flows through a pipe of constant cross-sectional area and without the possibility of heat transfer, the pressure decreases along the length of the pipe as a result of friction and the specific volume of the fluid increases because the pressure decreases, up to a point of zero entropy change in the fluid. It is not possible for the fluid to flow in a pipe of constant cross-sectional area faster than the speed of sound in the fluid (choked flow). At a fixed temperature, the speed of sound in a single-phase fluid depends on only the fluid density and the elastic modulus. The situation is more complicated for gas/liquid flows and flow choking is difficult to characterize. In the simple analysis of averaging Mach numbers of each phase of a gas-liquid flow, the average Mach number of the two-phase mixture depends on the ratio of the density of gas to liquid phases. This approximation results in anomalously low sound speeds in gas/liquid mixtures when the liquid density is much larger than the gas density. For example, for a water volume fraction of 0.5 in an air/water dispersed mixture, the sound speed is approximated as 20 m/s at 20 °C (compared with 343 m/s in air at 20 °C [11]).

Another distinguishing feature of two-phase gas/liquid flow is the dependence of the flow regimes on the liquid surface tension. Fluid flow of a single phase exhibits only three regimes: laminar, turbulent, and transition between laminar and turbulent. The transitions depend on only the Reynolds number of the fluid flow which, in turn, depends only on the kinematic viscosity of the fluid. The appearance of waves and droplets in two-phase gas/liquid flow depends on the liquid surface tension, which characterizes the tendency of the liquid to coalesce. Two-phase flow regime maps are often parameterized by the liquid surface tension (typically referenced to the surface tension of water).

3.3.3   Modeling Gas/Liquid Flow in Pipes

Gas/liquid flows are modeled to predict flow rates and pressure drops in pipes. Typically, this is done to determine the pipe diameter needed to transport a given mass flow rate of liquid and vapor through the pipe. There are two types of approaches: first, simple models that treat two-phase flow as a single "phase," either with averaged velocities and fluid properties, or completely separated phases with gas and liquid flows treated independently. Second, there are models that account for gas and liquid interactions and the interactions are an important part of the analysis. The primary interaction is "slip," which is the relative velocity of the gas to the liquid in the flow. Slip is correlated as a function of the flow variables and fluid properties by both theory and experiment.

Two-phase fluid flows are treated either as adiabatic and allowing no heat transfer from the pipe to the surroundings, or diabatic and heat transfer is accounted for. In the diabatic case, phase change is possible within the pipe, but "local" equilibrium is usually assumed, to relate equilibrium vapor pressures to the enthalpy of vaporization. The non-equilibrium condition of evaporation is difficult to characterize.

Steady-flow mass conservation and mechanical conservation are relatively easy to express for two-phase gas/liquid flow in a pipe of constant diameter (in the case of constant gas phase density), but are rather difficult to analyze in practical situations.

3.3.4   Flow Regimes and Flow Maps of Two-Phase Flows

3.3.4.1   Overview of Various Flow Maps

As noted in Sect. 3.3.3, calculations of pressure drop and flow rates of gas/liquid mixtures generally rely on correlations involving the physical and transport properties of the two fluids, under various initial conditions of gas and liquid flow rates and pipe positions: horizontal, vertical, or tilted, and a particular given direction of initial flow of each phase. Under some circumstances, the two-phase flow may be considered incompressible and, under this condition, the volumetric flow in a pipe may be considered constant. The assumption cannot hold when choked flow conditions are approached, and, in most cases, flow choking cannot be treated in the same manner as the flow of a single phase in a pipe. Moreover, it cannot be assumed that vapor cannot condense, or liquid cannot evaporate, when a two-phase mixture approaches critical flow at high speed, even under the assumption of adiabatic flow.

The numerous possible flow regimes of two-phase gas/liquid flow in a pipe are considerably more complicated in two-phase flows that admit the possibility of heat transfer to the surroundings (diabatic flow) than in the case of adiabatic flow regimes. The onset of flow boiling or condensation in the case of diabatic flow is particularly difficult to predict, as well as the situation of sudden dry-out in two-phase pipe flow under the possibility of heat transfer to the surroundings. A sudden dry-out in a pipe carrying a two-phase mixture can present a significant hazard in many industrial applications, such as when a flowing fluid or two-phase mixture must remove heat from a source (such as a boiler) or, otherwise, burn up the pipes or the system. The possibility must be carefully considered by design engineers.

Correlations of two-phase flow pressure changes and momentum transfer in pipes do not provide a picture of the flow situation in pipes, and, for this two-phase flow pattern, maps have been developed for numerous two-phase gas/liquid flows in many flow situations and pipe configurations such as horizontal, vertical, and tilted. These pattern maps should be consulted in any design configuration involving two-phase gas and liquid flow, in addition to the correlations, because it is important to understand what flow pattern behavior can be exhibited under specified conditions.

Flow pattern maps generally present regimes of two-phase flow under steady flow conditions and away from the transitions that must be present in practical piping applications, such as the entrance and exits, elbows and tees, and pipe expansion and contraction. Special considerations must be given to these fittings and every pipe segment.

High-speed digital computer processing has made two-phase flow visualization possible for many types of flow situations, and there are many bases chosen for these simulations. These provide no understanding of flow phenomena and cannot always be accepted as reproductions of reality. That is why flow pattern maps need to be consulted in engineering practice.

Flow pattern maps provide the foundation for correlations and calculations of two-phase flows, and especially for two-phase flow accompanied by heat transfer. This is why two-phase flow calculation methods that account for possible structuring of two phases are considered more reliable than calculations that do not account for these flow pattern maps.

The development of flow pattern maps began with observations of gas/liquid (usually air/water) flows in pipes of typical sizes, about 2–4 cm in diameter, and these were parameterized to account for pipe roughness and, in some cases, transitions. Many review articles appeared in the 1970s on flow pattern maps covering a variety of flow conditions and geometry. Since the 1990s, there has been considerable interest in two-phase flow with "microscale" channels, which have been assigned a hydraulic diameter smaller than 3 mm and built with hydraulic diameters of the order of 3–300 μm. The interest was inspired by the tremendous heat transfer possibly achieved with the application of microscale channel diameter heat exchangers, as well as chemical reactors, and has led to a distinct category of two-phase flow pattern maps. Within such microchannels, surface tension forces typically dominate all other forces acting upon the liquid phase. Because heat transfer rates are high in microchannels, mass transfer rates are also high, and fouling is a serious obstacle.

The now "classical" flow pattern maps of two-phase gas/liquid flows in channels of macroscopic hydraulic diameters fall into two categories: empirical and semi-empirical, which has some theoretical foundation. Both types have been prepared for co-current flow in horizontal and vertical channels. Empirical flow maps are available for inclined channels. All of these maps find use in steam power plant design. Few flow maps for counter-current flows of gas and liquid have been prepared. Counter-current flow in a vertical channel of gas rising against a falling liquid would encounter flooding, or hold-up, of the weight of the liquid at some upward gas velocity.

Well-known experimental flow pattern maps include the Baker map [12] of horizontal co-current gas/liquid flow, and the Hewitt and Roberts map of vertical upward co-current flow. Both involve the gas and liquid mass velocities GG and GL $(\text{kg/m}^2 \text{ s})$. The Hewitt and Roberts map parameterizes the fluid densities, the Baker map normalizes the liquid properties to the viscosity and surface tension of water. The Baker map was prepared with data measured on flowing hydrocarbon fluids.

One of the first theoretical treatments of two-phase flow and the development of a pattern map based on this was prepared by Taitel and Dukler [13] for horizontal gas/liquid co-current flow and can be compared with the Baker map.

3.3.4.2   Classical Flow Maps of Two-Phase Regimes: A Summary of Applications and Limitations

Maps that provide graphical interpretation of two-phase fluid flow regimes have been prepared mostly for adiabatic flows of a liquid together with an immiscible gas in co-current flow conditions for horizontal tubes, and for channels of macroscopic (6–30 mm) dimensions. Fewer have been prepared for vertical upward or downward flow of the phases or at various angles with respect to the horizontal tubes, or in channels of "micro" dimensions (<3 mm). Some have interpreted observed regimes with theory of momentum transfer such as the mechanistic model of Taitel and Dukler [13].

Fewer two-phase flow maps have been prepared under conditions that allow for heat transfer to the surroundings (diabatic flow), with attempts to correlate convective heat transfer. Diabatic two-phase flow pattern maps were prepared from

Fig. 3.8 Schematic diagram of flow patterns appearing in two-phase flow of gas and liquid for co-current flows in a vertical channel. (This figure was published in [14] and has been reused here)

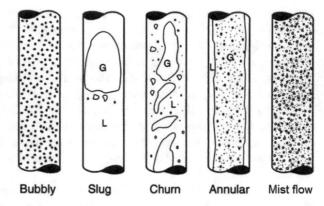

Bubbly   Slug   Churn   Annular   Mist flow

experimental studies of refrigerants and water/steam in the case of co-current flows in horizontal channels of both micro- and macro-scales. Attempts to correlate heat transfer have been limited and are not sufficient to design a microchannel heat exchanger that could replace a heat exchanger with channels of macroscopic dimensions. The improved heat transfer in microchannels offers the possibility of reducing size and weight of heat exchangers in many different applications.

Figure 3.8 shows a schematic diagram of flow patterns appearing in two-phase flow of gas and liquid for co-current flows in a vertical channel. In vertical channels, at low gas concentrations, the flow is bubbly and, as the gas velocity increases, the bubbles become dispersed. At larger void fraction of the gas phase, slugs of liquid appear, resulting in fluctuation of pressure. When the liquid forms near or at the wall, it can fall by gravity, even though the net upward flow is positive. Churn flow is the most unstable of the flow regimes at high gas upward velocities. Resulting chaotic behavior of a slug of liquid is difficult to predict and is a potentially hazardous condition in two-phase flows that must remove heat from something hot, such as a nuclear reactor or boiler. Annular flow appears at high gas velocities and heat transfer, if present, must result from evaporation of the film near the wall. Most heat transfer studies of the two-phase flow regime involve calculation of the possible heat transfer and/or evaporation of the annular film. Finally, at very high gas velocities and void fraction, a mist of the annular film at the wall is entrained in the vapor phase, and a very hot channel wall results.

The different vertical flow regimes are summarized in the Hewitt and Roberts flow pattern map for vertical upward flow, verified for air/water mixtures, and steam and liquid water (see Fig. 3.9). It is a widely consulted map in boiler and reactor design.

Patterns of horizontal flows are examined similarly to patterns of vertical flows. Figure 3.10 shows a schematic diagram of flow patterns appearing in two-phase flow of gas and liquid for co-current flows in a horizontal channel. At low speeds of gas, gravity will keep the liquid at the bottom of the channel, resulting in a stratified flow. The gravitational force acting on the liquid phase can be overcome by kinetic forces at high flow rates, causing stratified flows to revert to annular flows. At very high

Fig. 3.9 The different vertical flow regimes (Hewitt and Roberts flow pattern map [9]) for vertical upward flow

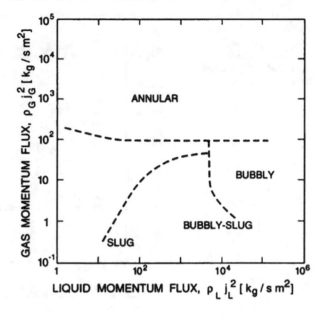

Fig. 3.10 Schematic diagram of flow patterns appearing in two-phase flow of gas and liquid for co-current flows in a horizontal channel. (This figure was published in [14] and has been reused here)

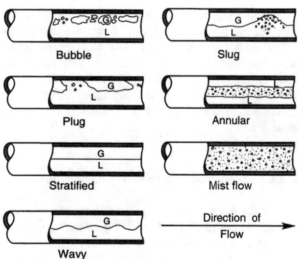

flow rates, the annular film is thinned by the shear of the vapor core and all the liquid is entrained as droplets in the vapor phase. This results in a mist flow.

Maps of the patterns of horizontal lows have been derived by many investigators and have been given theoretical treatments by numbers of authors. Mandhane et al. [7] present one such flow pattern map, as a function of the gas and liquid volumetric fluxes j_G and j_L (total mass flow of gas and liquid per fluid density per cross-sectional area of channel). Figure 3.11 shows maps of the patterns of horizontal two-phase flow of gas and liquid.

Fig. 3.11 The different vertical flow regimes for horizontal flow of gas and liquid. (This figure was published in [8] and has been reused here)

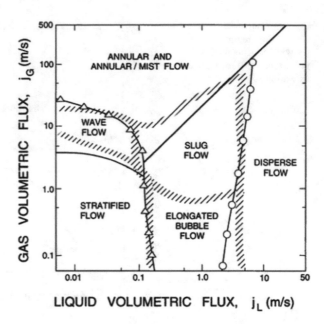

In both cases of vertical and horizontal gas/liquid flows, regions of flow instability are of particular interest, in that pressure fluctuations can be severe, resulting in unpredictable gas and liquid flow rates at particular lengths of channel. Such instabilities would appear at transitions from one regime to another, and especially so in the region of slug flow of gas and liquid, for coalescence and break-up of larger bubbles in the channel. Such instabilities and their prediction are important in the evaluation of critical heat flux in tubes that carry water and steam heated in a reactor core.

3.3.4.3   Industrial Application of Flow Maps

By far the most important application to date of two-phase flow studies has been nuclear reactor studies and design for safety, and, in particular, modeling of a loss of coolant accident. Most of the two-phase flow map studies as referenced in the previous section have been analyzed under an assumption of a constant void fraction and constant superficial velocities of two-phases. The most important consideration in nuclear reactor modeling and design is evaporation of the liquid and non-constant void fraction.

Considering the case of subcooled water entering by external pressure inside a vertical tube subjected to a constant heat flux (such as inside a nuclear reactor or furnace), bubbles begin to appear as the water begins to boil, and, as the boiling increases, the bubbles coalesce to form slugs of liquid that are churning as the bulk of the liquid rises. As more of the liquid vaporizes to steam, an annular region will form, still with a film of liquid on the side of the tube. Up to this point, the tube wall will remain at a temperature consistent with the saturation temperature according to the pressure drop arising from friction in the liquid phase. The critical point reached is dry-out of the liquid film, at which point the tube wall temperature can rise sharply. The vapor will be a mist of steam from the evaporated thin film, and, finally, vapor flow.

In the case of horizontal flow, similar boiling patterns are observed as in the case of vertical flow, except the liquid will tend to stratify along the bottom of the tube due to gravity, at low liquid speeds. At higher liquid speeds, the stratification can disappear. As in the case of vertical flow, there is a critical point along the tube at which a film of liquid can dry out and the tube wall temperature will rise sharply. For a given length of tube that carries the water/steam mixture, the point (or region of a tube) where this will occur is a desired calculation objective.

A use of the flow-pattern maps of horizontal and vertical flows of gas/liquid mixtures would be the evaluation of relative speeds of liquid and vapor at particular points along the tube to examine the type of flow to be expected at a particular location. For water entering into a vertical or horizontal tube as a subcooled liquid, the amount of heat flux needed to boil the water at a given length could be calculated. When saturation is achieved, bubbles begin to form; the void fraction of bubbles is related to the amount of water evaporated. The particular region of flow of the two-phase mixture is known from a two-phase flow pattern map related to the relative speeds of vapor and liquid.

Carrying out such a calculation involves evaluation of the pressure drop of the two phase mixture flowing in the tube. Calculation of frictional pressure drop of the flow of a fluid of single phase is well known. Calculation of the frictional pressure drop of a two-phase flow is considerably more complicated, and the frictional pressure drop of a two-phase flow is typically much larger than the pressure drop of a single-phase flow (at constant mass flow), apparently the result of additional friction due to bubbles.

The total pressure drop of the two-phase flow is the result of the static head, the momentum head of the fluid acceleration, and the friction head. Static and momentum head are evaluated in the same manner as a single-phase flow under certain assumptions, such as the approximation of a void-fraction averaged density of fluid mixture.

The frictional pressure drop can be evaluated by the application of a two-phase flow multiplier to a single-phase flow pressure drop,

$$(dp/dx)_{\text{two phase}} = \Phi^2 (dp/dx)_{\text{single phase}} \tag{3.2}$$

The Φ^2 factor can be evaluated by various methods, such as the Martinelli-Nelson [15] correlation, which is a function of mixture vapor quality, x. Figure 3.12 shows the variation of factor Φ versus pressure at different vapor quality.

The pressure drop for two-phase flow of gas and liquid flows may be estimated by the product of Φ^2 from Fig. 3.12 and the equivalent value from a single-phase flow pressure drop.

Fig. 3.12 The variation of factor Φ versus pressure at different vapor quality. (This figure was published in [15] and has been reused here)

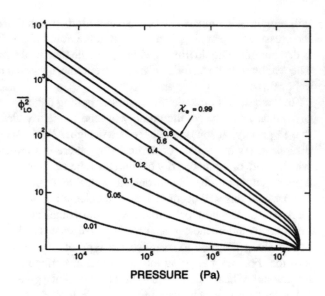

3.4 Exercises

3.4.1 Ex. 1: Minimum Fluidization Condition

Particles having a size of 0.10 mm, a sphericity of 0.86, and a density of 1200 kg/m^3 are to be fluidized using air at 25 °C and 202.65 kPa abs pressure. The void fraction at minimum fluidizing conditions is 0.43. The bed diameter is 0.60 m and the bed contains 350 kg of solids. Calculate the minimum height of the fluidized bed, the pressure drop at minimum fluidizing conditions, and the minimum velocity for fluidization.

3.4.2 Ex. 2: Particle Terminal Velocity

A sphere of density 2500 kg/m^3 falls freely under gravity in a fluid of density 700 kg/m^3 and viscosity of 0.5×10^{-3} Pa s. Given that the terminal velocity of the sphere is 0.15 m/s, calculate its diameter.

3.4.3 Ex. 3: Gas/Liquid Flow and Flow Regimes

Apply a multiphase CFD modeling software package, such as ANSYS Fluent [16], to study the two-dimensional, isothermal, horizontal, co-current flow of an air/water system at 25 °C in a 2 cm diameter glass tube. Begin the simulation at $j_g = 0.1$,

$j_l = 1.0$ m/s, then with j_l fixed, increase j_g from 0.1 to 100 m/s and verify the transitions from plug to annular flow as given in Fig. 3.6.

References

1. Smolders K, Baeyens J (2001) Gas fluidized beds operating at high velocities: a critical review of occurring regimes. Powder Technol 119:269–291
2. Geldart D (1973) Types of gas fluidization. Powder technol 93:285–292
3. Yang WC, Knowlton TM (1993) L-valve equations. Powder Technol 77(141):49–54
4. Abed R, Reeves J, Knowlton TM, Findlay J, Chan I (1995) L-valve feeding of a mixture of fine and coarse particles into a bubbling fluidized bed. Large JF, LaGuerie, C (eds) Proceedings of the eighth conference on fluidization, Engineering Conference International
5. Issangya AS, Karri SBR, Knowlton TM, Cocco R (2011) Effect of gas bypassing in deep beds on cyclone Dipleg Operation. Knowlton TM (ed) Proceedings of the tenth conference in circulating fluidized bed, Engineering Conference International, ISBN 978-1-4507-7082-5, pp 5–6
6. Govier GW, Aziz K (1972). The flow of complex mixtures in pipes. Van Nostrand Reinhold Company, New York
7. Mandhane JM, Gregory GA, Aziz, K (1974) A flow pattern map for gas-liquid flow in horizontal pipes. Int J Multiphase Flow 1(4):537–553
8. Ghiaasiaan, SM (2008) Two-phase flow, boiling, and condensation in conventional and miniature systems. Cambridge University Press, New York
9. Hewitt GF, Roberts DN (1969) Studies of two-phase flow patterns by simultaneous x-ray and flash photography. Atomic Energy Research Establishment, Harwell, UK
10. Whalley PB (1987) Boiling, condensation, and gas-liquid flow. Clarendon Press, Oxford
11. Shoham O (2006) Mechanistic modeling of gas-liquid two-phase flow in pipes. ISBN: 978-1-55563-107-9, Society of Petroleum Engineers, Richardson, TX
12. Baker O (1954) Simultaneous flow of oil and gas. Oil Gas J 53:185–195
13. Taitel Y, Dukler AE (1976) A model for predicting flow regime transitions in horizontal and near horizontal gas–liquid flow. AIChE J 22:47–55
14. Collier JG, Thome JR (1994) Convective boiling and condensation. Oxford University Press, New York
15. Martinelli RC, Nelson DB (1948) Prediction of pressure drops during forced circulation boiling of water. Transactions ASME 70(6): 695–702
16. ANSYS (2014) ANSYS Fluent theory guide. ANSYS Inc.

Chapter 4
Polymerization Process Intensification Using Circulating Fluidized Bed and Rotating Fluidized Bed Systems

Hamid Arastoopour, Dimitri Gidaspow, and Robert W. Lyczkowski

4.1 Introduction

Gas phase olefin polymerization is now widely achieved in fluidized bed reactors. In fluidized bed reactors, small catalyst particles (20–80 μm diameter) are introduced into the bed, and, when exposed to the gas flow (monomer), polymerization occurs. At early stages of polymerization, the catalyst particles fragment into a large number of small particles, and then the polymer particles grow continuously reaching a typical size of 1000–3000 μm diameter. The gas phase monomer diffuses through the boundary layer around the catalyst particle and through its pores to reach the active sites where the polymerization takes place. In a few seconds, the pores of the catalyst support will fill up with the polymer. The support then ruptures into many fragments, often called micrograins, microparticles, or primary particles. Once the particles fragment, the volume of polymer inside the particle continues to grow as the monomer diffuses to the active sites. This process of expansion continues until the polymer particles exit the reactor, at which point they would have reached about 2000 μm diameter. Because olefin polymerization is highly exothermic, the heat generation rate can be very high inside the particles, which may lead to the appearance of hot spots. This occurs when the heat of polymerization cannot be efficiently removed through the porous polymer matrix and particle boundary layer to the bulk phase of the reactor. One of the major advantages of the fluidized bed reactor process is that the solid polymer particles are vigorously mixed in the reactor by the fast flowing fluidizing gas, and thus the heat of reaction can be removed more effectively [1, 2]. Heat transfer issues should be addressed effectively to maintain a reasonably high production rate and prevent agglomeration.

A successful, innovative design analysis of this process should not only account for the heat removal and kinetics of polymerization, but also include the particle mixing and particle size distribution in the reactor. Despite significant application of fluidized bed polymerization reactors, they have shown limited flexibility in achieving high gas throughput mainly because of the heat removal limitation. To decrease

© Springer Nature Switzerland AG 2022
H. Arastoopour et al., *Transport Phenomena in Multiphase Systems*, Mechanical Engineering Series, https://doi.org/10.1007/978-3-030-68578-2_4

the very high temperature rise in fluidized bed polymerization reactors such as UNIPOL (the Union Carbide polymerization process) and intensify the process such that the polymer production rate increases significantly, two different process concepts are discussed in this chapter. This chapter investigates the feasibility of using circulating fluidized bed (CFB) and rotating fluidized bed [3–7] polymerization reactors for increasing the polymer production rate using the computational fluid dynamics (CFD) approach.

4.2 Circulating Fluidized Bed (CFB) Reactor for Polymerization

4.2.1 Introduction

This section is an extension of Gidaspow's patent relating to fluidized bed polymerization reactors [8], such as the UNIPOL reactor, or the later reactors containing cooling tubes. It shows that the polymer production rate can be increased by two or more orders of magnitude over that obtained in the UNIPOL reactor due to the orders of magnitude decrease in the adiabatic temperature rise. The CFD design of various-sized polyethylene reactors is described. In all cases, the small catalyst particle concentrations are high at the walls, showing that electrostatics is not the only mechanism for sheet formation.

An excellent review of the production processes for polymerization reactors was given by Xie et al. [9]. The reactor pressure has been reduced from as high as 3000 atm (309 MPa) to as low as 8 atm (0.8 MPa) [10]. For gas phase polymerization reactors, a typical operating pressure is 25 atm (2.5 MPa) and temperature is above 60 °C. The reactor is approximately 15 m high, with a diameter of 5 m for an $L/D = 3$. The catalyst diameter is about 80 μm or smaller, with the product polymer withdrawn from the reactor having a diameter of 100–5000 μm diameter. The catalyst residue remains in the polymer, and therefore its activity must be large enough to produce more than 50 kg/g of catalyst. The high activity catalysts are described in U.S. Patent 4,255,542 [11].

The rate of reaction is proportional to the monomer concentration and the concentration of polymer sites that decrease with reaction time [12]. The activation energy is about 5 kcal/mol for ethylene polymerization by chromium oxide catalysts. A typical effective first order rate constant is as high as 8 s^{-1}. Hence, at near constant temperature for a gas velocity of 1 m/s and a rate constant of 1 s^{-1}, the reactor height should be 1 m for a good conversion. Heat removal restricts the conversion to only 2% for both the well-stirred and the uncooled fluidized bed bubbling reactors [13]. The heat of reaction is removed by recycling.

Two types of fluidized bed bubbling reactors are used today, both developed at Union Carbide in the 1970s [14]. The first was built with an expanded section on top and is the one most modeled today [15]. It suffers from severe sheet formation that forces frequent reactor shutdown [16]. The second type of reactor with internal cooling tubes modeled by Chen et al. [17] may solve the sheet formation problem. However their paper presents too few details to be useful for this purpose.

4.2.2 Description of the Process

Figure 4.1 shows a simplified UNIPOL reactor [18]. Ethylene, hydrogen, and catalyst are fed into this bubbling bed reactor. In this type of reactor, the temperature rise will be 1600 °C [15]. This high temperature rise restricts the conversion of ethylene to only 2%. Sheet formation often limits the operation time. In 2016, Kashyap modeled such a reactor for Fluidization XV [19]. Similar simulations since that time have shown that the catalyst concentration will be the highest at the walls of the reactor.

Figure 4.2 shows a typical rate of reaction for production of polyethylene based on the data of Yermakov and Zakharov [12]. The rate of reaction is proportional to the monomer and surface concentrations. The rate decreases slowly after initiation. For CFD simulations, this slow decrease cannot be included in the calculations due to the long time needed to perform the calculations, typically 1 day for a 20 s run. Hence, only one typical reaction rate was chosen. It is possible to include more than

Fig. 4.1 Simplified UNIPOL reactor. (This figure was originally published in [20] and has been reused with permission)

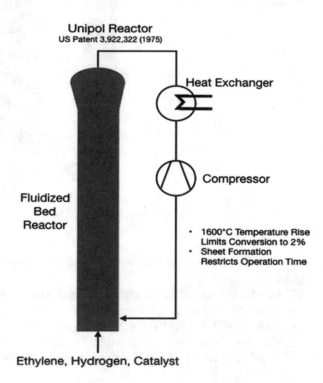

Unipol Reactor
US Patent 3,922,322 (1975)

Heat Exchanger

Compressor

Fluidized Bed Reactor

- 1600°C Temperature Rise Limits Conversion to 2%
- Sheet Formation Restricts Operation Time

Ethylene, Hydrogen, Catalyst

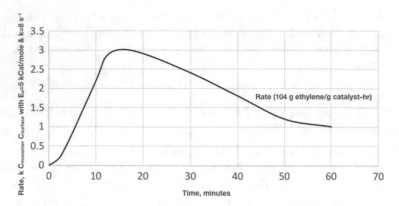

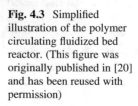

Fig. 4.2 Typical rate of reaction for polyethylene polymerization by chromium oxide catalyst on silicon oxide support at 80 °C. Catalyst activated in vacuum at 400 °C. (This figure was originally published in [20] and has been reused with permission)

Fig. 4.3 Simplified illustration of the polymer circulating fluidized bed reactor. (This figure was originally published in [20] and has been reused with permission)

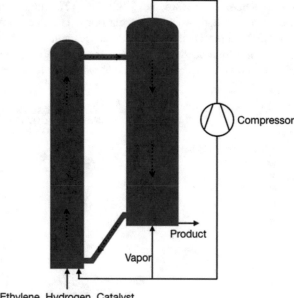

one such rate, a fast rate and a slow rate, to include the effect of partially deactivated catalyst.

Figure 4.3 is a simplified illustration of the polymer CFB reactor. It is the next-generation extension of the bubbling fluidized bed reactors shown in Figs. 1 and 2 in U.S. Patents 3,922,322 [14] and 4,255,542 [11]. In the proposed CFB reactor, the heat of reaction is removed by the circulating polymer particles rather than by the cooling tubes described in [11]. This major modification allows an order of magnitude larger polymer production per unit reactor volume and better temperature control.

The monomer, such as ethylene, with the diluent, such as hydrogen, together with suspended catalyst particles, enter the riser portion of the reactor and flow up the riser, as indicated in Fig. 4.3. To fluidize the particles in the riser, additional flow is supplied through the recycle system with the power produced by the compressor. From the riser, the polymer particles and the unreacted gas and the unreacted catalyst particles enter the downer. The hot particles, indicated by P in Fig. 4.3, are removed from the downer. They remove all of the heat of reaction. This invention removes the need for the cooling tubes in the older generation reactors. Gases are used to keep the downer fluidized. The portion of the polymer particles that is not removed is returned to the riser.

4.2.3 Steady State Energy Balance for a Fluidized Bed Riser

Multidimensional, multiphase energy balances are derived in Chap. 1 in Gidaspow's 1994 book [21]. The energy balance for a riser with conversion of monomer ΔY and heat of reaction ΔH is,

$$\rho_g v_g \varepsilon_g c_{pg} (T_{bed} - T_{in}) + \rho_s v_s \varepsilon_s c_{ps} (T_{bed} - T_{in}) = \rho_g v_g \varepsilon_g \Delta H \Delta \quad (4.1)$$

net rate of energy outflow + solids outflow = rate of energy generation

where the differences between the catalyst and the gas temperatures are small. This balance is similar to that used by Dhodakar et al. [13]. The main difference is that in the riser reactor, particles enter and leave the reactor. This difference is huge because the solids density, ρ_s, is of the order of magnitude of two to three times larger than the density of gas. Therefore the first term in the above equation can be neglected. Hence the rise in temperature becomes,

$$T_{bed} - T_{in} = ((\Delta H \Delta Y)/(c_{ps}\varepsilon_s)) \left(\rho_g/\rho_s\right) \left(v_g \varepsilon_g / v_s \varepsilon_s\right) \quad (4.2)$$

adiabatic riser temperature rise = conventional adiabatic temperature rise
× gas to solids density × slip ratio

The volume fractions and the velocities of the gas and the particles are of the same order of magnitude, but the ratio of the gas to solids densities is two to three orders of magnitude smaller. Hence the temperature rise in the riser is two to three orders of magnitude smaller than that in the stirred reactor or in the bubbling bed reactors. Instead of 800 °C in [13], the temperature rise is only of the order of 8 °C. Similarly, the temperature rise in the proposed design is reduced by two or more orders of magnitude from that of the UNIPOL bubbling fluidized bed. There the temperature rise is 1609 °C for ethylene and 850 °C for propylene [15].

4.2.4 High Production Rate

The proposed polymerization reactor, unlike the one shown in U.S. Patent 4,255,542 [11] assigned to Union Carbide and modeled by Chen et al. [17] does not need cooling tubes. The hot polymer particles withdrawn from the reactor remove all the heat of reaction. The steady state energy balance under adiabatic conditions is,

$$\Delta H \times \text{rate of reaction} = \text{polymer production rate} \times C_p \, \Delta T \qquad (4.3)$$

The polymer production rate is,

$$\text{polymer production} = A \, \rho_{monomer} v_{monomer} \Delta Y_{monomer} \qquad (4.4)$$

The polymer production rate equals the rate of catalyst injection described in [11], which states that the production rate can be controlled simply by increasing the rate of catalyst injection. This is true only when the temperature rise is small, as in the present invention. This patent also describes the catalysts and their method of preparation.

The low temperature rise in the riser allows the production rate to be increased by more than an order of magnitude. In the fluidized bed reactors used today, the production rate is limited by the low conversion per pass, approximately 2%. In Eq. (4.4), ΔY can be allowed to be nearly one, increasing the polymer production by 50 times for the same size reactor.

The CFD simulation for the polymerization reactor is similar to that published for the production of diesel fuel from synthesis gas [22] and to those in [21]. The method of solution and the computer code are described in [23].

4.2.5 CFD Design of a Large Ethylene Reactor

4.2.5.1 High Velocity

To prevent the formation of sheets at reactor walls due to electrostatics, it is best to have high wall velocities. This can be easily achieved using a Westinghouse-type reactor geometry, with a cone at the entrance of the reactor [23]. However, many preliminary CFD simulations have shown that the same effect can be achieved with a high velocity central jet entering a simple pipe reactor, with other jets supplying air to keep the particles fluidized [24]. Such a geometry is used as shown in Fig. 4.4, which shows the time-averaged axial gas velocity. The velocity of the central jet is about 1500 cm/s and the wall velocity is almost 10 m/s, providing a good sweeping effect. The jets contain gas having a mixture of 0.82 ethylene and 0.18 hydrogen weight fraction, respectively. The two entering solids are 0.1 cm diameter polymer particles with a volume fraction of 0.1, and 0.0008 cm catalyst particles with a

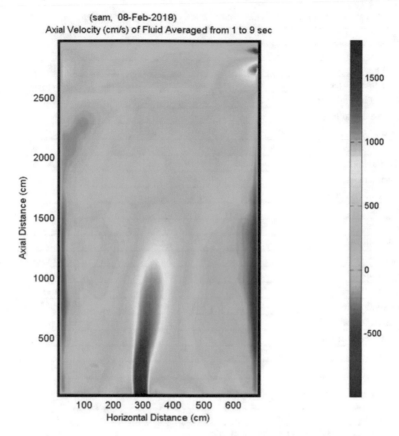

Fig. 4.4 Typical time-averaged gas velocity in the riser reactor with a high velocity central jet. (This figure was originally published in [20] and has been reused with permission)

volume fraction of 0.005 cm. The details of the simulation are shown in Table 4.1. The multiphase CFD code was an extension of that described in great detail for the KRW fluid bed geometry [23]. It is similar to the MFIX code [25] and the commercial ANSYS Fluent computer program [26]. The earliest CFD simulations for a gas phase polymerization reactor were conducted [27].

The conservation of mass, momentum, energy, and species equations are very similar to those used for the CFD design of a slurry bubble column reactor for making diesel fuel using the Fischer-Tropsch reaction [22]. These equations with a pressure drop only in the gas phase are well posed as an initial value problem. They are shown in Table 4.2.

Figure 4.5 shows the time-averaged volume fraction of solid 1, the polymer, in the center of the reactor. The jet region up to 15 m high has a slightly higher polymer concentration than the rest of the reactor higher up.

The first 2 m entrance section contains a very high polymer concentration. Figure 4.5 makes it clear that two regions exist in this reactor: a dense bottom region and a dilute top region.

Table 4.1 Input data for large polymerization reactor geometry simulation

Height	30 m
Width	7 m
Inlet jet width	0.512 m
Number of inlet jets	5
Outlet height: right wall	1.02 m
Splash-plate width at 26 m	3.68 m
Solids and gas phase properties	
Polyethylene density	0.9 g/cm^3
Catalyst diameter	0.008 cm
Catalyst density	0.9 g/cm^3
Inlet gas weight fractions	
Hydrogen	0.18
Ethylene	0.82
Inlet conditions	
Pressure	30.1 atm (3.1 MPa)
Temperature	333 K
Central jet velocity	18 m/s
Side jet velocities	6 and 3 m/s, each
Polyethylene weight fraction	0.1
Catalyst weight fraction	0.0005
Grid sizes	
Time step	5×10^{-6} s
Axial	106×0.28 m
Radial	42×0.167 m
Initial conditions	
Filled region with polymer and catalyst at inlet	Up to 11.5 m
Gas velocities in filled region	1.8 m/s
Gas velocities in top gas region	0.6 m/s

A comparison to the computed fluidization flow regimes [23, 28] shows flow in the turbulent fluidization regime. The turbulent flow regime computed in [29] was for 54 μm diameter fluid cracking catalyst (FCC) particles, with an average velocity of 3.25 m/s and a flux of 99 kg/m^2 s. In the two-story unit at IIT, a sharp transition was observed between the dense and the dilute regimes for similar conditions.

Figure 4.6 shows the computed time-averaged catalyst weight fraction. The catalyst concentration is much higher at the wall, similar to the high catalyst concentration in the core-annular regime in a single particle riser for flow of FCC particles [28] or the platelet concentration in blood vessels [30]. The blue dots near the bottom in Fig. 4.6 represent two cooling tubes placed into the reactor for safety in the event that hot spots form. The high catalyst wall concentration was consistently found in all polymerization reactors, in risers, and in bubbling bed reactors, such as the UNIPOL reactor. Hence, the mechanism for the high wall concentration is due not only to the electrostatics.

Table 4.2 The hydrodynamics model

Continuity equations

Gas phase

$$\frac{\partial}{\partial t}\left(\varepsilon_g \rho_g\right) + \nabla \cdot \left(\varepsilon_g \rho_g \vec{v}_g\right) = \dot{m}_g$$

Solids phases ($k = 1, 2$)

$$\frac{\partial}{\partial t}\left(\varepsilon_k \rho_k\right) + \nabla \cdot \left(\varepsilon_k \rho_k \vec{v}_k\right) = \dot{m}_k$$

Momentum equations

(a) Gas phase momentum

$$\frac{\partial}{\partial t}\left(\varepsilon_g \rho_g \vec{v}_g\right) + \nabla \cdot \left(\varepsilon_g \rho_g \vec{v}_g \vec{v}_g\right) =$$

$$-\nabla P + \sum_{k=1}^{N} \beta_{gk}\left(\vec{v}_k - \vec{v}_g\right) + \nabla \cdot 2\varepsilon_g \mu_g \nabla^s \vec{v}_g + \varepsilon_g \rho_g \vec{g}$$

(b) Solids phases, $k = (1, \ldots, N)$

$$\frac{\partial}{\partial t}\left(\varepsilon_k \rho_k \vec{v}_k\right) + \nabla \cdot \left(\varepsilon_k \rho_k \vec{v}_k \vec{v}_k\right) = \beta_{gk}\left(\vec{v}_g - \vec{v}_k\right)$$

$$+ \sum_{l=1}^{N} \beta_{kl}\left(\vec{v}_l - \vec{v}_k\right) - G_s \nabla \varepsilon_k + \nabla \cdot 2\varepsilon_k \mu_k \nabla^s \vec{v}_k + \varepsilon_k \rho_k \vec{g} \qquad k = 1, 2$$

Energy equations

Gas phase

$$\frac{\partial}{\partial t}\left(\varepsilon_g \rho_g H_g\right) + \nabla \cdot \left(\varepsilon_g \rho_g H_g \vec{v}_g\right) = \left(\frac{\partial P}{\partial t} + \vec{v}_g \cdot \nabla P\right)$$

$$+ \sum_{k=1}^{N} h_{vk}\left(T_k - T_g\right) + \nabla \cdot \left(K_g \varepsilon_g \nabla T_g\right)$$

Solids phases

$$\frac{\partial}{\partial t}\left(\varepsilon_k \rho_k H_k\right) + \nabla \cdot \left(\varepsilon_k \rho_k H_k \vec{v}_k\right) = h_{vk}\left(T_f - T_k\right) + \nabla \cdot \left(K_k \varepsilon_k \nabla T_k\right)$$

Species balances

Gas phase

$$\frac{\partial}{\partial t}\left(\varepsilon_g \rho_g Y_{ig}\right) + \nabla \cdot \left(\varepsilon_g \rho_g Y_{ig} v_g\right) = \varepsilon_s \rho_s r$$

where r has units of kg ethylene/kg catalyst s

Rate of reaction

Rate of reaction = $5 \, C_{ethylene} \, C_{catalyst}$

(for a high rate of reaction)

Constitutive equations

$$\varepsilon_g + \sum_{k=1}^{N} \varepsilon_k = 1$$

$$T_k = T_0 + \frac{H_k - H_0}{c_k}$$

Equations of state

Gas phase ideal gas law

(continued)

Table 4.2 (continued)

$\rho_g = \frac{P}{RT_g}$
Solids phase
$\rho_k = \rho_{sk} = \text{constant}$
Constitutive equations for stress
(a)
Fluid phase stress
$\left[\tau_f\right] = \varepsilon_f \mu_f \left(\left[\nabla \mathbf{v}_f + (\nabla \mathbf{v}_f)^T\right] - \frac{2}{3}\nabla \cdot \mathbf{v}_f[\mathbf{I}]\right)$
(b)
Solids phases stress $k = (1,\ldots,N)$
$\left[\tau_k\right] = \mu_k \left(\left[\nabla \mathbf{v}_k + (\nabla \mathbf{v}_k)^T\right] - \frac{2}{3}\nabla \cdot \mathbf{v}_k[\mathbf{I}]\right)$
Empirical solids viscosity and stress model
(a) Solids viscosity
$\mu_k = 5\varepsilon_k \text{P} \ (0.5 \text{ Pa s})$
(b) Solids stress
$\nabla P_k = G(\varepsilon_k)\,\nabla\,\varepsilon_k$
$G(\varepsilon_k) = 10^{-8.686\varepsilon_k + 8.577}\,\text{dynes/cm}^2$
Gas/solids drag coefficients $k = (1,\ldots,N)$
for $\varepsilon_f < 0.8$ (based on Ergun equation)
$\beta = 150\dfrac{\varepsilon_k^2 \mu_g \rho_k}{\left(\varepsilon_g d_k \phi_k\right)^2 (\rho_k - \rho_g)} + 1.75\dfrac{\rho_g \rho_k \varepsilon_k \lvert v_g - v_k \rvert}{\left(\varepsilon_f d_k \phi_k\right)(\rho_k - \rho_g)}$
for $\varepsilon_f \geq 0.8$ (based on empirical correlation)
$\beta = \frac{3}{4} C_D \dfrac{\rho_g \rho_k \varepsilon_k \lvert v_g - v_k \rvert}{d_k \phi_k (\rho_k - \rho_g)} \varepsilon_f^{-2.65}$
where $C_D = \begin{cases} \dfrac{24}{\text{Re}_k}\left[1 + 0.15\,\text{Re}_k^{0.687}\right] & \text{for Re}_k < 1000 \\[2mm] 0.44 & \text{for Re}_k \geq 1000 \end{cases}$
$\text{Re}_k = \frac{\varepsilon_g \rho_g \lvert v_g - v_k \rvert d_k \phi_k}{\mu_g}$
Solids/solids drag coefficient
$\beta_{kl}_{\ k,l \neq f} = \frac{3}{2}\alpha(1 + e)\dfrac{\rho_k \rho_l \varepsilon_k \varepsilon_l (d_k + d_l)^2}{\rho_k d_k^3 + \rho_l d_l^3}\left|\vec{v}_k - \vec{v}_l\right|$
where $\alpha = \left(\frac{d_e}{d_k}\right)^{1/2}$ and $e = 0.95$
Gas phase heat transfer
$K_f = 8.65 \times 10^5 \left(\frac{T_f}{1400}\right)^{1.786}\,\text{W/m K}$
Gas/solids heat transfer, phase $k\,(= s)$
for $\varepsilon_f \leq 0.8$
$Nu_k = (2 + 1.1\text{Re}^{0.6}\text{Pr}^{1/3})S_k \ \text{Re} \leq 200$
$\quad = 0.123\left(\frac{4\text{Re}}{d_k}\right)^{0.183} S_k^{0.17} \ 200 < \text{Re} \leq 2000$
$\quad = 0.61\text{Re}^{0.67}S_k \ \text{Re} > 200$
for $\varepsilon_f > 0.8$
$Nu_k = (2 + 0.16\text{Re}^{0.67})S_k \ \text{Re} \leq 200$

(continued)

Table 4.2 (continued)

$=8.2\text{Re}^{0.6}S_k \ 200 < \text{Re} \le 2000$
$=1.06\text{Re}^{0.457}S_k \ \text{Re} > 200$

where

$$\text{Re} = \frac{\rho_f |\vec{v}_f - \vec{v}_k| d_k}{\mu_f}$$

$$S_k = \varepsilon_k \frac{6}{d_k}$$

$$Nu = \frac{h_{vk} d_k}{k_f}$$

Solids phase heat transfer

$$\frac{K_k}{K_f} = \left(1 - \sqrt{1 - \varepsilon_f}\right)\left[1 + \varepsilon_f \frac{\lambda_R}{\lambda}\right] + \sqrt{1 - \varepsilon_f}\left[\varphi \frac{\lambda_s^*}{\lambda} + (1 + \varphi)\frac{\lambda_{SO}^*}{\lambda}\right]$$

with

$$\frac{\lambda_{SO}^*}{\lambda} = \frac{2}{(N-M)} \left(\frac{B\left(\frac{\lambda_s^*}{\lambda} + \frac{\lambda_R}{\lambda} - 1\right)}{(N-M)^2 \left(\frac{\lambda_s^*}{\lambda}\right)} \cdot \ln\left(\frac{\left(\frac{\lambda_s^*}{\lambda} + \frac{\lambda_{SO}^*}{\lambda}\right)}{B}\right) \right)$$

$$- \frac{B-1}{N-M} + \frac{B+1}{2B}\left(\frac{\lambda_R}{\lambda} - B\right))$$

$$N - M = 1 + \frac{(\lambda_R/\lambda) - B}{\left(\lambda_s^*/\lambda\right)}$$

$$B = 1.25 \left(\frac{1-\varepsilon_f}{\varepsilon_f}\right)^{10/9} \text{(for spheres)}$$

$$\frac{\lambda_R}{\lambda} = \frac{0.0004C_k}{(2/\varepsilon_r - 1)}\left(\frac{T}{100}\right)^3 \cdot d_k$$

$$\frac{\lambda_s^*}{\lambda} = 12.227$$

$$\varphi = 7.26 \times 10^{-8}$$

$$C_k = 5.67 \times 10^{-8}$$

$$\varepsilon_r = 0.93 \text{ emission ratio}$$

Figure 4.7 shows the ethylene concentration in this reactor. In the center jet region, the ethylene concentration at this high velocity is near 0.8 weight fraction, near that of the inlet concentration. However, at the exit left wall, above the blue splash-cooling plate, the exit concentration is near zero. There is more reaction near the wall, due in part to the higher catalyst concentration. The polyethylene production rate is the product of average gas velocity times the area, the density, and the weight fraction. Here, gas velocity equals 2.64 m/s, area equals 39.5 m^2, and density equals 30 kg/m^3.

With complete conversion, the production rate equals 2430 kg/s. With the more realistic weight fraction change of 0.6, the production rate equals 1823 kg/s or 61 million tons per year. This is the entire ethylene yearly production in the United States. The reactor height can be estimated roughly from the average gas velocity of 2.63 m/s divided by the input effective first order rate constant of 5 s^{-1} times the catalyst concentration of roughly 0.02. Hence the reactor height needs to be 26 m high (Fig. 4.7).

Fig. 4.5 Average polymer volume fraction in the center of the reactor. (This figure was originally published in [20] and has been reused with permission)

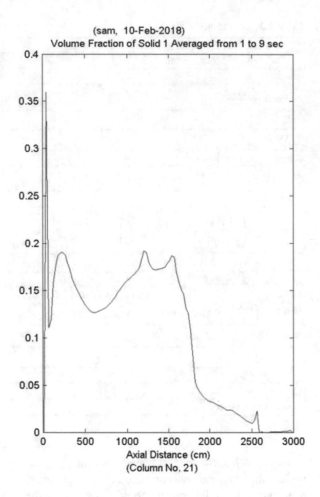

4.2.5.2   Low Velocity

Figure 4.8 shows the average polymer volume fraction with the inlet gas velocity reduced to 0.88 m/s and the particle size increased to 0.25 cm. At this low velocity, with the higher polymer size, the dense region volume fraction increases from approximately 0.2 to more than 0.6 at the bottom of the reactor. Figure 4.9 shows that we are again in the turbulent flow regime. At this low gas velocity, the time to a steady state increases from 1 to 10 s. Steady state was determined from plots of volume fractions, temperature, and ethylene concentrations as a function of run time. Figure 4.10 shows the ethylene weight fraction. Because almost complete conversion occurs in 10 m, the reactor is too tall for these conditions.

The outlet gas velocity at the left wall is approximately 25 m/s, with the polymer velocity going into the down comer at 16 m/s and gas temperature of 61 °C (334 K), a rise of only one degree even at this low velocity. The polymer production rate for this low velocity is 610 kg/s, again, an extremely high production rate compared to

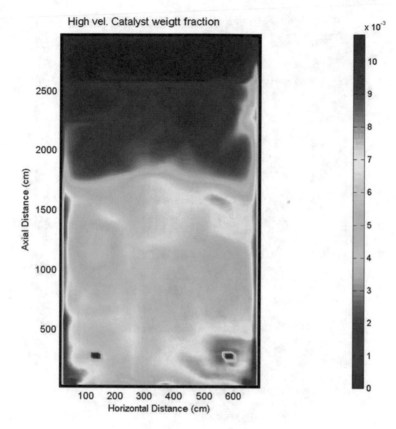

Fig. 4.6 Average catalyst weight fraction in the reactor. (This figure was originally published in [20] and has been reused with permission)

the commercial UNIPOL reactors. Figure 4.11 shows temperature peak formations in the reactor. While these temperature peaks are no more than 40 °C, this low velocity and high rate of reaction are close to the limit of operation. The lower inlet temperature of 60 °C (333 K) is due to the cooling splash-plate in the riser.

4.2.6   CFD Design of Smaller Reactors

To validate the proposed design of the large polymerization reactor, it is necessary to perform pilot plant tests with smaller reactors. A number of CFD simulations have been made for smaller reactors. Some of the results with an 80 cm diameter reactor are presented here. For high velocity flow, the ethylene weight fraction shown in Fig. 4.12 is very similar to that shown in Fig. 4.7 for the 6 m diameter reactor. Figure 4.13 shows the same core-annular flow as in Fig. 4.6. Figure 4.14 shows the

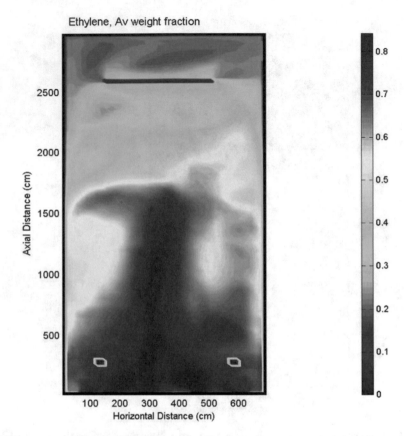

Fig. 4.7 Average ethylene weight fraction in the reactor [20]. (This figure was originally published in [20] and has been reused with permission)

variance of solid 2, the catalyst, and velocity. This is a measure of the turbulence. The turbulent granular temperature is approximately the square of the velocity. Figure 4.15 shows the outlet temperature and its variance, starting with the initial temperature of 60 °C (333 K).

The two-story IIT circulating fluidized bed with a splash-plate used to prevent asymmetrical flow in the 7.5 cm diameter riser [31] can be used for validation of the high production reactor. In the past, it was used for flow of Geldart Group D particles of a similar size as used here for the polymerization reactor [32]. Solids slugging was observed in this small diameter reactor. Similar slugging was computed here in this high L/D reactor, particularly during start-up through 8 s. There were other differences in flow behavior causing incomplete conversion. To obtain complete conversion, the inlet jet velocities were reduced. For the central jet velocity of 3.5 m/s, near complete conversion was achieved, assuring a high production of ethylene.

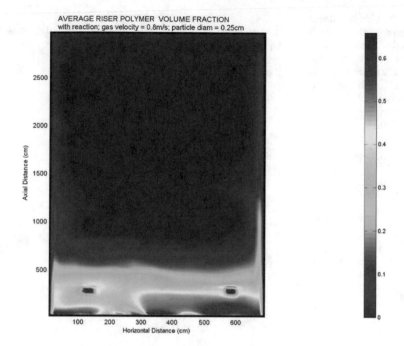

Fig. 4.8 Average polymer volume fraction with lower velocity, 0.8 m/s and higher particle size, 0.25 cm diameter. (This figure was originally published in [20] and has been reused with permission)

4.2.7 Conclusion

The major conclusions reached in this section are summarized below:

1. It is shown that the temperature rise in gas phase circulating fluidized beds is two to three orders of magnitude lower than that in bubbling fluidized beds. This conclusion is obtained from approximate energy balances and from CFD simulations.
2. This small temperature rise allows the circulating fluidized beds to operate with an order of magnitude higher monomer conversions. In the bubbling fluidized beds, such as in the UNIPOL reactor, the monomer conversion is only about 2%.
3. The high monomer conversion increases the polymer production by two to three orders of magnitude for the same size reactor.
4. CFD simulations for several different-sized polyethylene reactors confirm the high production rates.
5. The polyethylene reactor risers operate without bubbles, unlike the commercially used UNIPOL reactors. Formation of large bubbles in the UNIPOL reactors restrict their size due to shaking of the reactors.

Fig. 4.9 Turbulent flow regime computation for the low velocity flow in the reactor. (This figure was originally published in [20] and has been reused with permission)

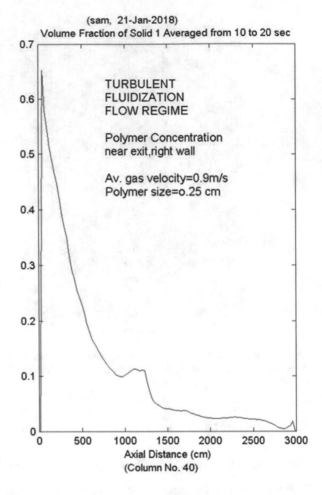

6. It is shown that the small catalyst particles move toward the walls for all of the studied reactors. Hence, sheet formation at the walls is due not only to electrostatics.
7. The ethylene reactors studied were all in the turbulent flow regime, with a dense polymer concentration at the bottom and a dilute region on top, similar to that found for UNIPOL reactors.
8. Operation of the reactors with a high velocity jet at the center produces a high downward flow at the reactor walls that minimizes the undesirable sheet formation observed in the UNIPOL type reactors.

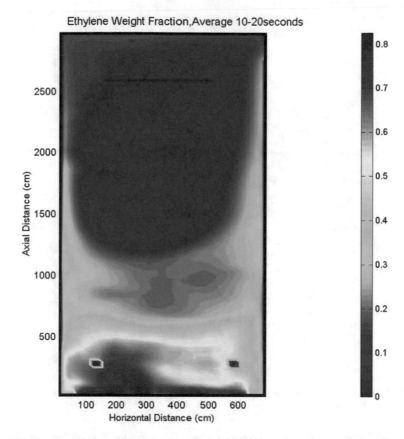

Fig. 4.10 Average ethylene weight fraction for the low velocity flow in the reactor. (This figure was originally published in [20] and has been reused with permission)

4.3   Rotating Fluidized Bed (RFB) Reactor for Polymerization

4.3.1   Introduction

Rotating fluidized bed reactors are a promising process to have better control of particle size distribution, particle separation, and increasing reactor efficiency. Owing to the high rotational acceleration (e.g., $10 \times$ g) that can be imposed in these kinds of reactors, the amount of throughput, i.e., monomer flow rate, can be increased significantly without concern for either temperature increase or change in the fluidization regime from the well-mixed condition to slugging. Thus the production rate and consequentially the polymerization yield will significantly increase. Despite significant application, conventional fluidized bed (bubbling) reactors have shown limited flexibility in achieving high gas throughput mainly because of the

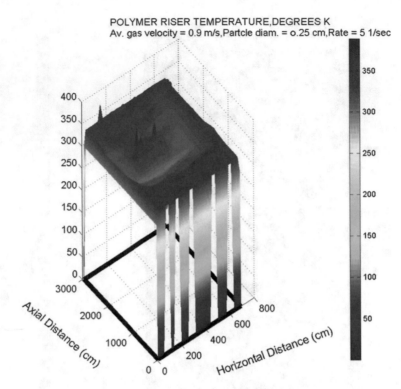

POLYMER RISER TEMPERATURE, DEGREES K
Av. gas velocity = 0.9 m/s, Partcle diam. = o.25 cm, Rate = 5 1/sec

Fig. 4.11 Typical peaks in temperature in the riser reactor. (This figure was originally published in [20] and has been reused with permission)

inability to provide suitable heat transfer rates. To overcome these disadvantages and enhance the efficiency of the fluidized reactors, the feasibility of using the rotating fluidized bed reactors in which polymer products continuously leave the reactor and carry a significant amount of thermal energy out of the rector, preventing temperature increase and in turn resulting in huge increase in polymer production is discussed. In this section, the feasibility study was performed using computational fluid dynamics (CFD), along with the population balance equation approach.

In conventional fluidized beds, fluidization occurs when the upward drag force balances the weight of the particles under gravity. Because the gravitational acceleration is a constant parameter, the minimum fluidization velocity is constant for particles with the same characteristic properties, i.e., particle diameter and density. Consequently these systems remain limited to moderate throughputs because of the relatively low velocity required to fluidize the catalyst and polymer particles. On the other hand, rotating fluidized beds (RFBs) provide added flexibility by allowing higher flow rates. Because the particles in an RFB are fluidized in a controlled centrifugal field, the control of particle residence time would be improved as opposed to the constant gravitational force of the conventional fluidized bed.

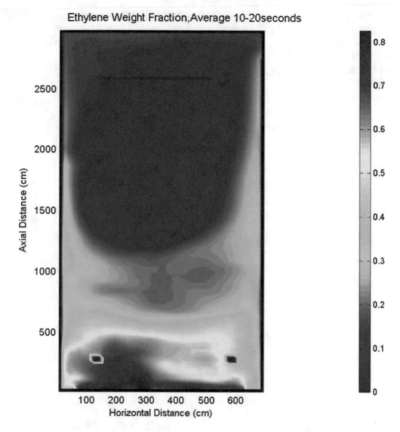

Fig. 4.12 Time-averaged weight fraction of ethylene in the 80 cm diameter reactor. (This figure was originally published in [20] and has been reused with permission)

In a rotating fluidized bed, the body force in a centrifugal bed becomes an adjustable parameter that is determined by rotational speed and basket radius. By using a strong centrifugal field much greater than gravity, the particles in the bed will withstand a high flow rate without formation of large bubbles. The gas/solids contact is also expected to improve at a higher aeration rate. In a rotating fluidized bed of uniform particles (mono-size solid particles), unlike in a conventional fluidized bed, the granules are fluidized layer-by-layer from the gas/solids interface toward the side-wall distributor, as the inlet gas velocity is increased. This is a very significant and interesting phenomenon and is important in the design of such fluidized beds [3, 6, 7]. There are several studies in the literature evaluating the effect of different parameters such as rotational speed on fluidization flow behavior, pressure drop, and elutriation of particles in RFBs [33–38]. Chen [3] was among the first who predicted theoretically the layer-by-layer fluidization of the mono-size solid particles, starting from the gas/solids interface, based on the one-dimensional local momentum balance. He also considered the variation of the voidage in the bed in order to clearly

Fig. 4.13 Time-averaged
catalyst concentration in the
80 cm diameter reactor.
(This figure was originally
published in [20] and has
been reused with
permission)

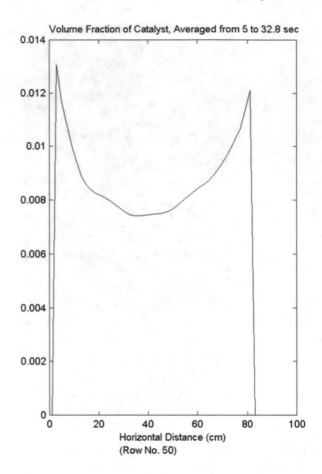

follow the transition from a packed bed to a fluidized bed. The advances in CFD and computational capability enabled us to simulate flow behavior and particle growth in a rotating fluidized bed. Ahmadzadeh et al. [7] were among the first who used CFD to obtain a fundamental understanding of gas/solids flow patterns in a rotating fluidized bed. They used a two-dimensional unsteady model to study the detailed flow patterns and particle concentrations, along with the quantities describing the overall behavior of the rotating fluidized beds such as pressure drop, bubble formation, and particle elutriation. Furthermore, the expected layer-by-layer fluidization behavior for beds with uniform particle size was simulated numerically. In order to rigorously account for the effect of particle growth during polymerization and agglomeration on the flow behavior, the population balance equations (PBEs) are solved along with the continuity and momentum equations for both the gas and solids phases using the method of moments [39–45].

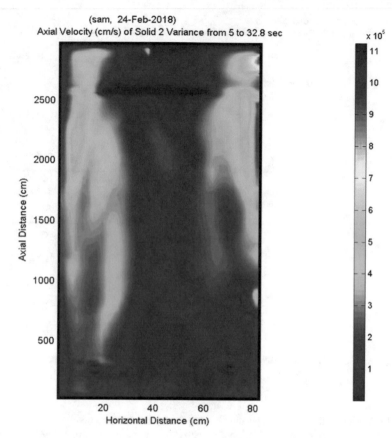

Fig. 4.14 Velocity variance of the catalyst in the reactor $(cm/s)^2$. (This figure was originally published in [20] and has been reused with permission)

4.3.2   Mathematical Modeling of a Rotating Fluidized Bed (RFB)

It is assumed that because of the efficient removal of the heat of reaction is achieved by polymers continuously exiting the reactor, the isothermal condition can be maintained throughout the bed. A rotating fluidized bed is essentially a cylindrical basket rotating about its axis of symmetry (Fig. 4.16).

Catalyst particles are introduced into the cylinder or placed at the bottom of the reactor and are forced to move toward the wall due to the large centrifugal force produced by the rotation. The gas phase is introduced in the inward direction from the side wall to fluidize the particles. Instead of having a fixed gravitational field as in a conventional bed, the body force in a centrifugal bed may be varied by changing the rotational speed of the bed. By using a strong centrifugal field much greater than gravity, the particles in the bed will be able to withstand a high aeration rate without

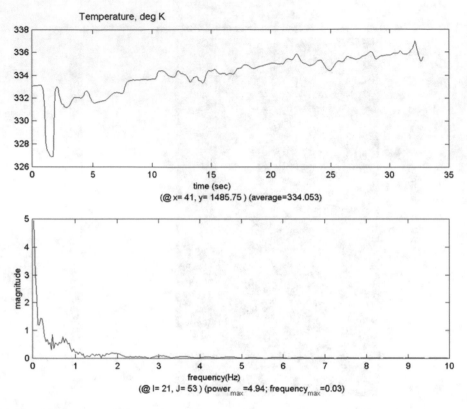

Fig. 4.15 Typical outlet reactor temperature leaving the riser reactor as a function of time (top) and its spectral distribution (bottom). (This figure was originally published in [20] and has been reused with permission)

Fig. 4.16 Geometry of the rotating fluidized bed used in this simulation [4]

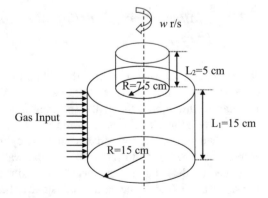

formation of large bubbles and elutriation. The gas/solids contact is also expected to improve at a higher aeration rate. In centrifugal or rotating fluidized beds, the wall serves as the distributor and the gas flows radially inward through the bed of granules. When the drag force on the granules balances the centrifugal force, the bed becomes fluidized. Minimum fluidization can be achieved at the desired gas flow rate by changing the rotating speed of the bed [6, 7]. The produced high temperature gas and polymers with lower density move toward the center of the RFB (at lowest gravitational force) and exit the reactor while removing a significant amount of heat from the reactor causing the polymerization to proceed at a significantly higher rate, while preventing the formation of high temperature regions in the reactor and polymer agglomeration. We assumed this fast heat removal results in approximately isothermal operation and we considered constant temperature in our numerical simulation.

The transient two-dimensional continuity and momentum equations for gas and polymer particles phases along with constitutive equations (similar to Table 4.2) were solved using the ANSYS computer code. The particle size distribution was defined by population balance equations and was linked with ANSYS CFD code and solved using the method of moments [4, 39, 45]. In olefin polymerization, the reaction starts with a catalyst of 20–50 μm diameter and the final polymer is about 1000–3000 μm diameter, and thus at any time there is particle size distribution in the bed. Successful CFD models for fluidized bed polyolefin reactors must be capable of describing the evolution of the particles. To reach this goal, the population balance equation (PBE) was used to predict the growth of the polymer particles in the rotating fluidized bed. All of the equations were solved in the axisymmetric cylindrical coordinates using the rotating frame of reference. The growth term in the population balance equation, the rate of change of particle diameter $G(d_s)$, is,

$$G(d_s) = d/dt \ (d_s) = rd_0^3/3\rho_s d_s^2 \qquad (4.5)$$

where d_s is the particle diameter, ρ_s is the particle density, d_0 is the initial catalyst particle diameter, and r_s is the intrinsic polymerization rate, where the intrinsic polymerization reaction rate is defined as the rate of propagation reaction,

$$r_s = k_p[C_r][M] \qquad (4.6)$$

where k_p is the propagation rate constant, M is the monomer concentration, and C_r is the catalyst concentration.

Neumann boundary conditions are applied at the outflow boundary for all of the variables except pressure, which is specified as atmospheric pressure. In addition, axisymmetric conditions are assumed around the vertical axis of symmetry. The no-slip boundary condition is used for the gas phase at the wall, with only the tangential velocity having a prescribed value proportional to the rotational speed. For the solids phase, the no-slip condition was used for the tangential velocity, but a non-zero or "partial-slip" velocity was defined for each of the radial and axial velocities. For the inlet boundary, the Dirichlet conditions are used for the tangential velocity and radial inflow velocity of the gas phase. Initially, the system is at rest and all velocities are zero. The rotational speed of the reference frame is defined such that no angular velocity is defined for either phase at the boundaries, contrary to the case of a stationary frame approach. In our simulations, due to symmetry, half of a two-dimensional cylindrical reactor with a radius of 15 cm and a height of 20 cm is considered.

The required computational time is very large due to the fact that the reaction is completed in 3 h of real time, while the fluid properties (velocity and pressure) are changing in less than a second. In other words, the maximum time step to obtain converged solution at every time step, when the momentum equations are linked with the population balance and continuity equations, is very small (i.e., 0.001 s).

Therefore solving the particle growth along with the flow field can be done using two different approaches. The first method is to decouple the PBEs from the flow equations because of the existence of two different time scales between the flow field (short time (e.g., 0.001 s)) and particle growth (long time, order of minutes). This means that first the flow calculation is performed. Next, the flow patterns are simulated by using the previously computed average particle diameter using PBEs. For the PBEs the time step can be chosen at approximately 10 s. Solving the flow field with small time step (e.g., 0.001 s) for a few seconds and freezing the PBEs allows us to update the flow field for the particle size distributions. It means that the drag force is calculated based on the average particle size in each numerical cell. The main issue with this approach is that when the PBEs are solved by freezing the flow field, the particles will grow in each individual numerical cell with no influence of drag and centrifugal forces.

The second method is to simultaneously solve the PBEs coupled with the flow hydrodynamics by assuming that reaction rate times the particle residence time in the reactor is constant. This enables use of a faster growth rate with lower residence time. This assumption will allow us to obtain a fundamental knowledge of the flow pattern in any kind of rotating fluidized bed where the particle size is changing. In the meantime, the effect of particle growth on the particle size distribution and, as a consequence, on the flow pattern can be predicted by calculating the drag force based on the average particle size in each numerical cell at each time step. The second approach was chosen in this study to solve the PBEs coupled with the conservation equations. It was assumed that the particles residence time is 50 s for particles to grow and reach to 1000 μm from an initial particle diameter of 200 μm, and the corresponding rate of growth was calculated accordingly.

4.3.3   Results and Discussion

The solids particles were placed in the vertical layer from the side-wall distributor with the initial volume fraction of 0.001. In this case, there is no flow and only the population balance equation was solved. The growth rate is based on 3 h of reaction

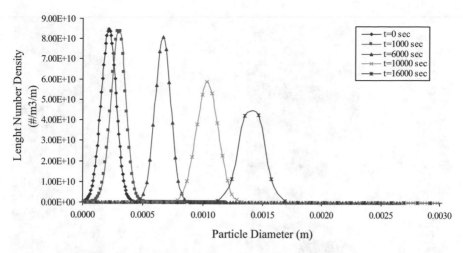

Fig. 4.17 Particle size distribution in a rotating bed, $\Omega = 30$ rad/s, with initial particle size of 200 μm diameter, for 3 h growth at a constant rate [4]

time. Figure 4.17 shows the change in the volume average particle size distribution as a result of the particle growth rate during 3 h of reaction time. As the particles grow and get bigger, the distribution becomes more flat.

Figures 4.18 and 4.19 show the growth of the particles as the reaction progresses in a case when there is a continuous flow of gas from the side-wall distributor and continuous injection of the 200 μm diameter particles from the location at the bottom of the reactor close to the side-wall distributor. In the early stage of growth, the particles move toward the side wall due to centrifugal force and form a layer by the wall, and the coverage area of the particulate phase increases as they grow. As gas flow is introduced after 12 s and increased to 0.75 m/s, the bed maintained a complete fluidization state, and the particle diameter reached approximately 350 μm after 17 s.

Figure 4.19 shows that, when a particle size distribution exists in the bed, the fluidization has a stripe.

There are some regions in which the drag is dominant because the particles are smaller and move toward the center and make that area dilute; meanwhile, the centrifugal force is pulling the particles toward the wall and creating areas with higher concentration (striped fluidization behavior). An important conclusion from these numerical results is that the fully grown particles that had higher residence time stayed close to the inlet-wall distributor and close to the top wall of the reactor. As a result, the optimum location to harvest these particles is at the top wall of the reactor and close to the inlet distributor. For more detailed modeling and additional results see Ahmadzadeh et al. [5].

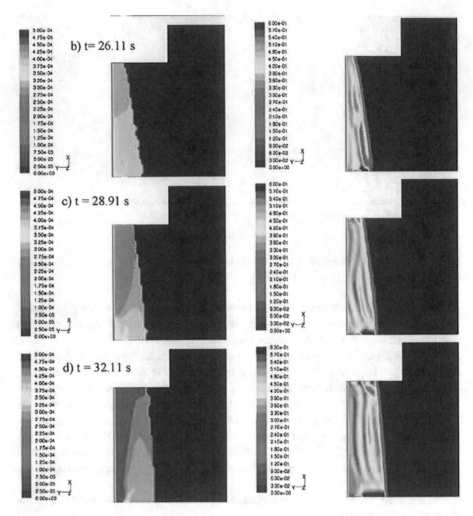

Fig. 4.18 Particle diameter distribution with its solid volume fraction contour in an RFB, with solid injection and rotational speed of 30 rad/s [4]

4.3.4 Conclusion

The population balance equation (PBE) model linked with CFD is an excellent tool for design and prediction of the evolution of the particle size distribution in a polymerization rotating fluidized bed reactor. The method of moments is also a very promising approach to obtain reliable and accurate numerical results for CFD/PBE linked models. Simulation results in this study revealed that the optimum

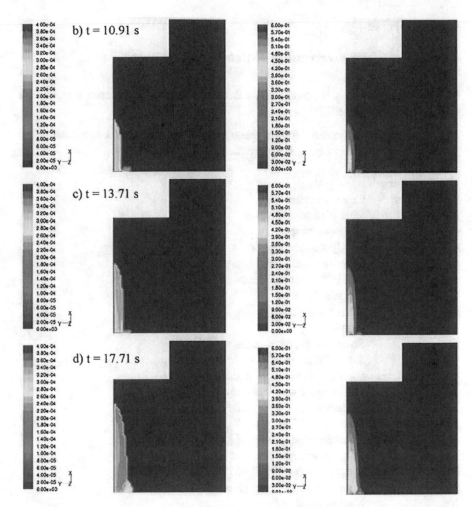

Fig. 4.19 Particle diameter distribution with its solid volume fraction contour in an RFB with solid injection. Gas inlet velocity = 0.75 m/s, and angular velocity = 30 rad/s [4]

location to harvest the fully polymerized particles should be at the top wall close to the inlet-wall distributor.

4.4 Exercise

4.4.1 Ex. 1: Polymerization Reactor with Downer

Complete simulations were presented in this chapter only for the riser portion of a circulating fluidized bed.

1. Extend the simulations for the complete circulating fluidized bed, including the downer, as shown in Fig. 4.3, for production of polyethylene.

Nomenclature

A	Cross sectional area
c_{pi}	Heat capacity of phase i
C_d	Drag coefficient
C_r	Catalyst concentration
d_k	Characteristic particulate phase diameter
d_0	Initial catalyst diameter
d_s	Particle diameter
E	Activation energy
g	Gravity
G	Solid compressive stress modulus
$G(d_s)$	Rate of change of particle diameter
g_o	Radial distribution function at contact
h_{vk}	Gas phase k heat transfer coefficient
H	Enthalpy
ΔH	Heat of reaction
K_g	Thermal conductivity of gas
k_p	Propagation rate constant
M	Monomer concentration
$\dot{m}_k$	Rate of generation of phase k
Nu_k	Nusselt number
P	Continuous phase pressure
P_k	Dispersed (particulate) phase pressure
r	Rate of reaction
r_s	Intrinsic polymerization rate
R	Gas constant
Re_k	Reynolds number for phase k
T	Temperature
ΔT	$T_{bed} - T_{in}$
t	Time
v_i	Hydrodynamic velocity in direction i
$\overline{v_i}$	Mean velocity in direction i
ΔY	Conversion of monomer
Y_{ig}	Weight fraction of species i in the gas phase

Greek Symbols

β	Interphase momentum transfer coefficient
ε_k	Volume fraction of phase k
φ_k	Particle sphericity
μ_k	Shear viscosity of phase k
θ	Granular temperature
ρ_k	Density of phase k
τ_k	Stress of phase k

References

1. Choi KY, Ray WH (1985) The dynamic behavior of fluidized bed reactors for solid catalysed gas phase olefin polymerization. Chem Eng Sci 40:2261–2279
2. Hutchinson RA, Chen CM, Ray WH (1992) Polymerization of olefins through heterogeneous catalysis. X. Modeling of particle growth and morphology. J Appl Polym Sci 44:1389–1414
3. Chen YM (1987) Fundamental of a centrifugal fluidized bed. AIChE J 33:722–72
4. Ahmadzadeh A (2016) Numerical simulation of olefin polymerization in a rotating fluidized bed. Ph.D. thesis, Illinois Institute of Technology
5. Ahmadzadeh A, Arastoopour H (2008) Three dimensional numerical simulation of a horizontal rotating fluidized bed. Powder Technol 183:410–416
6. Ahmadzadeh A, Arastoopour H, Teymour F (2004) Fluidization behavior of rotating fluidized beds. In: Arena L, Chirone R, Miccio M, Salatino P (eds), Fluidization XI, ISBN 0-918902-52-5, Engineering Conferences International, pp 667–674
7. Ahmadzadeh A, Arastoopour H, Teymour F (2003) Numerical simulation of gas and particle flow in a rotating fluidized bed. Ind Eng Chem Res 42:2627–2633
8. Gidaspow D (2018) Provisional US Patent Application, application number 62/615,798, 1/19/2018
9. Xie T, McAuley KB, Hsu JC, Bacon DW (1994) Gas phase ethylene polymerization: production processes, polymer properties, and reactor modeling. Ind Eng Chem Res 33:449–479
10. Choi KY, Ray WH (1985) Recent developments in transition metal catalytic olefin polymerization-a survey: I ethylene polymerization. JMS Rev Macromol Chem Phys C25 (1):1-55
11. Brown GL, Warner DF, Byon JH (1981) Exothermic polymerization in a vertical fluid bed reactor system containing means therein and apparatus thereof, US Patent 4,255,542, 10 March 1981
12. Yermakov Y, Zakharov V (1975) One-component catalysts for polymerization of olefins. Adv Catal 24:173–219
13. Dhodapkar S, Jain P, Villa C (2016) Designing polymerization reaction systems. CEP, February:1–25
14. Roger D, Laszlo H, Pierre M (1975) Process for dry polymerization of olefins. US Patent 3,922,322, 25 Nov 1975
15. Rokkam RG, Fox RO, Muhle ME (2011) Computational modeling of gas-solids fluidized bed polymerization reactors, Chapter 12. In: Pannala S, Syamlal M, O'Brien, TJ (eds) Computational gas-solids flows and reacting systems: theory, methods and practice, engineering science reference, Pennsylvania, pp 373–397
16. Hendrickson G (2006) Electrostatics and gas phase fluidized bed polymerization reactor wall sheeting. Chem Eng Sci 61:1041–1064
17. Chen XZ, Luo ZH, Yan WC, Lu YH, Ng IS (2011) Three-dimensional CFD-PBM coupled model of the temperature fields in fluidized bed polymerization reactors. AIChE J 57:3351–3366

18. Mayank K (2016) Application of multiphase flow CFD in the gas phase polymerization process. In: Chaouki J, Berruti F, Cocco R (eds) Fluidization XV, ECI Symposium Series, 2016 http://dc.engconfint.org/fluidization-xv/65
19. Mayank K (2016) Multiphase flow CFD capability development for gas phase polymerization processes. Presented at US Department of Energy, NETL, July 11 (2016)
20. Gidaspow D (2019) High production circulating fluidized bed polymerization reactors. Powder Technol 357:108–116
21. Gidaspow D (1994) Multiphase flow and fluidization. Academic Press, New York
22. Gidaspow D, He Y, Chandra V (2015) A new slurry bubble column reactor for diesel fuel. Chem Eng Sci 134:784–799
23. Gidaspow D, Jiradilok V (2009) Computational techniques. Nova Science, New York
24. Gidaspow D (2018) Design of the next generation polymerization reactor using CFD. AIChE Midwest Conference, IIT, Chicago, IL.
25. Syamlal M (1998) MFIX documentation: Numerical techniques. DOE/MC-31346-5824 NTIS/DE98002029. National Technical Information Service, Springfield, Virginia
26. ANSYS fluent theory guide, release 15.0 (Nov 2013) Canonsburg, PA
27. Gobin A, Neau H, Simonin O, Llinas JR, Reiling V, Selo JL (2003) Fluid dynamic numerical simulation of a gas phase polymerization reactor. Int J Numer Meth Fluids 43:1199–1220
28. Gidaspow D, Bacelos MS (2018) Kinetic theory based multiphase flow with experimental verification. Rev Chem Eng 34:299–318
29. Jiradilok V, Gidaspow D, Damronglerd S, Koves WJ, Mostofi R (2006) Kinetic theory based CFD simulation of turbulent fluidization of FCC particles in a riser. Chem Eng Sci 61:5544–5559
30. Gidaspow D, Chandra V (2014) Unequal granular temperature model for motion of platelets to the wall and red blood cells to the center. Chem Eng Sci 117:107–113
31. Tartan M, Gidaspow D (2004) Measurement of granular temperature and stresses in risers. AIChE J 50(8):1760–1775
32. Mayank K, Gidaspow D, Koves WJ (2011) Circulation of Geldart D type particles: Part I-High solids fluxes measurements and computation under solids slugging conditions. Chem Eng Sci 66:1649–167
33. Kroger DG, Levy EK, Chen JC (1979) Flow characteristics in packed and fluidized rotating beds. Powder Technol 24:9–18
34. Kroger, DG, Abdelnour G, Levy EK, Chen, JC (1980) Centrifugal fluidization: effects of particle density and size distribution. Chem Eng Commun 5:55–67
35. Levy EK, Shakespeare, WJ, Tabatabaie-Raissi A, Chen JC (1981) Particle elutriation from centrifugal fluidized beds. AIChE J 77:86–95
36. Qian GH, Bagyi I, Pfeffer R, Shaw H, Stevens J (1999) Particle mixing in rotating fluidized beds: inference about the fluidized state. AIChE J 45(7):1401–1410
37. Qian GH, Bagyi I, Burdick IW, Pfeffer R, Shaw H (2001) Gas–solid fluidization in a centrifugal field. AIChE J 47(5):1022–1034
38. Marchisio DL, Pikturna JT, Fox RO, Vigil RD, Barresi AA (2003) Quadrature method of moments for population-balance equations. AIChE J 49:1266
39. Marchisio DL, Vigil RD, Fox RO (2003) Quadrature method of moments for aggregation-breakage processes. J Colloid Interface Sci, 258:322–334
40. Abbasi, E, Abbasian J, Arastoopour H (2015) CFD–PBE numerical simulation of CO_2 capture using MgO-based sorbent. Powder Technol 286:616–628
41. Abbasi E, Arastoopour H (2013) Numerical analysis and implementation of finite domain trial functions method of moments (FCMOM) in CFD codes. Chem Eng Sci 102:432–441

42. Strumendo M, Arastoopour H (2010) Solution of population balance equations by finite size domain complete set of trial functions method of moments (FCMOM) for inhomogeneous systems. Ind Eng Chem Res 49(11):5222–5230
43. Strumendo M, Arastoopour H (2009) Solution of bivariate population balance equations using the FCMOM. Ind Eng Chem Res 48(1):262–273
44. Strumendo M, Arastoopour H (2008) Solution of PBE by non-infinite size domain. Chem Eng Sci 63:2624–2640
45. Ahmadzadeh A, Arastoopour H, Teymour F, Strumendo M (2008) Population balance equations' application in rotating fluidized bed polymerization reactor. Chem Eng Res Des 8 (6):329–343

Chapter 5
Circulating Fluidized Beds for Catalytic Reactors

Hamid Arastoopour, Dimitri Gidaspow, and Robert W. Lyczkowski

5.1 Introduction

The report by Spath and Dayton [1] shows how syngas can be converted to liquid fuels, such as diesel or gasoline, and chemicals, such as methanol and ammonia for fertilizers, using catalytic reactors as shown in Fig. 5.1. These catalytic reactions are all highly exothermic. As shown in Chap. 4, the circulating fluidized bed is therefore ideal for the design of the large-scale reactors needed for the economic production of fuels and large-volume chemicals, such as ammonia.

Circulating fluidized bed (CFB) systems have been used in the literature for several processes such as fluid cracking catalyst (FCC) and CO_2 capture [2–5]. The advantages of CFB reactors over bubbling bed reactors are:

No bubble formation. While bubbles may help mixing, large bubbles shake the reactor and, hence, limit the reactor size. The absence of bubbles in risers has been known for a long time, but has not been taken advantage of in fluidized bed reactors, such as those used for silicon production, where generation of fines in the bubbles causes contamination of the highly purified silicon product (see Chap. 8).

Near isothermal operation. This is due to the reduction of the adiabatic reaction temperature by more than two orders of magnitude, as explained in Chap. 4. Hence, the rate of production of the chemicals increases by two orders of magnitude for almost the same reactor size [6].

Operation at increased velocity. The oil industry recognized this advantage half a century ago and switched from bubbling beds to CFBs to obtain high production rates.

The shale gas revolution has made the United States a great place to make chemicals. Hence there is a need to design larger, less costly reactors for making chemicals and liquid fuels. The multiphase computational fluid dynamics (CFD) approach has not yet been used effectively. The Chinese Academy of Sciences is

© Springer Nature Switzerland AG 2022
H. Arastoopour et al., *Transport Phenomena in Multiphase Systems*, Mechanical Engineering Series, https://doi.org/10.1007/978-3-030-68578-2_5

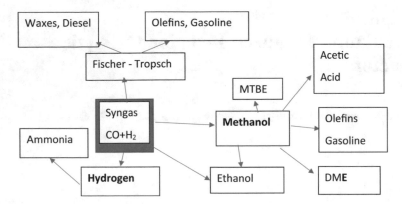

Fig. 5.1 Present and potential catalytic fluidized bed reactors for production of clean fuels and chemicals from syngas to be made from shale gas or coal. (Based in part on Spath and Dayton [1])

beginning to use the CFD approach for improvements of fluidized bed reactors [7, 8]. Gidaspow et al. [9] showed how to make syngas from coal using air in a CFB without the use of cyclones. Coal gasification is far more complex than oxidation of methane. Therefore, a shrinking core model usually has been used for coal gasification as a reasonable first approximation. Because it is often not accurate, certainly not for catalytic combustion, the first step is to review the assumptions.

5.2 Catalytic Rates of Reactions

Catalytic rates of reactions differ from gas phase reactions because they are proportional to the concentration of the catalyst. If there is no catalyst, there is no reaction. The larger the mass of the catalyst, the larger will be the rate of reaction, similar to that in the gas phase. A more than half-century-old example is the combustion of carbon deposited on the fluidized bed catalytic cracking catalyst presented in Pansing [10] in which the bubbling bed reactor is described by the rates of reaction as the sum of the resistances due to mass transfer and reaction.

Kashyap and Gidaspow [11] explained the origin of the very low Sherwood number in fluidized bed risers for the decomposition of ozone on a catalyst given by the reaction,

$$2O_3 = 3O_2 \tag{5.1}$$

The molar rate of reaction (moles i/m^3 s), r_i, for species i, ozone in this case, can be expressed,

$$r_i = KC_i\varepsilon_s \tag{5.2}$$

where K is the overall rate constant, and ε_s is the volume fraction of the catalyst. At steady state, the rate of mass transfer from the gas to the catalyst surface is equal to the rate of reaction at the catalyst surface,

$$k_{mass\ transfer}\ a_v\ (C_i - C_{i\ surface}) = k_{reaction}\ C_{i\ surface} \tag{5.3}$$

where $k_{mass\ transfer}$ is the mass transfer coefficient, m/s, and a_v is the surface area of the catalyst.

The overall rate constant, K, in Eq. (5.2) can be shown to be equal to the sum of the resistances,

$$1/K = 1/k_{mass\ transfer}\ a_v + 1/k_{reaction} \tag{5.4}$$

by first solving for the surface concentration in Eq. (5.3), an unknown,

$$C_{i\ surface} = [k_{mass}\ transfer\ a_v/(k_{mass\ ransfer}a_v + k_{reaction})]\ C_i \tag{5.5}$$

Then it is easy to see that K, the overall rate constant, is the expression given in Eq. (5.4).

The units of the mass transfer coefficient are m/s and those of the rate constants are s^{-1}. Equation (5.4) shows that, for rapid reactions, the mass transfer will control the overall rate of reaction.

The Sherwood number, Sh, is usually defined,

$$Sh = k_{mass\ transfer}\ d_p/D \tag{5.6}$$

where d_p is the diameter of the catalyst particle, and D is the diffusivity of species i. With diffusion controlling the rate of reaction, the minimum Sherwood number should equal 2. Measurements taken over decades have shown however that the Sherwood number in fluidized beds to be orders of magnitude lower. For fluidized bed risers, this is due to cluster formation. When the particle diameter is replaced by the cluster diameter, the Sherwood number becomes on the order of 2. Unfortunately, cluster diameters are a dynamic quantity, and it is difficult to obtain their precise values.

5.3 Shrinking Core Model and Rates in Conservation of Species

Consider the general gas/solid reaction,

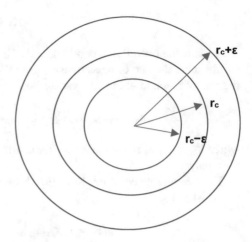

Fig. 5.2 Schematic of shrinking core radius

$$\gamma\, A(g) + S(s) = SA_\gamma \qquad (5.7)$$

where S is a solid phase, such as carbon, A is a gas phase species, γ is its stoichiometric coefficient, and SA_γ is the product of reaction. In spherical coordinates, the molar species balance for A is,

$$d/dr\left(r^2 dC_A/dr\right) = 0 \qquad (5.8)$$

where C_A is the molar concentration of gas species A. The boundary conditions are,

$$C_A(R_0) = C_0 \qquad (5.9)$$

$$C_A(r_C) = 0 \qquad (5.10)$$

It was assumed that, for the particle of radius, R_0, the surface concentration is C_0, and that, at the shrinking core radius, r_c, the gas species concentration vanishes.

A shock, or a stoichiometric balance, is now made around the shrinking core radius, r_c, in Fig. 5.2.

The species A balance in spherical geometry that results from this shock balance, neglecting accumulation in the gas phase compared to that in the solid phase, is as follows using Leibnitz's rule,

$$\gamma \mathrm{Lim}\, \varepsilon \to 0 \frac{d}{dt} \int_{r_c-\varepsilon}^{r_c+\varepsilon} C_s 4\pi r^2 dr = -\mathrm{Lim}\, \varepsilon \to 0 \ \left. \varepsilon v C_A 4\pi r^2 \right]_{r_c-\varepsilon}^{r_c+\varepsilon} \qquad (5.11)$$

where C_s is the molar concentration of the solid (equal to $\varepsilon_S \rho_S/M_S$), with ε_S the volume fraction of the solid phase, ρ_S its density, and M_S its molecular weight.

With the diluteness assumption already used in the spherical differential balance Eq. (5.8), the molar flux, $C_A v_A$, of the gas phase species is given by the simplified Fick's law for negligible bulk motion,

$$C_A v_A = -D \, dC/dr \tag{5.12}$$

Substitution of Eq. (5.12) into Eq. (5.8), using the boundary conditions given by Eqs. (5.9) and (5.10), results in the concentration gradient,

$$dC(r_c)/dr = C_0/\left(r_c - r_c^2/R_0\right) \tag{5.13}$$

Using Eqs. (5.12) and (5.11),

$$dr_c/dt = -(DC_0/\gamma_i C_{S0})[1/r_c(1 - r_c/R_0)] \tag{5.14}$$

Integration of Eq. (5.14) with the fraction W unconverted equal to,

$$W = (r_c/R_A)^3 \tag{5.15}$$

can be shown to give the well-known shrinking core formula,

$$1/2 \left(1 - W^{2/3}\right) - 1/3(1 - W) = (DC_0 t)/(R_0^2 \gamma_i C_{S0}) \tag{5.16}$$

This expression agrees with the experiments of Weisz and Goodwin for combustion of carbon deposited on several sizes of catalysts [12]. It gives the time for 85% burn-off, $T_{85\%}$,

$$T_{85\%} = 0.076 \left(R_0^2 \gamma_i C_{S0}/DC_0\right) \tag{5.17}$$

For a 0.2 cm particle consisting of 3% carbon, combusting with air having 20% oxygen, and using a diffusion coefficient $D = 0.01 \text{ cm}^2/\text{s}$, the time to 85% burn-off is 4 s.

Using Eq. (5.12), the molar flow per unit volume of particle becomes,

$$\left(3DC_0/R_0^2\right)(r_c/R_0)/(1 - r_c/R_0) \tag{5.18}$$

Multicomponent effects were neglected. Hence, the diffusion coefficient, D, is that of the component with the lowest value.

If we define r_i to be the molar flow, or rate per unit of particle and gas volume, as in [13], Eq. (5.13) gives,

$$r_i = (1 - \varepsilon) \left(3DC_0/R_0^2\right)(r_c/R_0)/(1 - r_c/R_0) \tag{5.19}$$

The gasification rate, m_i, from the solid species, S, mass balance,

$$\partial(\varepsilon_S\rho_S)/(\partial t) + \partial(\varepsilon_S\rho_S v_S)/(\partial x) = -m_i \tag{5.20}$$

shows that,

$$m_i = r_i\,M_S/\gamma \tag{5.21}$$

where the carbon particle, S, moves with the particle velocity, v_S, and M_S is the molecular mass of phase S.

5.4 Denn Shrinking Core Model

The Denn model [13] for coal gasification involves the sum of the resistances due to diffusion and reaction. In terms of the partial pressure in the gas phase for components i, p_i, and its equilibrium pressure, p_i^*, the rate is,

$$r_i = \varepsilon_S(p - p_i^*)/\left[d_p/6K_{p,i} + d_pRT\left(1 - W^{1/3}\right)/12DW^{1/3} + 1/(\eta WK_rC_C\right] \tag{5.22}$$

where d_p is the particle diameter, $K_{p,I}$ and K_r are overall rate constants, C_C is the initial concentration of carbon, and the effectiveness factor, η, is,

$$\eta = 1/\Psi(1/\tanh 3\Psi - 1/3\Psi) \tag{5.23}$$

with

$$\Psi = d_p/6[K_rC_CRT/D]^{1/2} \tag{5.24}$$

5.5 Combustion Reaction

The principal combustion reaction is that between carbon and oxygen,

$$C + \gamma_iO_2 = 2(1 - \gamma_i)CO + (2\gamma_i - 1)CO_2 \tag{5.25}$$

The stoichiometric coefficient, γ_i, depends upon the temperature and the type of coal used. We have used Kashyap and Gidaspow [11],

$$CO/CO_2 = 2(1 - \gamma_i)/(2\gamma_i - 1) = 10^{3.4} \exp(-12,400/RT) \qquad (5.26)$$

where T is the temperature of the solid in degrees K. The intrinsic reaction rate constant, K_r, is given by $K_r = 1.79 \times 10^6 \exp(-27,000/RT)$ mol/(mol C atm s).

The heat of reaction $\Delta H_{r,1298K}$ is $\Delta H_{r,1298K} = (-97.7)(2\gamma - 1) + (-28.714)$ $(2 - 2\gamma) + 2.40426$ kcal/(gmol K) and $p_{O2}{}^* = 0$.

5.6   Gasification Reactions

The principal gasification reaction is that which produces carbon monoxide,

$$C + \gamma_2 CO_2 = 2CO \qquad (5.27)$$

with the rate constant, $K_{r,2} = 930\exp(-45,000/RT)$ mol/(mol C s) and the equilibrium constant $K_{eq} = (p_{CO}{}^2/p_{CO2}{}^2) = 1.222 \times 10^9 \exp(-40,300/RT)$. The heat of reaction is $\Delta H_{298k} = 40.27356$ kcal/gmol. This carbon/carbon dioxide reaction proceeds very slowly compared with the combustion reaction.

The second reaction is that to produce methane,

$$C + 2H_2 = CH_4 \qquad (5.28)$$

with the rate constant,

$$K_3 = 8.36 \times 10^{-4} \exp(-1650/RT) \qquad (5.29)$$

The water/gas shift reaction is,

$$CO + H_2O = CO_2 + H_2 \qquad (5.30)$$

The rate of the catalytic water/gas shift reaction, in terms of the volume fraction, X, the solid catalyst volume fraction, εs, and its density, ρ_s, is,

$$\text{rate} = 0.775\exp\left[-8421.3/T_g\right]P^{0.5-P/250}\left(X_{CO}X_{H2O} - X_{Co2}X_{H2}/K_{wg}\right)\rho_s\varepsilon_s \quad (5.31)$$

gmol/(cm^3 s) with a rate constant,

$$K_{wg} = 0.0265 \exp(7860/RT_g) \qquad (5.32)$$

The heat of reaction for the water/gas shift reaction is −9.838 kcal/gmol.

This catalytic rate of reaction is identical to that in [14] for the U-GAS process for coal gasification developed by the Institute of Gas Technology. The water/gas shift reaction can be used to produce hydrogen for fuel cells for pollution-free cars. The expensive oxygen in the partial oxidation reaction can be replaced by carbon dioxide.

5.7 Circulating Fluidized Bed (CFD) Simulations for Synthesis Gas

The first CFD model for a circulating fluidized bed, with a riser and down comer, was published by Gidaspow et al. [9]. Results for the kinetic theory model are described in [15]. The non-CFD design, including rates of reaction, is described in [16].

Figure 5.3 shows a simplified full-scale Pyropower CFB, together with dimensions and flow conditions. In place of a cyclone used in such units, the particles are

Fig. 5.3 Simplified full-scale Pyropower CFB with dimensions and flow conditions [17]

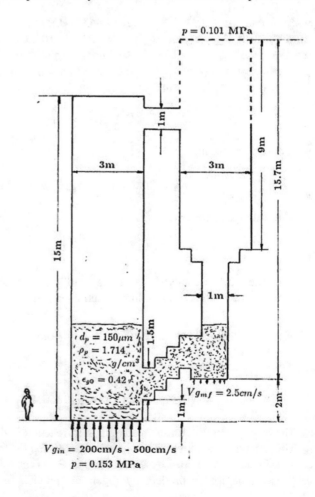

Fig. 5.4 Particle distribution (left) and solids velocity (right) at time = 10 s with inlet gas velocity $Vg_{in} = 500$ cm/s [17]

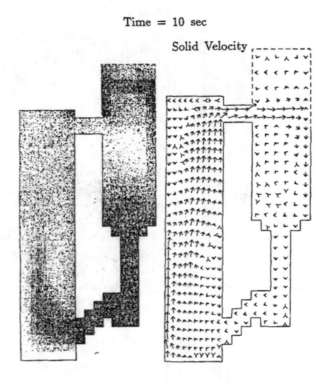

Time = 10 sec

Solid Velocity

separated by gravity, with a large settler above the down comer. Repeated CFD simulations have shown that the cyclones can be eliminated, producing a simpler design. The initial bed height was 4.5 m, and the porosity was 0.42, as indicated. The Denn shrinking core model [13], with the combustion, gasification, and water/gas shift reactions already discussed were used in the simulations.

Figure 5.4 shows the particle distribution and the velocities in the CFB at 10 s from start-up for the high inlet velocity of 5 m/s. The concentration in the down comer on the right is much more dense than that in the riser due to the high inlet velocity there. The concentration in the settler on top of the riser is dense, thus allowing the particles to drop into the down comer.

Figure 5.5 shows the solids mass flux distributions at two locations in the riser at the high inlet velocity of 5 m/s. The fluxes are high. There is down flow at some of the walls.

Figure 5.6 shows the mole fractions of CO_2, steam, and oxygen in the gasifier at 15 s. The oxygen from the inlet air reacted almost completely near the bottom of the reactor. Figure 5.7 shows the gas temperatures as a function of time at the height of 4.5 m at three radial locations.

Time-averaged values were obtained over a 10 s transient. Such results must be obtained as the first step for data analysis to get correct time-averaged values for the computed variables. Figure 5.8 shows the outlet gas compositions as a function of time. One can see that 10 s is a good number to begin obtaining time-averaged

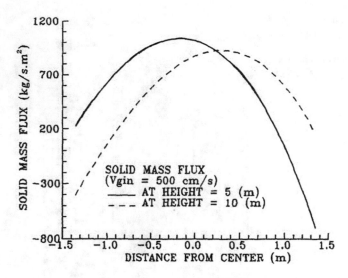

Fig. 5.5 Time-averaged radial profiles of solid mass flux with inlet gas velocity $Vg_{in} = 500$ cm/s [17]

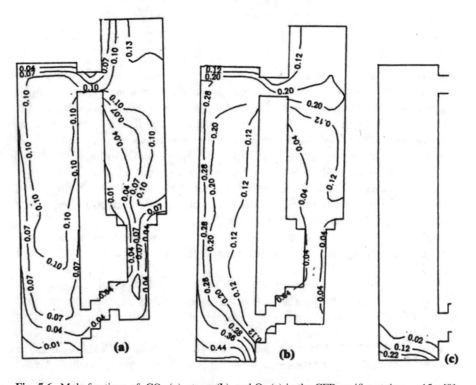

Fig. 5.6 Mole fractions of: CO_2 (**a**), steam (**b**), and O_2 (**c**) in the CFB gasifier at time $= 15$ s [9]

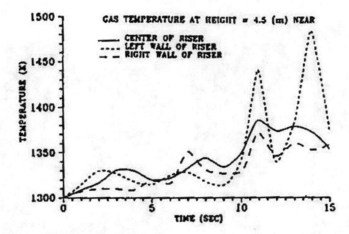

Fig. 5.7 Gas temperatures from time = 0 to 15 s [9]

Fig. 5.8 Outlet gas compositions from time = 0 to 15 s [9]

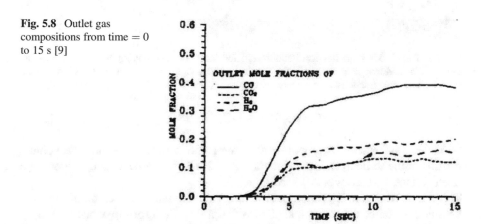

compositions and that 15 s is sufficient for time averaging. In this simulation, the outlet gas consists principally of carbon monoxide. More hydrogen is needed for synthesis gas.

5.8 CFB Simulations for Sulfur Dioxide Capture

Therdthianwong and Gidaspow [18] modeled the sulfur dioxide sorption in the Pyropower CFB reactor discussed above. The reaction of the sorbent, calcium oxide, with sulfur dioxide and oxygen is described by Basu and Frazer [16],

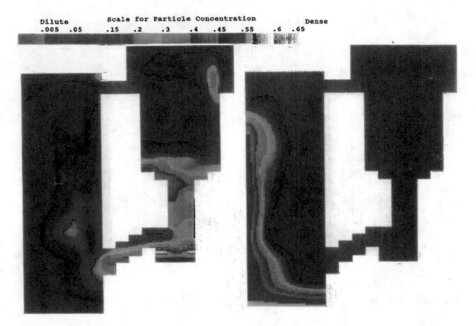

Fig. 5.9 Particle concentration distributions (left) and SO_2 concentrations (right) in the CFB for the asymmetric solid feed case at $Vg_{in} = 5$ m/s at time $= 20$ s. Color contours represent solid volume fractions and mole fractions, respectively [19]

$$CaO + SO_2 + \tfrac{1}{2}O_2 = CaSO_4 \tag{5.33}$$

The rate of reaction is proportional to the sulfur concentration and the volume fraction of the sorbent, similar to that in Eq. (5.2) for the ozone decomposition. The rate constant, K, used was 2018 mol/(cm³ s).

The rate of reaction is proportional to the sulfur dioxide concentration and the volume fraction of the sorbent, similar to that in Eq. (5.2) for the ozone decomposition. The rate constant, K, used was 2018 mol/(cm³ s). The particle diameter was 210 μm with a density of 1530 kg/m³. The inlet SO_2 concentration was 5000 ppm. The other initial and inlet conditions were the same as those for the synthesis gas discussed above.

Figure 5.9 shows the particle and SO_2 concentrations at 20 s for an inlet velocity of 5 m/s. The down comer is dense as expected. However, due to the symmetric inlet into the riser, the right wall has a higher sorbent concentration. This causes much better SO_2 sorption at the right wall. The total absorption is 97.52%, with an outlet SO_2 concentration of 123 ppm at the inlet velocity of 5 m/s.

The outlet concentration of SO_2 is 4 ppm for an inlet velocity of 3 m/s with 99.92% removal (see Fig. 5.10). A symmetric feed into the riser gives even better removal, but at the expense of a much more complex system.

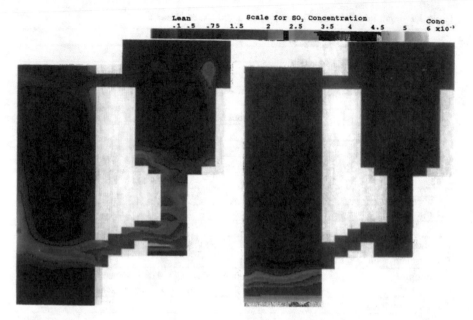

Fig. 5.10 Particle concentration distributions (left) and SO_2 concentrations (right) in the CFB for the asymmetric solid feed case at $Vg_{in} = 3$ m/s at time $= 20$ s. Color contours represent solid volume fractions and mole fractions, respectively [19]

5.9 Exercises

5.9.1 Ex. 1: Catalytic Conversion of Methane to Synthesis Gas by Partial Oxidation

Y.-H. Hu and Eli Ruckenstein [20] have given an excellent review of synthesis gas production. They point out that the development of hot spots is an issue that limits the design of such reactors. Multiphase CFD is a tool that can compute the temperature distribution in such reactors. Ding and Gidaspow [21] modeled the hydrodynamics of a reactor with a downward jet for oxygen injection built at Illinois Institute of Technology (IIT) that can be scaled up to commercial-size units. The objective of this exercise is to model such a reactor that is large enough to demonstrate the feasibility of this process. A larger-diameter reactor can then be designed for commercial production of synthesis gas.

The dimensions of the pilot plant reactor and the operating conditions are given in Table 5.1. The reaction models to be used are similar to those in the literature and are shown in Table 5.2. The effect of cohesion can be ignored. The commercial code ANSYS Fluent [22], the Department of Energy MFIX code [23], or the IIT code described in [24] can be used for the simulations in two dimensions.

Table 5.1 Input data for synthesis gas by partial oxidation of methane

Geometry	
Height	20 m
Width	0.4 m
Central jet width	2 cm
Central jet height	4 cm
Central jet location	2 m from bottom
Inlet conditions	
Velocity	
Bottom	30 cm/s
Jet	30 m/s
Temperature	
Bottom	920 K
Jet	530 K
Pressure	25 atm (25.3 kPa)
Composition, weight fraction	
Bottom	0.7 methane, 0.02 CO, 0.04 hydrogen 0.03 CO_2, 0.21 steam, 1.0 jet oxygen
Particle properties	
Diameter	45 μm
Density	3.0 g/cm^3
Initial conditions	
Temperature	1300 K
Pressure	25 atm (25.3 kPa)
Composition, weight fraction	0.4 methane, 0.3 hydrogen, 0.1 steam, 0.1 CO, 0.1 CO_2
Filled bed height	9 m
Solids volume fraction	0.6

The variables to be computed are the porosity, the pressure, the velocity components for the gas and the solids particles, the two temperatures for the gas and the solids, and the weight fractions of each component.

1. Run the simulation for about 15 s. Make a video of the variables to see the initial bubble formation. The first bubble is large. Time average the variables after a steady state has been reached.
2. Starting with the suggested initial conditions, compute the temperature, pressure, concentrations, and velocities as a function of time. The temperature is the critical variable because it changes only slowly due to the large heat capacity of the solids compared with that of the gas. Hence, its initial guess is critical. Commercial CFD codes cannot be run for a long time due to the large number of variables. Their convergence is normally only on pressure. But, in bubbling beds, the porosity in the free board is an order of magnitude lower than that in the lower bed section.

   This order of magnitude difference prohibits good convergence of concentrations. Errors in concentrations cause the CFD code to stop when all the variables

Table 5.2 Reaction rates for partial oxidation of methane

Steam reforming
$CH_4 + H_2O = CO + 3H_2$
See also Gidaspow et al. [9]
In the gas phase,
Rate of reaction (gmol/cm^3 s) = $k_1(C_{CH4} - C_{CH4eq})$
The first order rate constant, k_1 has units of s^{-1} and is,
$k_1 = 0.1\exp[-24,811(1/T_g - 1/1256.6)]$
In the solid catalyst phase the rate constant, k_{1s}, is,
$k_{1s} = 2.3\exp[-16,104(1/T_s - 1/1255.6)]$.
The equilibrium methane concentration expression in units of gmol/cm^3 is,
$C_{CH4,eq} = (P_{CO}P_{H2}{}^3)/(K_{eq}P_{H2O}RT)$ with an equilibrium constant K_{eq} is,
$K_{eq} = \exp[-(26,672 - 29.5T - 0.00045T^2)/T]$ in units of atm^2
The temperature, T, is in degrees Kelvin and the pressure, P, is in atmospheres. The total reaction rate is the sum of the gas and the catalyst rates. The rates have to be multiplied by their respective volume fractions.
The value of the heat of reaction at 298 K is 49.27 kcal/gmol
Partial oxidation
$CH_4 + O_2 = CO + H_2 + H_2O$
Rate = 23.1 C_{CH4}
The heat of reaction is −66.33 kcal/gmol
Water/gas shift reaction
$CO + H_2O = CO_2 + H_2$
The rate of the catalytic water/gas shift reaction, in terms of the volume fractions, X, the solid catalyst volume fraction ε_s, and its density, ρ_s, is,
rate = $0.775\exp[-8421.3/T_g]P^{0.5-P/250}(X_{CO}X_{H2O} - X_{Co2}X_{H2}/K_{wg})\rho_s\varepsilon_s$ gmol/cm^3 s) with a rate constant,
$K_{wg} = 0.0265\exp(7860/RT_g)$
The heat of reaction for the water/gas shift reaction = −9.838 kcal/gmol. This catalytic rate of reaction is identical to that in the paper by Gidaspow et al. for the U-GAS process for coal gasification developed by IGT [14].

are solved simultaneously. This limitation also prohibits the programming of complex reaction mechanisms.

3. Obtain a graph for time-averaged temperature distributions and oxygen concentrations near the inlet oxygen jet to see whether oxygen reacted properly to form synthesis gas. Explain the observed near-circular behavior.
4. Plot the time-averaged mole fractions of methane, water vapor, remaining small concentrations of oxygen, and carbon monoxide along the bed height for several radial positions. Where should the outlet of the reactor be located?
5. Find the rate of production of the synthesis gas. What reactor size should you increase to double the production rate? How else can you increase the production rate?

5.9.2   Ex. 2: Catalytic Conversion of Methane to Synthesis Gas in a Riser No Bubbles

1. Repeat Exercise 1 for a riser using the same size reactor. Program the outlet to be on the top left side of the reactor. Close the top of the reactor. Introduce the catalyst through the inlets at the same velocities as the inlet gases.
2. To increase the rate of synthesis gas production, increase the particle size to a diameter of 500 μm. For this particle size, the Thiele modulus shows that diffusion decreases the rates of reaction by about 30%. However, because much higher velocities produce much higher production rates, this reduction is tolerable. Find the velocities needed to obtain such high synthesis gas production rates.

5.9.3   Ex. 3: Catalytic Conversion of Methane to Synthesis Gas in a Circulating Fluidized Bed

1. Repeat Exercise 2 for a circulating fluidized bed using 500 μm diameter catalyst particles. Unlike conventional fluidized beds, such a fluidized bed needs no cyclones. For a picture of the downer and the riser, see [19].

5.9.4   Ex. 4: Parabolic Rusting Law

1. Consider diffusion-controlled rusting of iron. Show that, as an approximation, the rate of reaction in terms of moles of iron reacted per square meter per second is,

$$\text{rate} = [3DC_o \, \rho/8tM]^{0.5} \tag{5.34}$$

where D is the diffusion coefficient of oxygen, C_o is the molar oxygen concentration at the surface, M is the molecular weight of iron, and t is the time of exposure to rusting. Comment on the simplifying assumptions made in the derivation.

5.9.5   Ex. 5: CFD Scale-Up of a Fluidized Bed Coal Gasification Process, IGT U-GAS Process

In the 1970s, before the invention of CFD, the Institute of Gas Technology (IGT) developed and commercialized a fluidized bed coal gasification process using air. The scale-up was done using one-dimensional and black box models. Gidaspow, Ettehadieh, and Bouillard, in their paper, "Hydrodynamics of Fluidization: Bubbles and Gas Compositions in the U-GAS Process," AIChE Symposium Series

No. 241, Vol. 80, pp. 57–64, used a supercomputer to model this process. The cone geometry was approximated by three rectangular steps shown in Fig. 1 in the above-mentioned paper. It was assumed that the gasifier was initially filled with nitrogen and that it was at minimum fluidization. At zero time, the flows were increased to the indicated values. In the published paper, the results shown were for only a short time due to limitations of computer time. At least 10 s are needed to obtain a steady state. Today's work stations need about a day to compute 10 s of real time. Table 5.2 presents the reaction model.

1. Compute the time-averaged porosities, velocities, and the concentrations of oxygen, steam, nitrogen, carbon monoxide, and hydrogen.
2. What is the rate of coal gasification using the shrinking core model?
3. In the simulations, large bubbles are formed. Show how you can eliminate these bubbles.

Nomenclature

C_i	Molar concentration of species i
D	Diffusivity
H	Enthalpy
K	Overall rate constant
$k_{mass\ transfer}$	Mass transfer coefficient
$k_{reaction}$	Reaction rate constant
M	Molecular weight
m	Mass
m_i	Rate of mass production
p_i	Pressure of component i
p_i^*	Equilibrium pressure of component i
R_0	Outside radius
r	Radial direction
r_c	Core radius
r_i	Rate of reaction of species i
T	Temperature
v	Velocity
W	Weight fraction

Greek Symbols

γ_I	Stoichiometric coefficient
ε	Gas volume fraction, also limit variable in Eq. (5.10)
ε_s	Solid volume fraction
η	Effectiveness factor

References

1. Spath PL, Dayton DC (2003) Preliminary screening-technical and economic assessment of synthesis gas to fuels and chemicals with emphasis on the potential for biomass-derived syngas. NREL/TP-510-34929
2. Arastoopour H, Gidaspow D (2014) CFD modeling of CFB: from kinetic theory to turbulence, heat transfer, and poly-dispersed systems. 11th International Conference on Fluidized Bed Technology (CFB 11), Beijing, China, May 14–17
3. Arastoopour H, Gidaspow D, Abassi E (2017) Computational transport phenomena of fluid-particle systems. Springer, Cham, Switzerland
4. Benyahia S, Arastoopour H, Knowlton TM, Massah H (2000) Simulation of particles and gas flow behavior in the riser section of a circulating fluidized bed using the kinetic theory approach for the particulate phase. Powder Technol 112:24–335
5. Ghadirian E, Abbasian J, Arastoopour H (2019) CFD simulation of gas and particle flow and a carbon capture process using a circulating fluidized bed (CFB) reacting loop. Powder Technol 344:27–356
6. Gidaspow D (2018) Provisional U.S. patent application, confirmation No.8603, application number 62/615,798, 01/19/2018
7. Li J (2008) Plenary lecture presented at the 8th World Congress on Particle Technology, Orlando, Florida, April 22-26, 2008
8. Zhang J, Lu B, Chen F, Li H, Ye M, Wang W (2018) Simulations of a large methanol-to-olefins fluidized bed reactor with consideration of coke distribution. Chem Eng Sci 189:212–220
9. Gidaspow D, Bezburuah R, Ding J (1992) Hydrodynamics of circulating fluidized beds: kinetic theory approach. In: Potter OE, Niklin DJ (eds) Fluidization VII proceedings of the seventh engineering conference on fluidization, May 3-8 1992, Engineering Foundation, New York, pp 75–82
10. Pansing WF (1956) Regeneration of fluidized cracking catalysts. AIChE J 2:71-74
11. Kashyap M, Gidaspow D (2012) Measurements and computation of low mass transfer coefficients for FCC particles with ozone decomposition reaction. AIChE J 58:707–729
12. Weisz PB, Goodwin RD (1963) Combustion of carbonaceous deposits within porous catalyst particles I. diffusion-controlled kinetics. J Catalysis 2:397–404
13. Yoon H, Wei J, Denn MM (1978) A model for moving bed coal gasification reactors. AIChE J 24:885–903
14. Gidaspow D, Ettehadieh B, Bouillard JX (1984) Hydrodynamics of fluidization: bubbles and gas compositions in the U-GAS process. AIChE Symposium Series No. 241, Vol. 80, 57-64, American Institute of Chemical Engineers, New York
15. Gidaspow D (1994) Multiphase flow and fluidization. Academic Press, New York
16. Basu P, Frazer SA (1991) Circulating fluidized bed boilers. Butterworth-Heineman
17. Ding J (1990) A fluidization model using kinetic theory of granular flow. PhD thesis, Illinois Institute of Technology, Chicago, IL
18. Therdthianwong A, Gidaspow D (1993) Hydrodynamics and SO_2 sorption in a CFB loop. In: Avidan AA (ed) Circulating fluidized bed technology IV, pp 351–358
19. Jayaswal UK (1991) Hydrodynamics of multiphase flows separation, dissemination, and fluidization. PhD thesis, Illinois Institute of Technology, Chicago, IL
20. Hu YH, Ruckenstein E (2004) Catalytic conversion of methane to synthesis gas by partial oxidation and CO_2 reforming. Adv Catal 48:297–345

21. Ding J, Gidaspow D (1994) A semi-empirical model for fluidization of fine particles. Indian Chem Engr 36(4):139–150
22. ANSYS, Inc., FLUENT user's guide. Canonsburg, PA
23. Syamlal M, Rogers W, O'Brien TJ (1993) MFIX documentation theory guide, technical note. DOE/METC-94/1004, NTIS/DE94000087, U.S. Department of Energy, Office of Fossil Energy, Morgantown Energy Technology Center Morgantown, WV. National Technical Information Service, Springfield, VA
24. Gidaspow D, Jiradilok V (2009) Computational techniques: the multiphase CFD approach to fluidization and green energy technologies. Nova Science Publishers, New York

Chapter 6
Synthetic Gas Conversion to Liquid Fuel Using Slurry Bubble Column Reactors

Hamid Arastoopour, Dimitri Gidaspow, and Robert W. Lyczkowski

6.1 Introduction

The shale gas revolution has turned the United States into a great place to make chemicals. There is also a need to develop energy efficient technology processes to convert shale gas that is presently flared in North Dakota and carbon dioxide emitted by burning fossil fuels into storable liquid fuels, such as diesel.

The first step in the process is the production of synthesis gas by reaction of methane with steam. This highly endothermic reaction is carried out in heated packed tubular reactors. About three decades ago, Exxon developed a fluidized bed process to produce syngas by partial oxidation of methane. This more efficient process of producing syngas was not commercialized, probably due to the high cost of oxygen, about 40% of the cost of syngas. More recently, it was found that methane can be oxidized by carbon dioxide on 1% Ru/Al_3O_2 catalyst with rates of reaction much faster than those of the conventional steam reforming reaction used in the Exxon fluidized bed process. The discovery of this catalyst suggests that a fluidized bed reactor can be designed to make syngas with the carbon dioxide waste product in place of the expensive oxygen.

The second step in the production of liquids from shale gas is the conversion of syngas into liquids, such as diesel, using Fischer-Tropsch technology. The National Energy Technology Laboratory (NETL) has shown that the best reactors for this purpose are the slurry bubble column type. Commercially, they produce up to 24,000 bbl/day of diesel fuel. Traditionally, they are designed using hold-up correlations [1–3]. However, researchers at Illinois Institute of Technology (IIT) have been able to design them using the principles of conservation of mass, momentum, and energy for each of the phases and have developed a multiphase computational fluid dynamics (CFD) program for such designs [4–6].

This fundamental approach leads to the elimination of all cooling tubes in the reactor design [7] described in this chapter. Hence, the cost of novel reactors based on this design is much lower than for those used in the past. The cooling tubes can be

© Springer Nature Switzerland AG 2022   149
H. Arastoopour et al., *Transport Phenomena in Multiphase Systems*, Mechanical Engineering Series, https://doi.org/10.1007/978-3-030-68578-2_6

eliminated by overflow of the hot liquid product due to the reduction of the adiabatic temperature rise from 1600 °C to only 16 °C, with a liquid-to-gas density ratio of 100. The use of large catalyst particles greatly reduces the filtration problem encountered in such reactors in the past. Elimination of the cooling tubes also greatly reduces the attrition of the catalysts in the reactors of the present invention.

6.2 Diesel Fuel Reactor

Figure 6.1 is a schematic of the slurry bubble column reactor (SBCR) without cooling tubes [5, 7]. It is similar to Fig. 1 in [4], or to the right of Fig. 2 in [1] for making methanol from synthesis gas, except that, in this case, the internal heat exchangers have been eliminated.

The heat of reaction is now removed by filtered overflow of the hot liquid product. The cold or hot synthesis gas enters the reactor through three or more jets at high velocity to produce superficial velocities in the reactor of 0.05–1 m/s, depending upon the catalyst and reactor size. These jets produce high turbulence in the reactor, eliminating mass transfer resistances. The porous catalysts are manufactured to be larger than those used in the reactor designs of [1, 4]. The large catalyst particles, of the order of 500 µm diameter or more, are necessary to achieve high velocities in the reactor without blow-out of the catalyst. The low rate of reaction of the Fischer-Tropsch catalysts, on the order of 0.01 s^{-1}, permits the use of large catalyst sizes without a severe internal diffusion limitation. The use of large

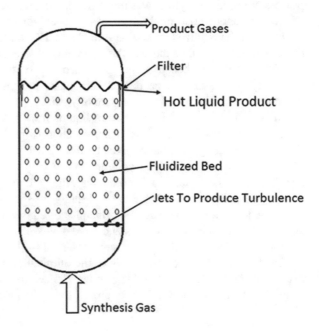

Fig. 6.1 Slurry diesel fuel reactor with no cooling tubes. (This figure was originally published in [5] and has been reused with permission)

Table 6.1 Input data for diesel fuel reactor: 500 µm and (0.5 cm) catalyst diameters

Geometry	
Height	30 m
Inventory	12 m
Width	7 m
Solid and liquid properties	
Density of catalyst	1490 kg/m^3
Density of diesel fuel	700 kg/m^3
Diameter of catalyst	500 µm (0.5 cm)
Droplet diameter	1000 µm (1 cm)
Inlet gas weight fractions	
CO	0.82
H$_2$	0.18
Grid sizes	
Lateral	$\Delta X = 0.167$ m
Axial	$\Delta Y = 0.283$ m
Number of grids	
Radial × axial	42 × 106 cells
Operation conditions	
Superficial gas velocity	0.05 m/s (0.12 m/s)
Inlet jet velocity (five jets)	0.20 m/s (0.5 m/s)
Pressure	3.01 MPa
Inlet gas temperature	300 K
Effective first order rate constant	0.00811 s^{-1}
Initial conditions	
Initial volume fraction of solid and liquid	0.29 each
Initial temperature	513 K

catalyst particles minimizes the filtration problems encountered in the past. The product gases and the unreacted synthesis gas are removed at the top of the reactor in the free board through multiple outlets. The free board is designed to be large enough to prevent the loss of catalyst and the liquid product. This design is done using a unique multiphase CFD program and theory [4–6].

A hydrodynamic model for the production of diesel fuel from synthesis gas in slurry bubble column reactors was developed [4–6]. The hydrodynamic approach to modeling gas/liquid/solid flow is based on the principles of conservations of mass, momentum, and energy for each phase. The liquid phase is considered to comprise droplets, as in the model in [4]. The conservation laws and constitutive equations are summarized in Table 1 in [5]. They are essentially the same as presented in Table 4.2 in Chap. 4 on polymerization. In this study, the particle viscosities and stresses have been approximated rather than computed from kinetic theory. Table 6.1 summarizes the input data. Two inlet velocities for the five inlet jets were studied, 0.2 and 0.5 m/ s, producing superficial gas velocities of 0.05 and 0.12 m/s as shown in Table 6.1.

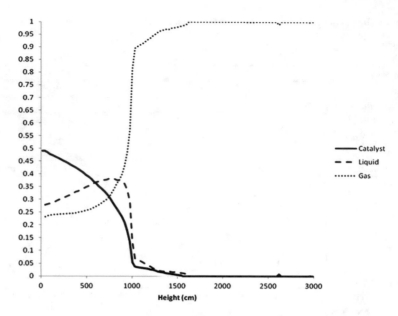

Fig. 6.2 Computed time-averaged (from 5 to 25 s) catalyst, liquid, and gas volume fractions for 7 m diameter Fischer-Tropsch reactor. (This figure was originally published in [5] and has been reused with permission)

Figures 6.2, 6.3, 6.4, 6.5 and 6.6 are for the case of the five inlet jet velocities equal to 0.2 m/s to produce a superficial velocity of 0.05 m/s. The catalyst particle diameter used is 500 μm and the droplet diameter used is 1000 μm as shown in Table 6.1. Figure 6.2 shows the computed time-averaged volume fractions, often called holdup, of the gas, liquid, and the 500 μm diameter catalyst at the center of the SBCR, with dimensions and properties given in Table 6.1. The computations showed no significant radial variation. Qualitatively, the computed volume fractions were similar to those measured in and computed for the Air Products methanol reactor [6]. The volume fraction of the catalyst is highest at the bottom of the reactor because its density is higher than that of the liquid.

Computations show that by making the catalyst porous enough to have the same density as the liquid, the catalyst will have a more uniform concentration. Such a catalyst will produce a more complete conversion of the synthesis gas into liquids. The liquid has a maximum volume fraction, again due to the lower density of the liquid as shown in Fig. 6.2. In the free board, the gas volume fraction is one. This figure shows that the free board need not be as large as shown.

Figure 6.3 shows the time-averaged axial velocity in cm/s for the 500 μm diameter catalyst in the SBCR. There is up flow near both walls and down flow at the center. The turbulence, as measured by the standard deviation of the catalyst velocity, is of the order of this velocity. It is high enough to produce good mixing. This velocity distribution is similar to experimental measurements done with large

Fig. 6.3 Axial velocity
(cm/s) of catalyst (time-
averaged from 0 to 38.8 s).
(This figure was originally
published in [5] and has
been reused with
permission)

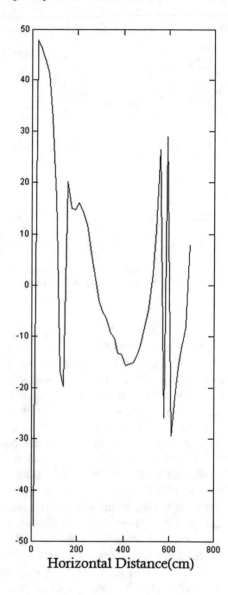

Horizontal Distance(cm)

particles at Illinois Institute of Technology using velocity image techniques, but
different from the Air Products SBCR, where there was basically up flow at the
center and down flow near the wall [5]. Computations with high velocities produce
the more common behavior with down flow at the walls.

Figure 6.4 illustrates the computed carbon monoxide concentration in the SBCR
as a function of time starting with an initial weight fraction of 0.82. After about 40 s,

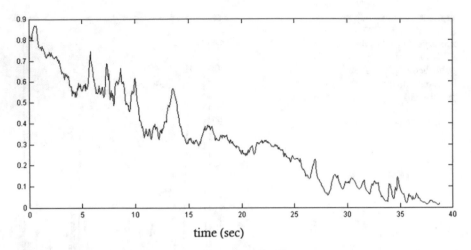

time (sec)

Fig. 6.4 Outlet weight fraction of CO (inlet and initial weight fraction CO = 0.82). (This figure was originally published in [5] and has been reused with permission)

there is nearly complete reaction. A plug flow approximation for the gas confirms this result. Figure 6.5 illustrates the computed hydrogen concentration in the SBCR as a function of time starting with an initial weight fraction of 0.18. Again, by 40 s, there is nearly complete reaction. Figure 6.6 shows the weight fraction of the gaseous product, water vapor as steam, starting with zero initial steam.

Figures 6.7, 6.8 and 6.9 are for the case the five inlet jet velocities increased to 0.5 cm/s to produce a superficial velocity of 0.12 m/s. The catalyst particle diameter used is 0.5 cm and the droplet diameter used is 1 cm as shown in Table 6.1. Figure 6.7 illustrates the axial velocity of the liquid droplets in the SBCR for a gas superficial velocity of 0.12 m/s.

Figure 6.8 shows the granular temperature for the same conditions as in Fig. 6.7. The granular temperature is defined as the variance of the liquid droplet velocity. High granular temperature means high turbulence. High turbulence means good mixing and reduced mass transfer resistance.

Figure 6.9 illustrates the rise in temperature of the 0.5 cm catalyst particles at the center of the SBCR starting with an initial temperature of 513 K. The inlet temperature and the reactor walls were at 300 K. After 20 s, the temperature rose to 535 K. A steady state was reached. The temperature rise was of the order of the expected adiabatic temperature rise. At the high reactor pressure and temperature, the adiabatic temperature rise is 65 K. The lower computed temperature rise in the reactor is due to wall cooling. With no liquid withdrawal, the catalyst temperature rises without limit.

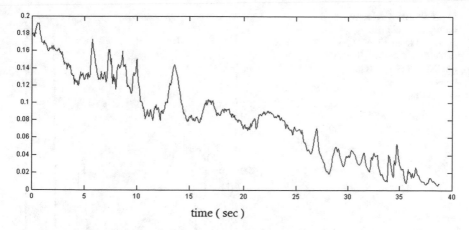

Fig. 6.5 Outlet weight fraction of hydrogen (inlet and initial = 0.18). (This figure was originally published in [5] and has been reused with permission)

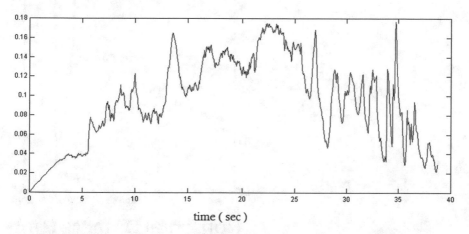

Fig. 6.6 Outlet weight fraction of product water vapor. (This figure was originally published in [5] and has been reused with permission)

6.3 Reactor Model for Fischer-Tropsch Kinetics

To minimize the water/gas shift reaction, the IIT reactor model for syngas reactor design is based on the use of cobalt catalyst, with the intrinsic kinetics [8],

$$-R_{co} = \frac{aP_{co}P_{H_2}}{(1 + bP_{co})^2} \tag{6.1}$$

Fig. 6.7 Axial velocity (cm/s) of liquid (time-averaged from 10 to 23.5 s). (This figure was originally published in [5] and has been reused with permission)

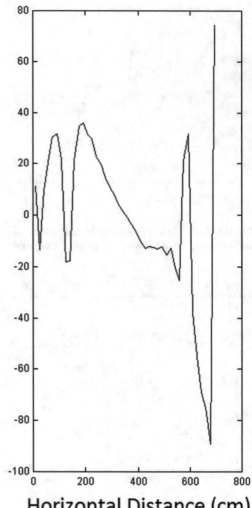

Horizontal Distance (cm)

Reactor temperature (°C)	a	b
240	75.76	11.61
220	53.11	22.26

The reaction rate, R_{co}, has the units, $\frac{mmol \cdot CO + H_2}{min \cdot g \cdot catalyst}$. The pressures, P_{CO} and P_{H_2}, are in MPa.

Fig. 6.8 Granular temperature of liquid (time-averaged from 10 to 23.05 s). (This figure was originally published in [5] and has been reused with permission)

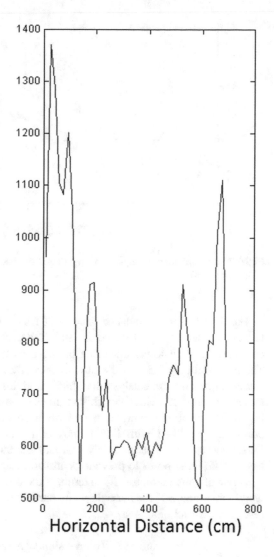

It is convenient to have the reactor model in terms of an effective first-order rate constant. For the hydrogen species balance, the rate of reaction, r, in terms of $\frac{\text{kg } H_2}{\text{kg } catalyst \cdot s}$ is related to R_{co} through its molecular weight. Then, the rate constant, k, in s^{-1} is related to r, the rate of reaction in the species balance through the ratio of the entering gas to solid catalyst volume,

$$k = r \times \left(\frac{\varepsilon_s \rho_s}{\varepsilon_g \rho_g}\right) \tag{6.2}$$

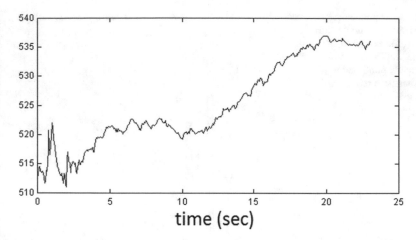

Fig. 6.9 Temperature of catalyst. (This figure was originally published in [5] and has been reused with permission)

For the CO species balance, $k = 0.0081$ s^{-1}. This rate constant agrees within a few percent with those of Post et al. [9] of the Shell Laboratory in Amsterdam, when corrected for their lower operating pressure by multiplying by 3.01/2.1 MPa. The Thiele modulus is 1.3 and the effectiveness factor is 98% for a 500 μm diameter catalyst. For a 0.5 cm catalyst, the Thiele modulus is about 6.5 and the effectiveness factor is about 40%. However, this assumes that the catalyst pores are all filled with liquid. The catalyst should be made with large and small pores, and only the small pores will be filled with liquid, thus eliminating the pore diffusion limitation. Furthermore, such a large catalyst should be encased in a thin metal or ceramic casing with large pores to prevent its attrition caused by catalyst collisions.

A rate of conversion for CO in terms of the difference between the inlet and outlet weight fractions, ΔY, the mass flow rate per unit area entering the reactor, ρv, and the reactor cross-sectional area, A, is,

$$\text{rate of CO conversion} = A \cdot \rho_{mixture} \cdot v \cdot \Delta Y \tag{6.3}$$

In this example, $\rho_{mixture} = 28.4$ kg/m^3 entering at 300 K and 3 MPa. The superficial velocity, $v = 0.05$ m/s, and ΔY is approximately 1.0. Since the stoichiometry shows that we produce one mole of diesel per mole of CO, we produce 0.5 kg of CH$_2$ per kg of CO. Hence, the diesel production rate is,

$$0.5 \, A \cdot \rho_{mixture} \cdot v \cdot \Delta Y \tag{6.4}$$

In the IIT model, $A = 38.5$ m^2·and $\rho_{mixture} = 28.4$ kg/m^3. Hence, the diesel production rate is 0.5×38.5 m$^2 \times 28.4$ kg/m$^3 \times 0.05$ m/s $= 22.4$ kg CH$_2$/s $= 21,200$ bbl/day.

The production rate is high for the 500 µm diameter catalyst particles due to the use of the large catalyst particle size. An even larger catalyst diameter will give a much larger production rate due to the permissible use of the larger flow rates. But beyond a 0.5 cm diameter, pore diffusion will limit the rate of diesel production. Computed production rates of up to 100,000 bbl/day were achieved in the 7 m diameter reactor without catalyst blow-out at the high gas superficial velocities on the order of 0.5 m/s. For such high gas velocities, the slurry bubble height has to be increased to 30 m or more to achieve high conversion, without gas recycle.

6.4 Potential High Production Reactor Simulation

To obtain a high production, such as a 100,000 bbl/day unit, without increasing the already high diameter, the syngas velocity can simply be increased, but this increase will be at the expense of large bubble formation. Due to the normally sudden change in the CFD simulations from an initial condition with a small flow rate to that of a high flow, the first bubble will be large and will not be shown here. The simulation shown here is for conditions of steady state operation. To speed up the simulations, no reactions are included.

In this simulation, the three inlet jet velocities are increased to 3.5 m/s. This increases the production rate, as given by a factor of 14 (Eq. 6.4). Hence, even with incomplete conversion, the goal of 100,000 bbl/day is easily achieved, without increasing the reactor height. Also, to minimize segregation, the catalyst density is decreased to 0.7 g/cm^3. The first bubble formed at 6 s and had a diameter of 6 m. The next series of bubbles are much smaller, as will now be discussed.

The second bubble at 20 s is shown in Fig. 6.10 in the vicinity of the axial height of 400 cm. It is small in comparison with the first bubble (not shown). At 22 s, as shown in Fig. 6.11, it has grown somewhat.

Figures 6.12 and 6.13 show how this bubble continues to grow.

Figure 6.14 shows that this bubble has split into two.

Figure 6.15 shows that these two bubbles begin to coalesce and that a third bubble begins to form near the bottom, emanating from two high-speed jets near the center.

Figure 6.16 shows complete coalescence of the two top bubbles into a one large bubble. Figure 6.17 shows further growth of the bubble nearing the top as well as the bubble below it.

Figure 6.18 shows that by 34 s, the large top bubble has burst and combined with the low free board. The bubbling process continues. As shown in Fig. 6.18 that the

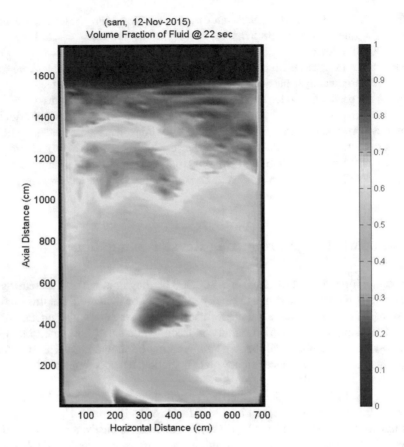

Fig. 6.10 Second bubble in the high production reactor

bubble below the top bubble shown in Fig. 6.17 has decreased in size. The plot at the top of Fig. 6.19 shows the gas volume fraction as a function of time to nearly 50 s.

The power spectrum plot in the lower figure clearly shows the formation of the bubbles having frequencies in the range below 1 Hz. The time-averaged gas volume fraction is shown in Fig. 6.20. It clearly shows the small free board at the top and much lower gas volume fraction at the bottom on the sides of the gas jets. Figure 6.21 shows the time-averaged liquid volume fraction, labeled solid 1, consisting of liquid droplets in the simulation. There is more liquid at the bottom and near the walls due to the two low-velocity gas jets needed to maintain the particles above minimum fluidization.

The time-averaged liquid axial velocity is shown in Fig. 6.22. The velocity in the central region is around 2 m/s, with a strong downward flow near the walls. The time-averaged gas velocity is shown in Fig. 6.23.

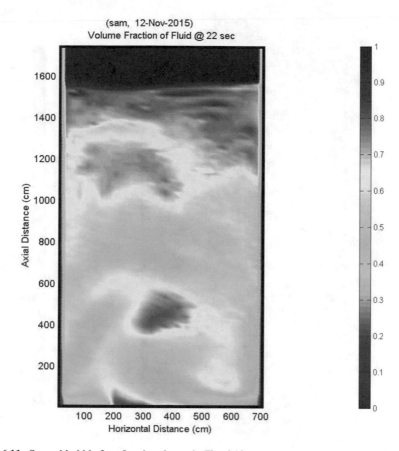

Fig. 6.11 Second bubble 2 s after that shown in Fig. 6.10

The time-averaged catalyst volume fraction, labeled solid 2, is shown in Fig. 6.24. Catalyst is present in all regions of the reactor, except in the free board, assuring that reaction will occur all over, thus producing a high yield of diesel product. The computed velocity variance of the catalyst is of the order of magnitude of $O(1)\ (\text{m/s})^2$ and is highest in the vicinity of the reactor walls. This suggests that the heat of reaction will not cause the development of hot spots. All the heat of reaction can be removed with the hot liquid product.

The remaining concern of this reactor design presented in this simulation is the formation of large bubbles that may cause excessive shaking of the reactor. These large bubbles can be completely eliminated by the use of a circulating fluidized bed. But such a more complex reactor, once designed, will have to be tested as a pilot unit.

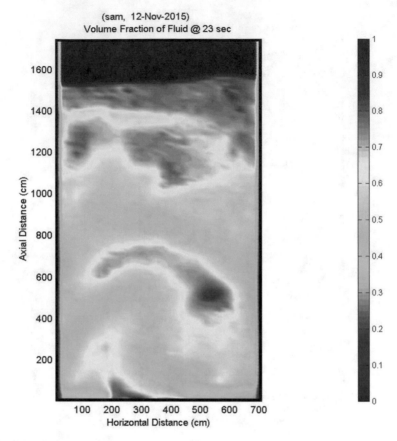

Fig. 6.12 Growth of second bubble

6.5 New Reactor Features

1. The new SBCR model demonstrates a method for converting synthesis gas, and a mixture of carbon monoxide and hydrogen, into liquids, such as diesel, molten waxes, or methanol, in which the heat of reaction is removed by the filtered, hot, liquid product.

   This new method of removing the heat of reaction eliminates the cooling tube bundles normally put into the SBCR and therefore leads to a much simpler and far less expensive design. In this process, the porous catalyst, such as the Fischer-Tropsch cobalt catalyst, is manufactured to be larger than 100 μm diameter, but less than 0.5 cm, to prevent filtration problems but have an effectiveness factor of larger than 40%. The reactor is designed using a modified CFD program, such as that used in [4]. The CFD program uses the principles of conservation of mass,

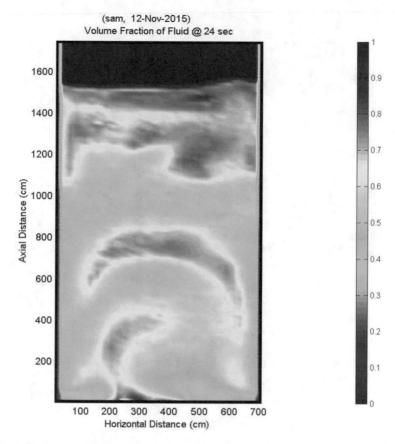

Fig. 6.13 Further growth of second bubble and detachment of third

momentum, and energy for each of the gas, liquid droplet, and catalyst particle phases rather than empirical hold-up correlations used in almost all previous SBCR designs [1]. The multiphase CFD program design had been previously validated with Air Products pilot plant data for a methanol SBCR having cooling tubes [1].

2. The new SBCR model demonstrates a method for removing the heat of reaction for the manufacture of any other liquid product requiring heat removal due to an exothermic reaction. Production of liquid fuels from biomass, such as wood chips, is an example of a method requiring such a Fischer-Tropsch reactor without cooling tubes.

3. The new SBCR model demonstrates a method for removing the heat of reaction operating with large catalyst particles and at high superficial gas velocities on the order of 0.5 m/s to achieve high liquid product production rates, on the order of 100,000 bbl/day. The SBCR may contain one or more tubes for temperature control.

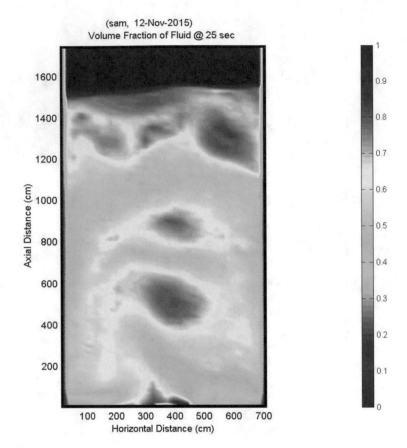

Fig. 6.14 Splitting of second bubble and formation of third bubble

6.6 Exercises

6.6.1 Ex. 1: Computed Bubble Coalescence Explanation and Flow Regimes

Figure 6.15 shows the computed bubble coalescence of the two bubbles shown in Fig. 6.14 for the slurry bubble column reactor described in Sect. 6.4. Gidaspow explains the origin of bubbles in fluidized beds [10]. He shows that the approximate void propagation equation is,

$$\frac{\partial \varepsilon}{\partial t} + C\frac{\partial \varepsilon}{\partial x} = 0 \tag{6.5}$$

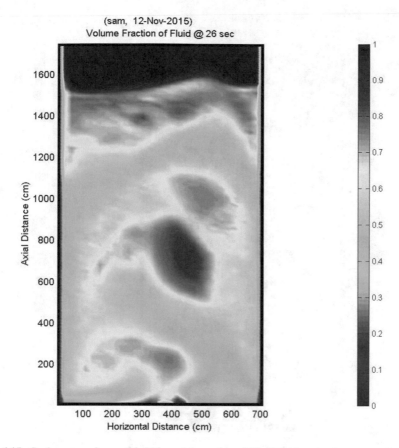

Fig. 6.15 Coalescence of second bubble and formation of third bubble

where the propagation velocity, C, for particle Reynolds numbers greater than 1000 in turbulent flow is given in his Eq. (6.6),

$$C = \left[\frac{\Delta \rho g d_p}{1.75}\right]^2 (1 - \varepsilon)\varepsilon^{0.5} \tag{6.6}$$

1. Explain the phenomenon of bubble coalescence using the above approximate equations.
2. For high diameter reactors and high gas velocities, the flow regime map of Shah et al. [11] drawn schematically below shows that we will be in the coalesced bubble regime. Is this observation consistent with the bubble coalescence theory in part (1)? For small diameter reactors, the Shah et al. map below shows that, for high velocities, we will be in the slugging regime. Explain this phenomenon.

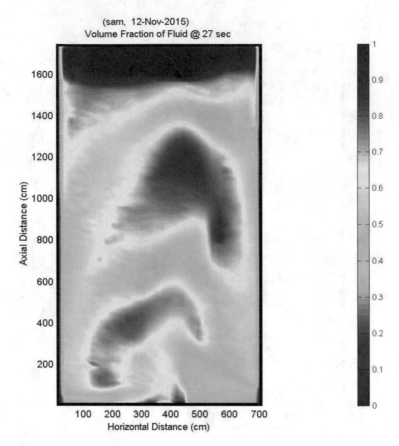

Fig. 6.16 Coalescence of second bubble into a huge bubble

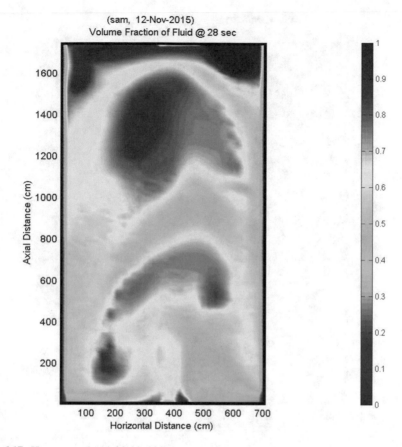

Fig. 6.17 Huge second and third bubbles

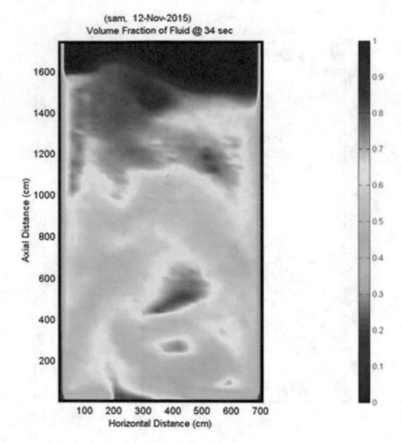

Fig. 6.18 Merge of second bubble with the free board

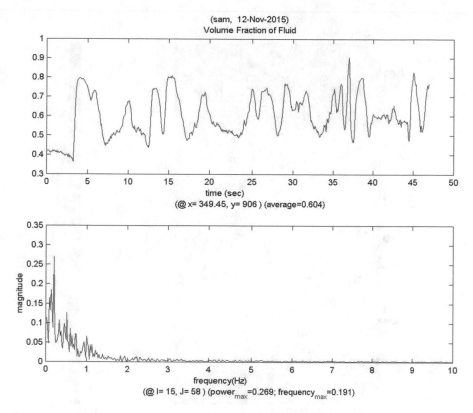

Fig. 6.19 Gas volume fraction variation and its spectrum starting from the initial condition

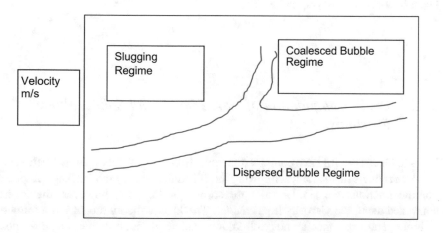

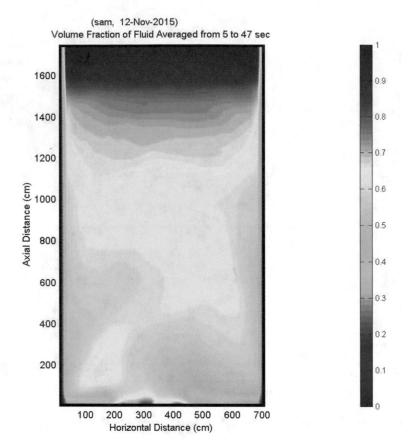

Fig. 6.20 Time-averaged gas volume fraction

6.6.2 Ex. 2: Computation of Gas, Liquid, and Solid Volume Fractions

Wu and Gidaspow [6] have computed the gas, liquid, and solids volume fractions in the Air Products methanol reactor using the IIT kinetic theory model. The computed volume fractions are similar to those shown in Fig. 6.2; however, they were computed using the viscosity input model. The kinetic theory model has a number of input parameters that are not well-known and are difficult to measure. The most disputed parameter is the restitution coefficient. It should be close to unity to give the measured core-annular regime discussed in Chap. 2, Sect. 2.4.3, Core-Annular Flow Regime Explanation. Hence, a comparison of the two models is useful.

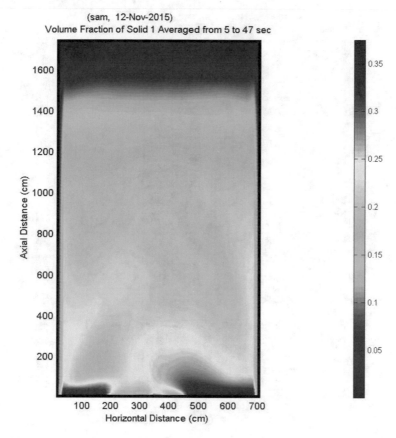

Fig. 6.21 Time-averaged liquid volume fraction

1. Repeat the simulations of Wu and Gidaspow [6] for the Air Products reactor using the viscosity input model used in this chapter for isothermal conditions with no reaction. Compare the results to the Air Products measurements for:

   (a) The liquid volume fractions
   (b) The gas volume fractions
   (c) The catalyst

6.6.3 Ex. 3: Computation of Gas Hold-Up (Effect of Pressure)

Krishna and Sie [3] studied the effect of pressure on the gas volume fraction, called gas hold-up. Their Fig. 11 shows an increase with pressure. For gas/solids flow, it was known long ago that, at high pressure, fluidization is smoother.

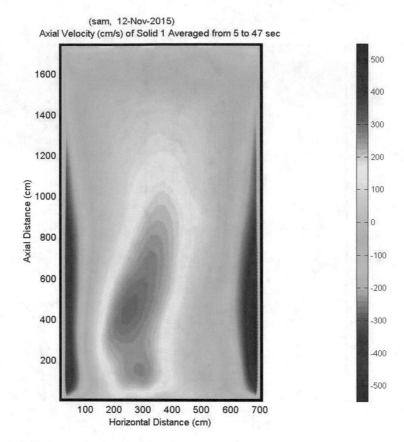

Fig. 6.22 Time-averaged liquid axial velocity

1. Compare their result of increase of the gas volume fraction with pressure to their experimental data. Use the hydrodynamic model shown in Table 1 in Gidaspow et al. [5], which is essentially the same as that shown in Table 4.2 in Chap. 4 on polymerization.
2. Explain this phenomenon.
3. How good are their measurements?

6.6.4 Ex. 4: Elimination of Bubbles

Large bubbles increase mixing and help to eliminate hot spots in fluidized bed reactors. But they also produce large vibrations that may cause equipment failure.

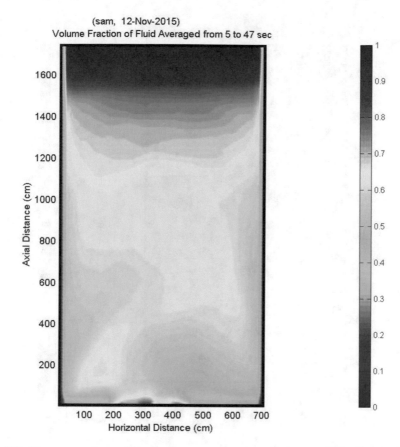

Fig. 6.23 Time-averaged gas axial velocity

Such accidents had occurred in the storage of nuclear waste in large tanks where large hydrogen bubbles had formed and had caused leakage. Such large bubbles have been computed using the IIT CFD model as shown in this chapter. Bubbles can be eliminated using circulating fluidized bed reactors.

1. Repeat the simulation for the high production reactor presented in this chapter for a circulating fluidized bed. To model a riser in place of a bubbling bed, it is only necessary to replace the inlet gas with flow of gas, particles, and liquid at equal velocities.

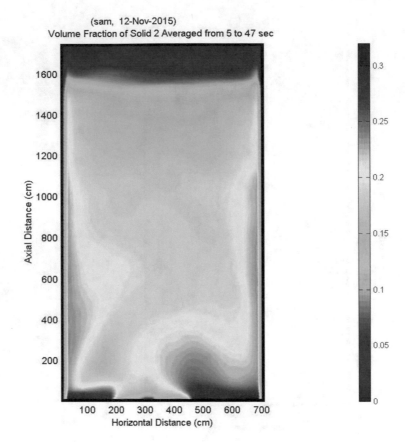

Fig. 6.24 Time-averaged catalyst volume fraction

References

1. Gamwo IK, Gidaspow D, Jung J (2007) Slurry bubble column reactor optimization, Chapter 17. In: Ultra clean transportation fuels, Ogunsola OI, Gamwo IK (eds), ACS Symposium Series Vol. 959, American Chemical Society, Washington, DC, Distributed by Oxford University Press, pp. 225–252
2. Krishna R, Sie ST (2003) Design and scale-up of the Fischer-Tropsch bubble column slurry reactor. Fuel Process Technol 64:73–105
3. Sehabiague L, Lemoine R, Behkish A, Heintz YJ, Sanoja M, Oukaci R, Morisi BI (2008) Modeling and optimization of a large-scale slurry bubble column reactor for producing 10,000 bbl/day of Fisher-Tropsch liquid hydrocarbons. J Chin Inst Chem Eng 39:169–179
4. Gamwo I, Gidaspow D, Jung J (2009) Design of slurry bubble column reactors: novel technique for optimum catalyst size selection. Contractual origin of the invention, U.S. patent 7,619,011 B1, 11/17/2019
5. Gidaspow D, Yuting HY, Chandra V (2015) A new slurry bubble column reactor for diesel fuel. Chem Eng Sci:134:784–799

6. Wu Y, Gidaspow D (2000) Hydrodynamic simulation of methanol synthesis in gas-liquid slurry bubble column reactors. Chem Eng Sci 55:573–587
7. Gidaspow D (2016) Slurry bubble column reactor without cooling tubes. U.S. patent application number 14/998,647, filing date 01/29/2016
8. Yates IC, Satterfield CN (1991) Intrinsic kinetics of the Fischer-Tropsch synthesis on a cobalt catalyst. Energy and Fuels 5:168–173
9. Post MFM, Van't Hoog AC, Minderhoud JK, Sie ST (1989) Diffusion limitations in Fischer-Tropsch catalysts. AIChE J 35(7):1107–1114
10. Gidaspow D (1994) Multiphase flow and fluidization, continuum, and kinetic theory description. Academic Press Inc., San Diego, California
11. Shah YT, Kelkar BG, Godbole SP, Deckwar W-D (1982) Design parameters estimations for bubble column reactors. AIChE J 28:353–379

Chapter 7
Application of Multiphase Transport to CO_2 Capture

Hamid Arastoopour, Dimitri Gidaspow, and Robert W. Lyczkowski

7.1 Introduction

Climate change, in part due to emission of fossil energy powered power plants, has created a world-wide interest in capturing carbon dioxide. Presently, fossil fuels are still the major source used for electric power generation and most of the industrial CO_2 emission results from their combustion and gasification. By cost-effective capturing of the CO_2 before it is emitted to the atmosphere and then storing it, fossil fuels can continue to be used without affecting economic growth, while reducing carbon emissions to the atmosphere. Therefore, carbon capture and sequestration (CCS) is one of the key technologies needed to mitigate carbon dioxide emission from industrial sources and power plants. Chemical and physical sorption of CO_2 from flue gases is a challenging process due to the low pressure and low concentration of CO_2 in flue gas, which requires high volumetric flow rates of flue gas to be processed. In addition, these processes generally use sorbents to capture the CO_2 and these sorbents need to be regenerated and used continuously in the process. The regeneration of sorbents is an energy demanding process that reduces the overall efficiency of the power plant. Therefore, developing more efficient and economically feasible processes for CO_2 removal is the focus of this chapter. It has been shown in the literature that carbon dioxide is captured using liquid amine scrubbing, a now commercial technology [1, 2]. However, commercially available processes operate at low temperatures and atmospheric pressure, imparting a severe energy penalty on the system and, consequently, their use could significantly increase the costs of electricity production. Therefore, development of regenerative processes based on solid sorbents offers an attractive alternative option for carbon capture at competitive costs.

The design technology reviewed by Rochelle [1] is standard in the chemical industry. The design is done using black box mass and energy balances with simulation codes such as ASPEN Plus [3]. Equilibrium of CO_2 with methyl ethyl amine is assumed. Rates of reactions are sometimes added. Empirical Murphree

© Springer Nature Switzerland AG 2022   177
H. Arastoopour et al., *Transport Phenomena in Multiphase Systems*, Mechanical Engineering Series, https://doi.org/10.1007/978-3-030-68578-2_7

efficiencies of 40–100% are used in the models. A good example of such a design is the stripper design for CO_2 capture by aqueous amines described by Oyenekan and Rochelle [4]. In their model, the rate of reaction equals the mass transfer coefficient times the difference between the equilibrium and the operating partial pressures of CO_2. A similar rate expression was used for sorption and stripping of CO_2 using dry sorbents. Rochelle [1] shows that, to capture carbon dioxide from fossil energy power plants, 15–30% of the power will be consumed due to the use of steam from the power plant and compression of carbon dioxide.

In this chapter, it is shown that this loss can be significantly reduced using solid sorbents and bubbling or circulating fluidized bed (CFB) systems with a regeneration system that may be performed at lower pressure and energy supplied by waste heat and partial recovery of the heat of reaction. The use of CFBs ensures continuous CO_2 removal processes based on the dry sorbent concepts [5]. In the chemical and energy industries, CFB reactors are among the most important devices. One of the well-established applications of CFB reactors is in fluid catalytic cracking (FCC) applications with more than seven decades of history and more than 400 units in operation worldwide today [6]. Gasification of coal and biomass, synthesis of olefin from methanol, and chemical looping are among the relatively new applications of CFB reactors.

Using solid particles for CO_2 capture in a CFB reactor allows continuous carbon dioxide removal in a relatively compact unit. The basic configuration of a CFB reactor consists of a riser where the particles are transported by the gas flow, a cyclone to separate gas and solid at the top of the riser, a standpipe (down-comer) to return the separated solid to the riser inlet, and a flow controlling device (e.g., L-valve) to control the solid flow. Depending on the application, other vessels or reactors can be added to the system. For instance, in processes that include a regenerative sorbent or catalyst, a second fluidized bed reactor can be added between the down-comer and the L-valve to serve as a regenerator reactor. Figure 7.1 shows a schematic diagram of a CFB system for the CO_2 capture process [7, 8].

Shimizu and coworkers [9] were among the first groups who proposed the concept of a dual fluidized bed for carbon dioxide capture using CaO solid sorbents for post-combustion applications. Following their work, applications of CFB for carbon dioxide capture were studied by several researchers [10–13]. Despite the long history of operation that has led to a wealth of design and operational experience with these systems, confidence to design and build commercial plants without significant levels of pilot-scale testing at various intermediate scales is still lacking. This is due to the lack of complete understanding of the origin and nature of the inherently complex flow structures observed in these devices, and uncertainties as to how they change upon scale-up [14]. Another challenge in the deployment of these novel technologies is the fact that the majority of the promising technologies for CO_2 capture are still in the lab or pilot scales. To successfully scale up these processes, a powerful tool, such as computational fluid dynamics (CFD), is needed to fill the gap between the lab/bench scale and the large scales needed for demonstration. IIT researchers believe that CFD has been proven to be a cost-effective tool for conducting virtual experiments, prototype testing, and parametric studies. Analysis

Fig. 7.1 Schematic diagram of a typical circulating fluidized bed for the CO_2 capture process [8]

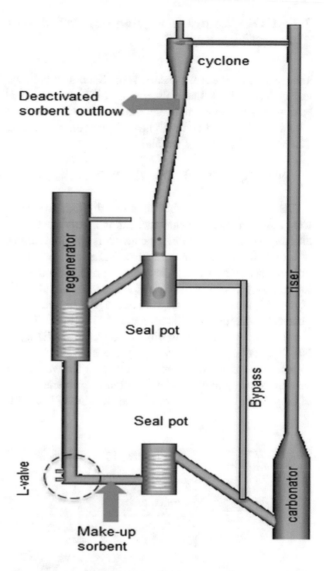

using CFD complements and reduces physical testing, and it can result in significant time and cost savings [15]. Therefore, CFD and rigorous numerical modeling are essential to shed light on the complex behavior and flow structure in these systems and provide a reliable option to study and design the proposed fluidization process required for continuous CO_2 removal using solid sorbents in a systematic and economically feasible way.

7.2 CO$_2$ Capture Using Sodium or Potassium Carbonate Solid Sorbents

The sorbent is essentially baking soda. Such a sorbent was developed by Gidaspow and Onischak [16]. Later, the group at RTI International [17] reported depositing baking soda on pellets and demonstrated that CO$_2$ can be removed from flue gases in a bubbling fluidized bed with heat exchangers to remove the high heat of reaction. The reaction involved is as follows,

$$Na_2CO_3 + CO_2 + H_2O = 2NaHCO_3 \ \Delta H_{298K} = -129 \ kJ/gmol \ CO_2 \qquad (7.1)$$

The RTI International bubbling bed dry carbonate prototype unit consists of the CO$_2$ absorber and sorbent regenerator involving two mechanical conveyors/heat exchangers. The purpose of the conveyors was to minimize the pressure drop for use of a draft fan rather than a compressor. Such unconventional design proved to be poor for heat removal. The temperature rise in their system is large due to the very low ratio of gas to particle densities, as explained in Chap. 4 on polymerization. Figure 7.2 shows a flow diagram of the RTI International technology unit [18]. The physical properties of their sorbent are as follows. The mean particle diameter was about 70 μm with a bulk density of 1.0 g/cm^3, a BET surface area of 100 m^3/g, and a Na$_2$CO$_3$ content of 10–40 weight percent [17].

The rates of reactions are very rapid, as shown here, leading to smaller reactors than those using amines. The diffusion coefficients in the amine reactors are orders of magnitude lower than those in the gas phase. Thus, 24 m high amine reactors are

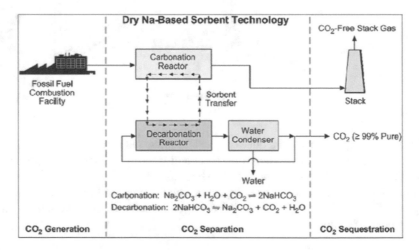

Fig. 7.2 Dry Na-based sorbent technology [18]

typical units. A bubbling fluidized bed is unfortunately not a good reactor to use as explained in the following sections.

7.2.1 Conceptual Design of Fluidized Bed Systems Based on the CFD Approach

It has been known for many decades that, for the design of efficient processes, entropy production must be minimized (Fitzmorris and Mah [19], Bejan [20], and Prigogine [21], who received the Nobel Prize for such investigations). Unfortunately this approach has only recently been used in practice due to difficulties in its implementation. To capture CO_2 from flue gases and to produce pure CO_2, both the absorber and the regenerator must be taken together as a system. In order to obtain an efficient absorber-regeneration system, the absorption must be as close as possible to equilibrium. The initial CFD design by Kongkitisupchai and Gidaspow [22, 25] showed that one stage was insufficient to remove CO_2 from flue gases to the desired purity. Such a restriction is similar to the well-known calculation of the number of stages in distillation. Boonprasop et al. [23] were the first to calculate the number of stages needed for the separation. Since the rates of reaction are high, the rate of CO_2 sorption was controlled by mass transfer through clusters.

The equilibrium diagram in Fig. 7.3 shows that two inexpensive sorbents can be used to capture CO_2 [16, 18]. Sodium carbonate used by RTI International is less expensive, but cannot be used to remove CO_2 from the atmosphere owing to its higher vapor pressure. The potassium carbonate has a sufficiently low decomposition pressure to remove CO_2 from the atmosphere using the IIT sorbent based on thermodynamic analysis [24].

7.2.1.1 Bubbling Beds and Plug Flow Approximation

A bubbling bed for CO_2 capture was used with some success at RTI International, as described in Sect. 7.1. Chaiwang et al. [26] modeled similar reactors using the IIT CFD computer program. Such reactors are simpler than the preferred circulating beds. IIT researchers found that, with a proper design and the use of a mixture of large and small particles, bubbles can be eliminated. This discovery is important for the design of fluidized beds for silicon production for solar collectors to be discussed in Chap. 8. It was also found that bubbling beds and plug flow model are good first-order approximations for calculating the reactor height. It is as follows,

$$(C_{out} - C_{eq})/C_{in} = \exp\ (-kL/V) \qquad (7.2)$$

where C is the species concentration, k the first order rate constant, L is the reactor height, and V is velocity. The above approximation allows a rapid estimate of reactor

Fig. 7.3 Equilibrium
curves for sodium and
potassium carbonate/
bicarbonate systems. (This
figure was originally
published in [22] and has
been reused with
permission)

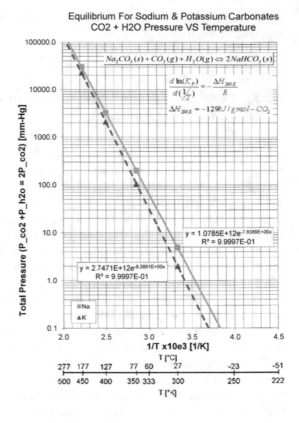

height. Hence, the concentration of species equations do not need to be solved for most simulations. Such a simplification permits a more rapid fluidized bed reactor design.

7.2.1.2   CO$_2$ Capture with Reduced Pressure in a Downer of a CFB

Figure 7.4 shows the fluidized bed riser and the downer system at a reduced pressure for sorbent regeneration. Figure 7.5 shows the sorber-regenerator CO$_2$ capture fluidized bed process and the temperature and partial pressure equilibrium for sodium carbonate/bicarbonate. The loss in thermodynamic availability is only 1.5 kJ/g mol CO$_2$ [22, 25] compared to the minimum energy of separation of 7.3 kJ/g mol CO$_2$. The equilibrium decomposition pressure of the carbonate limits the percent CO$_2$ removal. Figure 7.6 shows the two-stage riser-sorber with a down-comer regenerator process. These results are summarized in a later study using thermal sorbent regeneration in the next section.

Fig. 7.4 Carbon dioxide capture system with reduced pressure regeneration based on CFD design. (This figure was originally published in [22] and has been reused with permission)

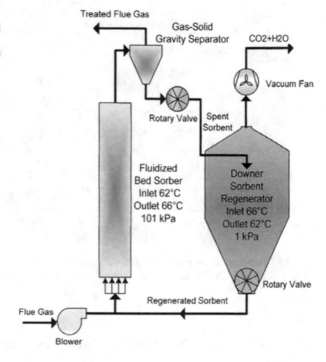

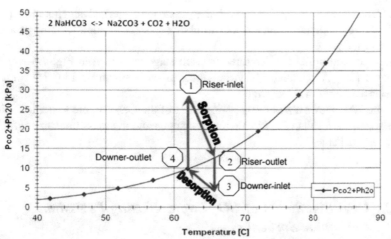

Fig. 7.5 Sorber-regenerator CO$_2$ capture fluidized bed process. Temperature and partial pressure equilibrium for sodium carbonate/bicarbonate. (This figure was originally published in [22] and has been reused with permission)

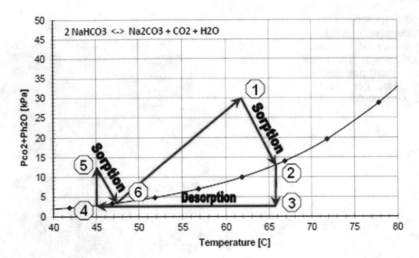

Fig. 7.6 Two-stage riser-sorber with a down-comer-regenerator for large percent CO_2 capture. (This figure was originally published in [22] and has been reused with permission)

7.2.1.3 CO_2 Capture in a Multistage Sorber with Thermal Regeneration

Figure 7.7 shows the multistage sorber system with thermal regeneration [23] in which IIT researchers computed the number of stages needed to remove a given percent of CO_2.

The inlet and outlet temperatures in each stage as shown in Fig. 7.8 were also computed [26]. A schematic of the cooling tubes is shown in Fig. 7.9 for a 1-m diameter reactor with three stages.

Figure 7.10 shows a plot of operating lines of the CO_2 capture system. Table 7.1 summarizes the results. The kinetic rate constants and the mass transfer coefficients are shown in Table 7.2.

7.2.2   CFD Simulation of CO_2 Capture Using Potassium Carbonate Sorbent

7.2.2.1   Introduction

The alkali-metal-based solids could be a promising sorbent for efficient and cost-effective CO_2 removal from combustion gases and they possess excellent features like superior attrition resistance and high CO_2 sorption capacity [28]. Potassium-based solid sorbent was used successfully to capture CO_2 using a circulating fluidized bed (CFB) system. Use of CFB ensures the continuous CO_2 removal

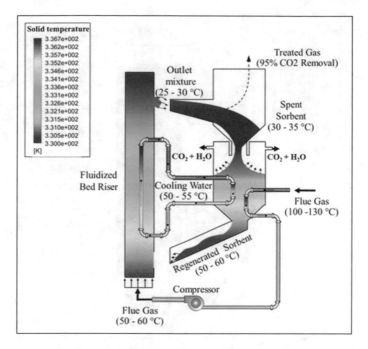

Fig. 7.7 Multistage sorber system with thermal regeneration. (This figure was originally published in [23] and has been reused with permission)

process from dilute flue gases at laboratory and bench scales [29, 30]. To scale up this CFB process, a state-of-the-art design tool based on CFD simulation is needed.

In this section, CFD simulation for the CO_2 capture process in the riser part of a CFB using a potassium-based solid sorbent is presented and experimental data [5] were used to validate the simulation results.

7.2.2.2 Numerical Analysis

For this simulation, the experiments of Yi et al. [29] were used, which include a circulating fluidized bed consisting of a riser as the carbonator and a bubbling fluidized bed as the regenerator. In this study, the focus was on the simulation of the 2.5 cm ID and 6 m height riser carbonation reactor. The operating condition was atmospheric pressure at a constant temperature of 80 °C in the riser. The flue gas inlet velocity was 2 m/s with 12% CO_2 (dry basis) and 12.3% H_2O composition. The solid circulating rate is set to 21 kg/m^2 s. The potassium-based sorbent has a particle density of 2394 kg/m^3 with average particle diameter of 98 μm. CFD simulation of this work is based on a two-dimensional Eulerian-Eulerian approach in combination with the kinetic theory of granular flow (see Chap. 2 for governing and constitutive

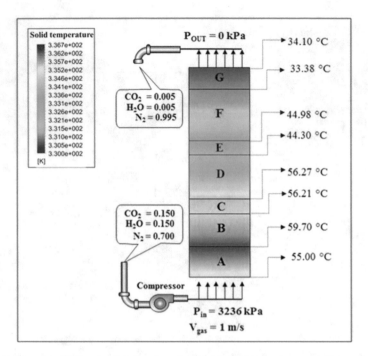

Fig. 7.8 Inlet and outlet temperatures in a multistage sorber. (This figure was originally published in [27] and has been reused with permission)

equations). To convert the real geometry to a reduced two-dimensional domain, the solid mass flux was kept constant as the basis for calculation. The assumptions of this numerical simulation include: the isothermal condition for the process, ideal gas equation of state for gas phase, and uniform particle size with constant size and density. The ANSYS/Fluent computer code was used to obtain the numerical solution for the governing and constitutive equations.

In addition, the k-ε turbulent model was used to take care of turbulent fluctuations of the gas phase. Initially there was no solid in the riser and the concentration of CO_2 was zero as well.

A second order discretization scheme was used to discretize the governing equation throughout the domain with 34×1200 uniform rectangular cells. Grid independence tests were performed and showed that our numerical results are grid independent.

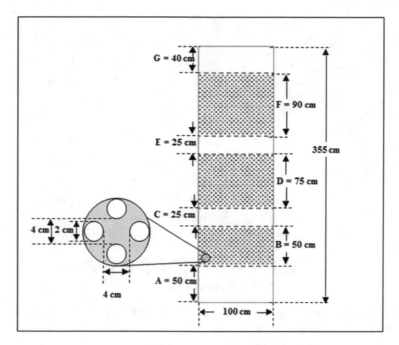

Fig. 7.9 Schematic diagram of a riser with cooling tubes for a three-stage system. (This figure was originally published in [27] and has been reused with permission)

7.2.2.3 Reaction Kinetic Model

There is little available on the kinetics of the carbonation reaction of K_2CO_3 in the literature [16, 31]. In this study, the deactivation model (DM) proposed by Park et al. [31], which includes gas volumetric flow rate in the kinetic model, is chosen. The concept of this model is based on an analogy between deactivation of catalyst particles by coke formation and deactivation of sorbent particles by carbonation. In the DM model, the effect of the formation of a product layer on the surface of the sorbent particles that results in an additional diffusion resistance is lumped into a reducing activation factor with an exponentially deactivation rate,

$$r = kC_{co2}a \tag{7.3}$$

where a is the activity of the sorbent and defined,

$$a = exp\left[\frac{[1 - exp(\tau k_s(1 - exp(-k_d t)))]}{1 - exp(-k_d t)} exp(-k_d t)\right] \tag{7.4}$$

where τ is called surface time and is defined as the ratio of available pore surface to the volumetric flow rate of flue gas. k_s, k_d, and k are three adjustable parameters of

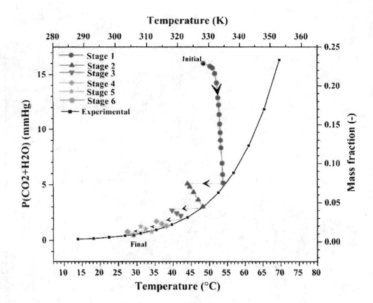

Fig. 7.10 Operating lines for the CO_2 capture system. (This figure was originally published in [27] and has been reused with permission)

Table 7.1 Number of stages vs. percent CO_2 capture [23]

Stage	Gaseous condition			DT (°C)	% CO₂ removal
	[CO₂] mass fraction	[CO₂] mole fraction	Temp. (°C)		
1	0.2284	0.1500	52.50	+6.00	67.316381
2	0.0746	0.0480	47.50	+4.86	83.16
3	0.0385	0.0246	42.50	+3.79	88.54
4	0.0262	0.0167	37.50	+3.5	91.32
5	0.0198	0.0126	32.50	+3.25	94.32
6	0.0130	0.0083	27.50	+2.96	96.27
Exhausted	0.0085	0.0054	30.46	–	–

Table 7.2 Kinetic constants [23]

Stage	k_{eff} (s⁻¹)	k_{eq} (s⁻¹)	k_{mass} (s⁻¹)	Sh
1	1.951	19.190	2.172	0.372
2	1.230	356.559	1.235	0.211
3	0.731	861.093	0.732	0.125
4	0.532	1899.465	0.532	0.091
5	0.801	4214.787	0.801	0.137
6	0.801	9042.221	0.801	0.137

the model and are called surface reaction constant, deactivation constant, and rate constant, respectively.

7.2.2.4   Simulation Results and Comparison with Experimental Data

The simulation results for CO_2 removal were compared with the experimental data provided by Yi et al. [29]. The calculated CO_2 removal percentage at the outlet of the riser is 58%, which is very close to the reported 54% removal in the experimental data. However, the experimental data have been reported over a time period of 20 h and the simulation has been run just for a few minutes. This difference could be a potential source of difference between simulation and experimental data. Simulation results also showed that the increasing inlet gas flow rate decreases the CO_2 removal percentage, which is in line with the experimental data. Table 7.3 shows our simulation results and their comparison with experimental data for time-averaged differential pressure at four different elevations.

In general, the simulation agrees well with the experimental data. The small deviation could be because the difference between the simulation time and the experiment time period is huge and, at such short simulation time, the simulation is not able to capture the possible accumulation effect. Furthermore, the deviation could be because the two-dimensional simulation is not appropriate to capture the wall effect on the flow that exists due to the very slender geometry of the riser.

7.3 CO_2 Capture by MgO-Based Sorbents Using a Circulating Fluidized Bed (CFB)

7.3.1   Introduction

Biomass and fossil fuels continue to play a significant role in the total energy picture. Advanced power generation technologies such as integrated gasification-combined cycle (IGCC) are among the leading contenders for power generation conversion in the twenty-first Century because such processes offer significantly higher efficiencies and superior environmental performance compared to fossil fuel and biomass combustion processes. The key is finding a sorbent to absorb CO_2 at high

Table 7.3 Differential pressure at four elevations along the riser

Location (m)	Differential pressure (mm H_2O) KIER experiments [28]	Differential pressure (mm H_2O) Simulation
0.52	100	75
2.27	200–500	125
4.07	100–210	230
5.87	70	120

temperature of about 350–450 °C. This condition makes MgO-based sorbents ideal for such operations. The MgO-based sorbents can absorb CO_2 to form magnesium carbonate and can easily be regenerated back to MgO [32–34]. The cyclic pre combustion CO_2 capture processes utilizing dry MgO-based sorbents proceed via the following reactions,

$$MgO + CO_2 = MgCO_3 \text{ (CO}_2 \text{ absorption reaction)} \tag{7.5}$$

$$MgCO_3 = MgO + CO_2 \text{ (regeneration reaction)} \tag{7.6}$$

$$CO + H_2O = CO_2 + H_2 \text{ (water/gas shift reaction)} \tag{7.7}$$

In addition, using the circulating fluidized bed (CFB) process for chemical looping of MgO-based sorbents for CO_2 capture and regeneration ensures a continuous carbon dioxide removal process in a relatively compact unit. In this section, chemical looping of MgO-based sorbents for CO_2 capture in IGCC that involves a process similar to the National Energy Technology Laboratory (NETL) carbon capture CFB unit experimental setup is used [7]. Gas containing CO_2 enters the bottom of the riser sorption reactor and mixes with fresh sorbent. The MgO-based sorbent particles mix with the coal gases and absorb CO_2 through chemical reaction. The CO_2-laden particles flow up the riser, turn, and flow into the cyclone. CO_2-lean gas is separated from particles in the cyclone and exits the system, and the CO_2-laden particles pass through a loop-seal and enter the regenerator where CO_2 is released from the sorbent particles by heating the spent sorbent with high-temperature steam. The CO_2-lean gas exits the carbon capture unit system and the regenerated sorbent particles continue through the loop to the next loop-seal. The regenerated sorbent particles pass through the loop-seal to the riser and the process continues. Figure 7.1 shows the schematic diagram of the NETL carbon capture unit experimental setup.

7.3.2 Numerical Analysis and Simulation of Entire CFB Loop

There are several publications in the literature on the numerical modeling of CO_2 capture in the riser and regenerator reactors of a circulating fluidized bed system [22, 35, 36]. However, to simulate the entire CFB loop using the computational fluid dynamics (CFD) approach requires comprehensive modeling of gas/particle interactions as well as particle/particle interactions, which are continuous and valid at different flow regimes. In addition, simulating these systems requires significant computational resources and use of high performance computing units. This study investigates two-dimensional numerical modeling of the entire CFB using the CFD approach and the state-of-the-art constitutive relations developed for interfacial interaction expressions between the phases in the system. The CFD simulation of this work is based on a two-dimensional Eulerian-Eulerian approach in combination with the kinetic theory of granular flow (see Chap. 2 for continuity, momentum, and constitutive equations). In addition, the k-ε turbulent model has been used to take care of turbulent fluctuations of the gas phase. To reduce required computational time, it was assumed that the convection and conduction terms in the granular temperature equation were neglected. The CFB loop in this simulation is the same as the ones that were developed and used by NETL as a bench-scale model for the CO_2 capture processes. An MgO-based solid sorbent was used to separate CO_2 from carrying gas (usually syngas) in a carbonator reactor. The carbonated sorbent was then regenerated at high temperature and returned to the carbonator to be used again. The concentrated CO_2 stream leaves the system to be sequestered or converted to fuels or useful products.

The energy minimization multi-scale (EMMS) approach was used to model the drag force between phases and the solid frictional pressure and frictional viscosity equations in the dense part used were the one discussed in Chap. 2, Sects. 2.7 and 2.8. The Johnson and Jackson [37] partial slip boundary conditions for particles at walls have been used in this study. The values of the particle/wall restitution coefficient, particle/particle restitution coefficient, and specularity coefficient are assigned as 0.7, 0.85, and 0.1, respectively, for the MgO-based sorbent used in this study. A cut-cell mesh with 611,224 cells was used in ANSYS/Fluent (v13.0) to perform the simulations. For information regarding the details of numerical simulation and system operating conditions, see Ghadirian et al. [38]. In this study, the rate of carbonation and regeneration reactions used were the ones developed using experimental data and the variable diffusivity shrinking core model [32, 33, 39].

7.3.3 Numerical Simulation of Full CFB Loop for CO_2 Capture and Sorbent Regeneration

To simulate CO_2 sorption and regeneration using the CFB loop, the carbonation and regeneration heterogeneous reactions were added to the two-dimensional CFD governing equations [32, 33, 39]. The carbonation reaction was assumed to take place only in the carbonator at a constant temperature of 380 °C, while the regeneration reaction was assumed to take place only in the regenerator at constant temperature of 500 °C. The system pressure was set to 50 atm, which is appropriate for carbon capture units in an IGCC power plant. The initial sorbents in the regenerator and loop-seal in front of regenerator were assumed to be 65% carbonated, while all other parts of the system were filled with the fresh sorbent. The inlet gas in the regenerator was pure steam and, in the carbonator inlet, the CO_2 mole fraction was 14.26% and the rest was steam [40].

Figure 7.11 shows the solid circulating rate at the standpipe of the loop. The average value of the solid circulating rate was 0.043 kg/s at 50 atm pressure. At this pressure, the gas density is considerably higher in comparison with the simulation at

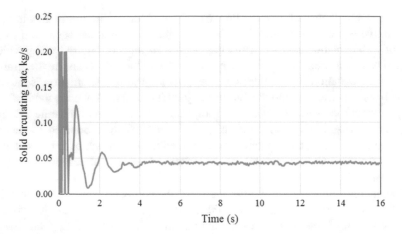

Fig. 7.11 Solid circulating rate at the standpipe as a function of time. (This figure was originally published in [38] and has been reused with permission)

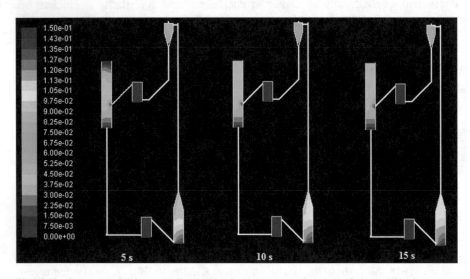

Fig. 7.12 Contours of CO_2 mole fraction in the system as a function of time. (This figure was originally published in [38] and has been reused with permission)

atmospheric pressure. This means higher drag force exerted by gas on the solid phase, and, in turn, change on the solid sorbents circulation rate.

Figure 7.12 shows the contours of the CO_2 mole fraction at different times. The red color represents about 15% CO_2 and the blue color represents almost zero CO_2 concentration. As can be seen, the mole fraction of CO_2 is almost identical after 10 s suggesting a statistical steady state operation of the system.

As the carbonation proceeded in the carbonator, the rate of carbonation was reduced due to higher sorbent conversion; meanwhile, fresh sorbents entered the carbonator from the lower loop-seal and resulted in a higher reaction rate. These competing phenomena resulted in small variations in the CO_2 mole fraction in the system.

7.4 Use of Carbon Dioxide

Carbon dioxide can be used to make chemicals [41, 42]. The oxidation of methane using carbon dioxide in place of the more expensive oxygen produced by the very energy intensive liquefaction of air was first discovered in 1992 at Cambridge University [43]. The catalytic reaction is,

$$CH_4 + CO_2 = 2CO + 2H_2 \qquad (7.8)$$

The product is synthesis gas from which liquid fuels, like diesel, and a large number of other chemicals can be made, as described in this book. Catalytic conversion of carbon dioxide into fuels also provides an attractive utilization of CO_2. Recently, Asadi et al. [44] demonstrated catalytic electrochemical CO_2 conversion to carbon monoxide in an ionic liquid.

Sequestration of carbon dioxide may be achieved by creation of additional forest areas that will be used as a CO_2 sink, or by injection to underground reservoirs such that the CO_2 cannot escape from the rock formation into which it is injected and be stable during the time that CO_2 is trapped in the pores of the underground formations [45–47].

7.5 Exercises

7.5.1 Ex. 1: Order of Magnitude Design of CO_2 Capture Riser/Loop

Chapter 5, Circulating Fluidized Beds for Catalytic Reactors, describes the CFD design of IIT researchers. Chapter 4 on polymerization shows that such catalytic reactors with a high heat of reaction operate at nearly constant temperature due to a high heat capacity of the solid particles compared to that of the gas, unlike the bubbling reactors built by RTI International. To take advantage of this phenomenon, IIT researchers made a loop CFD simulation using a Pyropower-type reactor similar to that described in Chap. 5 (see Chalermsisuwan et al. [48]). The effective first-order rate constant was $2\ s^{-1}$, as shown in Table 7.2. The reactor diameter and the height were 3 m and 15 m, respectively. The inlet CO_2 and water inlet for the flue gas were both 15 mol%. The flue gases were cooled to 55 °C, as in the amine reactor designs.

1. Using a plug flow model, estimate the reactor height needed for near complete conversion, discuss the difference between your result and the assumed 15 m diameter reactor height.
2. What is the maximum CO_2 conversion for potassium and sodium carbonate sorbents at 55 °C?
3. Estimate the number of stages needed to obtain 95% CO_2 removal assuming plug flow.

7.5.2 Ex. 2: CFD Design of CO_2 Capture Loop

1. Repeat the CFD simulations in the article by Boonprasop et al. [23] for a large 7 m diameter reactor. Compute the rate of CO_2 removal from the flue gases cooled to 55 °C, as in Exercise 1.

7.5.3 Ex. 3: CFD Design of an Amine Sorber

Oyeneken and Rochelle [4] used the conventional simulation code ASPEN to design their reactors. There has been no attempt to use CFD models to possibly improve their designs. The input into their model is similar to that used for sorber design discussed in this chapter. It involves mass transfer coefficients and rates of reactions proportional to the difference of CO_2 partial pressures.

1. The temperature rise in their sorber is not high. Explain.
2. Formulate a CFD model.
3. Compare the CFD results to those in Rochelle [1].

References

1. Rochelle GT (2009) Amine scrubbing for CO_2 capture. Science 325:1652–1654
2. Bottoms RR (1930) Process for acidic gases. U.S. Patent 1,783,901, Dec 2, 1930
3. Arastoopour H (2019) The critical contribution of chemical engineering to a pathway to sustainability. Chem Eng Sci 203:247–258
4. Oyenekan BA, Rochelle GT (2006) Energy performance of stripper configurations for CO_2 capture by aqueous amines. Ind Chem Eng Res 45:2457–2464
5. Yi CK, Jo SJ, Seo Y, Lee JB, Ryu CK (2007) Continuous operation of the potassium-based dry sorbent CO_2 capture process with two fluidized-bed reactors. Int J Greenhouse Gas Control 1:31–36
6. Chen YM (2011) Evolution of FCC — past present and future — and the challenges of operating a high-temperature CFB system. In: Knowlton TM (ed) Proceeding of the tenth international conference on circulating fluidized beds and fluidization technology, CFB-10, ECI, New York, pp. 58–85
7. Shadle L, Spenik J, Monazam RE, Panday R, Richardson G (2013) Carbon dioxide capture in fully integrated fluidized process using polyethylenimine (PEI) sorbent. DOE Report

8. Abbasi E (2013) Computational fluid dynamics and population balance model for simulation of dry sorbent based CO_2 capture process. Dissertation, Chicago (IL): Illinois Institute of Technology.
9. Shimizu T, Hirama T, Hosoda H, Kitano K, Inagaki M, Tejima K (1999) A twin fluid-bed reactor for removal of CO_2 from combustion processes. Chem Eng Res Des 77:62–68
10. Abanades CJ, Grasa G, Alonso M, Rodriguez MN, Anthony EJ, Romeo LM (2007). Cost structure of a post combustion CO2 capture system using CaO. Environ Sci Technol 41:5523–5527
11. Abbasi E, Arastoopour H (2011) CFD simulation of CO_2 sorption in a circulating fluidized bed using deactivation kinetic model. In: Knowlton TM (ed). Proceeding of the tenth international conference on circulating fluidized beds and fluidization technology, CFB-10, ECI, New York, pp. 736–743
12. Breault RW, Huckaby ED (2013) Parametric behavior of a CO_2 capture process: CFD simulation of solid-sorbent CO_2 absorption in a riser reactor. Appl Energy 112:224–234
13. Nikolopoulos A, Nikolopoulos N, Charitos A, Grammelis P, Kakaras E, Bidwe AR, Varela G (2012) High–resolution 3-D full-loop simulation of a CFB carbonator cold model. Chem Eng Sci
14. Sundaresan S (2011) Reflections on mathematical models and simulation of gas-particle Flows. In: Knowlton TM (ed), Proceeding of the tenth international conference on circulating fluidized beds and fluidization technology, CFB-10, ECI, New York, pp. 21–40
15. Benyahia S, Arastoopour H, Knowlton TM, Massah H (2000) Simulation of particles and gas flow behavior in the riser section of a circulating fluidized bed using the kinetic theory approach for the particulate phase. Powder Technol 112(1):24–33
16. Gidaspow D, Onischak M (1975) Process for regenerative sorption of CO_2. U.S. Patent 3,865,924 (Feb 11, 1975)
17. Nelson TO, Coleman LJI, Green DA, Gupta RP (2009) The dry carbonate process: carbon dioxide recovery from plant flue gas. ScienceDirect Energy Procedia 1:1305–1311
18. Nelson TO, Coleman LJI, Green DA, Gupta RP, Figueroa JD (2008) The dry carbonate process: carbon dioxide recovery from power plant flue gas. Presentation at the seventh annual conference on carbon capture & sequestration, Pittsburgh, Pennsylvania, May 5-8 2008
19. Fitzmorris RE, Mah RSH (1980). Improving distillation column design using thermodynamic availability analysis. AIChE J 26:265–273
20. Bejan A (1982) Entropy generation through heat and fluid flow. Wiley, New York
21. Prigogine I (1955) Thermodynamics of irreversible processes. Charles C. Thomas, Springfield
22. Kongkitisupchai S, Gidaspow D (2013) Carbon dioxide capture using solid sorbents in a fluidized bed with reduced pressure regeneration in a downer. AIChE J 59:4519–4537
23. Boonprasop S, Gidaspow D, Chalermsinsuwan B, Piumsomboon P (2017). CO_2 capture in a multistage CFB: Part I: number of stages. AIChE J Founders Tribute to Roy Jackson issue, 63:5267–5279
24. Worek WM, Lavan Z (1982) Performance of a cross-cooled desiccant dehumidifier prototype. J Solar Energy Eng 104:187–196
25. Kongkitisupchai S, Gidaspow D (2012) Carbon dioxide capture with solid carbonate sorbents in a fluidized bed riser and regeneration in a reduced pressure downer. Presentation at the NETL 2012 conference on multiphase flow science, Morgantown, WV, May 22-24, 2012
26. Chaiwang P, Gidaspow D, Chalernsinsunwan B, Piumsombon P (2014) CFD design of a sorber for CO_2 capture with 75 and 375 micron particles. Chem Eng Sci 105:32–45
27. Boonprasop S, Gidaspow D, Chalermsinsuwan B, Piumsomboon P (2017) CO_2 capture in a multistage CFB: Part II: riser with multiple cooling stages. AIChE J, Founders Tribute to Roy Jackson issue, 63:5280–5289.
28. Lee SC, Chae HJ, Lee SJ, Park YH, Ryu CK, Yi CK, Kim JC (2009) Novel regenerable potassium-based dry sorbents for CO_2 capture at low temperatures. J Mol Catal B: Enzym 56:179–184

29. Yi CK, Jo SJ, Seo Y, Lee JB, Ryu CK (2007) Continuous operation of the potassium-based dry sorbent CO_2 capture process with two fluidized-bed reactors. Int J Greenhouse Gas Control 1:31–36

30. Park YC, Jo SJ, Ryu, CK, Yi CK (2009) Long-term operation of carbon dioxide capture system from a real coal-fired flue gas using dry regenerable potassium-based sorbents. Energy Procedia 1:1235–1239

31. Park SW, Sung DH, Choi BS, Lee JW, Kumazawa H (2006) Carbonation kinetics of potassium carbonate by carbon dioxide. J Ind Eng Chem 12(4):522–530

32. Abbasi E, Hassanzadeh A, Zarghami S, Arastoopour H, Abbasian J (2014) Regenerable MgO-based sorbent for high temperature CO_2 removal from syngas: CO_2 capture and sorbent enhanced water gas shift reaction. Fuel J 137:260–268

33. Zarghami S, Ghadirian E, Arastoopour H, Abbasian J (2015) Effect of steam on partial decomposition of dolomite. Ind Eng Chem Res 5398–5406

34. Zarghami S, Hassanzadeh A, Arastoopour H, Abbasian J (2015) Effect of steam on the reactivity of MgO-Based sorbents in precombustion CO_2 capture processes. Ind Eng Chem Res 54(36):8860–8866

35. Abbasi E, Abbasian J, Arastoopour H (2015). CFD–PBE numerical simulation of CO_2 capture using MgO-based sorbent. Powder Technol, 286:616–628

36. Ghadirian E, Abbasian J, Arastoopour H (2017) Three-dimensional CFD simulation of an MgO-based sorbent regeneration reactor in a carbon capture process. Powder Technol 318:314–320

37. Johnson PC, R. Jackson (1987) Frictional-collisional constitutive relations for granular materials, with application to plane shearing. J Fluid Mech:67–93

38. Ghadirian E, Abbasian J, Arastoopour H (2019) CFD simulation of gas and particle flow and a carbon capture process using a circulating fluidized bed (CFB) reacting loop. Powder Technol 344:27–35

39. Abbasi E, Hassanzadeh A, Abbasian J (2013) Regenerable MgO-based sorbent for high temperature CO_2 removal from syngas: 2. Two-zone variable diffusivity shrinking core model with expanding product layer. Fuel 2013 105:128–34

40. Arastoopour H, Abbasian J (2019) Chemical looping of low-cost MgO-based sorbents for CO_2 capture in IGCC. In: Breault RW (ed.) Handbook of chemical looping technology, ISBN: 978-3-527-34202-0 Wiley –VCH. Verlag, Gmbh and Co., Germany pp. 435–459

41. Tullo AH (2016) Dry reforming puts CO_2 to work. Chem Eng News, 30

42. Extavour M, Bunje P (2016) CCUS: Utilizing CO_2 to reduce emissions. Chem Eng Prog 52–59

43. Vernon PDF, Green MLH, Cheetham AK, Ashcroft AT (1992) Partial oxidation of methane to synthesis gas, and carbon dioxide as an oxidizing agent for methane conversion. Catalysis Today 13:417–426

44. Asadi M, Kim K, Liu C, Addepalli A, Abbasi P, Yasaei R, Phillips P, Behranginia A, Cerrato, JM, Haasch R, Zapol P, Kumar B, Klie RF, Abiade J, Curtiss LA, Salehi-Khojin A (2016). Nanostructured transition metal dichalcogenide electrocatalysts for CO_2 reduction in ionic liquid. Science 353(6298):467–470

45. Orr FM Jr (2009) Online geological storage of CO_2. Science 325:1656–1658

46. de Chalendar, JA, Garing C, Benson, SM (2018). Pore-scale modelling of Ostwald ripening. J Fluid Mech 835:363–392

47. Arastoopour H (2019). The critical contribution of chemical engineering to a pathway to sustainability. Chem Eng Sci 203:247–258

48. Chalermsinsuwan B, Piumsomboon P, Gidaspow D (2010) A CFD design of a carbon dioxide sorption CFB. AIChE J 56(11):2805–2824

Chapter 8
Fluidized Bed Reactors for Solar-Grade Silicon and Silane Production

Hamid Arastoopour, Dimitri Gidaspow, and Robert W. Lyczkowski

8.1 Introduction

Production of electricity using solar collectors is determined to a large extent by the price and energy required to produce pure silicon. Inexpensive metallurgical grade silicon is produced from sand by an established technology. It can be converted to solar grade using a closed loop design, based on the following reversible reaction,

$$4SiHCl_3(g) = Si(s) + 3SiCl_4(g) + H_2(g) \qquad (8.1)$$

Metallurgical grade silicon is converted to gaseous trichlorosilane ($SiHCl_3$) using silicon tetrachloride ($SiCl_4$) and hydrogen at high temperature and high pressure. Mui [1] measured the reaction kinetics of this reaction in his study of the hydrochlorination of $SiCl_4$. This rate successfully was used to scale such fluidized bed reactors using CFD in [2]. Such reactors were constructed by AE Polysilicon Corporation in Fairless Hills, PA, in 2010.

The trichlorosilane product is then purified via filtration and distillation processes. The trichlorosilane is then used to produce solar-grade silicon by the reaction shown in Eq. (8.1) in a fluidized bed reactor. The hydrogen and silicon tetrachloride are then recycled to react with the metallurgical grade silicon. However, a very important difference between these two processes is the reaction temperature. The hydrochlorination of $SiCl_4$ with hydrogen is carried out at temperatures of the order of 500 °C, while the deposition of silicon is carried out at much higher temperatures. Mui [1] used a copper catalyst in his study.

Trichlorosilane is used in the silicon refining industry in the production of high purity silicon via the conventional Siemens process [3, 4]. Figure 8.1 shows a sketch of this process. Silicon is deposited on silicon rods heated to about 1150 °C by high voltage current controlled by complex switchgear and transformers. This batch process is very energy intensive due to heat loss by radiation. It requires 70–120 kWh/kg Si. Such a high energy loss was checked using the data reported

© Springer Nature Switzerland AG 2022
H. Arastoopour et al., *Transport Phenomena in Multiphase Systems*, Mechanical Engineering Series, https://doi.org/10.1007/978-3-030-68578-2_8

Fig. 8.1 Schematic of
solar-grade silicon
production reactor via the
conventional Siemens
process

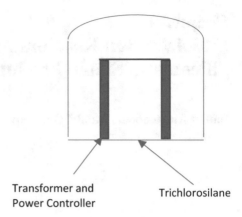

Transformer and   Trichlorosilane
Power Controller

by Hsu et al. [5]. Fluidized bed reactors were reported to use only 10 kWh/kg Si. The
minimum energy of separation needed to produce solar-grade silicon is only
0.3 kWh/kg Si. This energy savings and the continuous operation are the reasons
for interest in fluidized bed deposition reactor designs. Unfortunately, decomposition of silicon in the bubbles causes product contamination by chlorine [6]. Bubbles
can be eliminated by operating in the region between the minimum bubbling velocity
and the minimum fluidization velocity. An even better alternative is to use a
circulating fluidized bed.

The National Renewable Energy Laboratory and others [7] explored possible
silicon production using silane decomposition at 850–900 °C. However, its production is more complicated and, therefore, expensive. Hence, there is a need to design
better deposition reactors. One such design presented in this chapter has the potential
to be scaled up easily to obtain high rates of silicon production. Electrically heated
silicon deposition reactors [5] cannot be scaled up due to heat transfer limitations.
The CFD reactor simulation presented in this chapter has no size limitation because
the energy needed for trichlorosilane decomposition and preheating is supplied by
high voltage silicon rods, similar to those shown in Fig. 8.1 for the Siemens process.

8.2   Innovative Technology Description

Figure 8.2a shows a sketch of a polysilicon deposition reactor with internal heating.
Silicon is produced by vapor phase deposition of purified trichlorosilane by the
reaction given by Eq. (8.1), with rate of reaction,

$$k \frac{\varepsilon_g \rho_g}{MW_{SiHCl_3}} \left(Y_{SiHCl_3} - Y_{SiHCl_3,eq} \right) \text{ with } k = Ae^{-E/RT} \tag{8.2}$$

where ε_g is volume fraction, ρ_g is density, MW_{SiHCl_3} is molecular weight, and Y is
weight fraction. The reactant $SiHCl_3$ is produced by the reverse reaction shown in

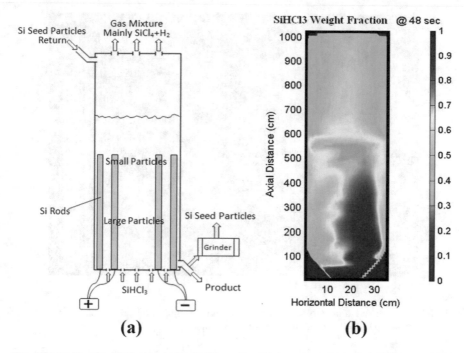

Fig. 8.2 (a) Sketch of IIT proposed polysilicon deposition reactor with internal heating, and (b) computed trichlorosilane concentration

Eq. (8.1) at 20 atm ($\approx$2 MPa) [1]. Figure 8.2b shows the computed trichlorosilane concentration using the IIT computer program [2] and the rate constant measured at AE Polysilicon Corporation [1]. In this simulation, heat was added internally using a measured fluidized bed resistivity equal to 1400 kΩ cm. The production rate is,

$$0.05183 A v \rho (Y_{SiHCl_3,In} - Y_{SiHCl_3,Out}) \tag{8.3}$$

where A is the cross-sectional area of the fluidized bed, v is the gas velocity, and ρ is its density. The particle diameter used was 850 μm diameter for the computed trichlorosilane concentration shown in Fig. 8.2b. The reactor operated in the bubbling bed regime. Unfortunately, nanoparticles are produced in the bubble phase. These nanoparticles contaminate the products.

By numerical simulation, from the theory and experiments of Gidaspow and Chaiwang [8], it was shown that, by using larger particles, operation can occur in the regime without bubbles. Figure 8.3 shows the published comparison of experimental and computed bubbles similar to that in the silane deposition reactor [5].

Figure 8.4 shows that using 5 mm diameter glass beads and the mixture of 3.5 and 5 mm diameter beads, allows operation in the regime without bubbles. Furthermore, the production rate will be much higher due to the higher reactant velocity as shown in Eq. (8.3).

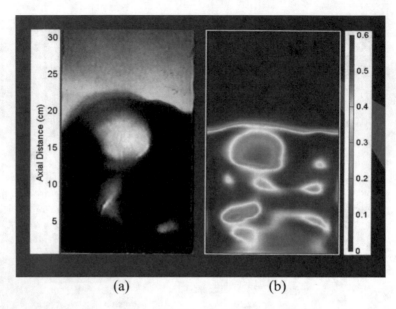

Fig. 8.3 Comparison of (**a**) experimental bubbles, and (**b**) computed bubbles obtained for 530 μm glass beads at $U/U_{mf} = 2.5$, using IIT computer program [2]

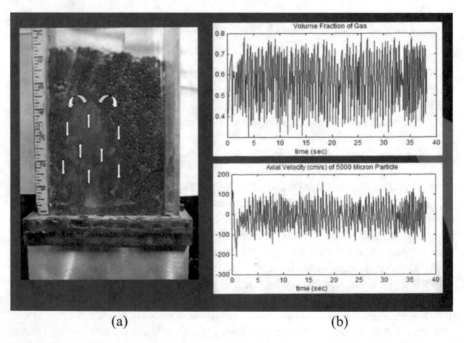

Fig. 8.4 (**a**) Bubble-less fluidization experiment at IIT, and (**b**) computed gas volume fractions and axial particle velocities

Fig. 8.5 Comparison of computed minimum bubbling velocity, $v_{mb\ jet}$, for the Westinghouse 3 m bed [9]

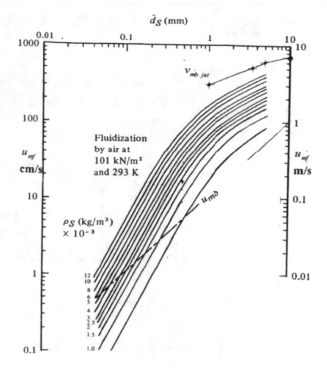

Figure 8.5 shows how to determine the minimum bubbling and the minimum fluidization velocities of particles as a function of particle diameter.

8.3   Exercises

8.3.1   Ex. 1: Hydrochlorination of $SiCl_4$

Very pure silicon for solar cells and computers is produced commercially by thermal decomposition of $SiHCl_3$. The product is $SiCl_4$. To close the loop, $SiHCl_3$ is made using inexpensive impure silicon made from sand. Mui [1] experimentally determined the kinetics of this reaction for the Department of Energy. This rate is given in the text. The following figures are taken from his report.

1. The objective of this exercise is to verify and to understand the reaction rate to design a workable reactor for production of inexpensive silicon for clean, unlimited energy using solar cells. Figure 8.6 (Fig. II in [1]) shows the fluidized bed reactor he used for obtaining his data.

To withstand the high operating pressure, the reactor was made of steel. The temperature was controlled by nichrome wires wrapped around the tube of 2 in. diameter. The reactor was filled with 862 g of 32 mesh silicon powder to a height of

FIGURE II   THE FLUIDIZED-BED REACTOR AND GRID DESIGN

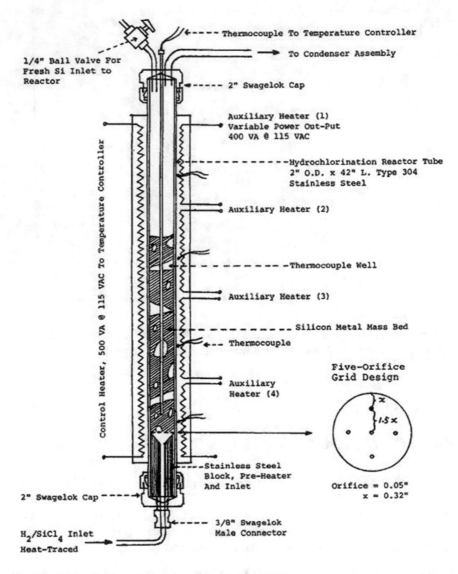

Fig. 8.6 Schematic diagram of Mui experiment. (Fig. II [1])

Table 8.1 Table XXIV [1]

TABLE XXV SUMMARY OF RATE CONSTANTS FOR THE HYDROCHLORINATION OF SiCl$_4$ AT 500 °C AS A FUNCTION OF REACTOR PRESSURE					
Reactor pressure		H$_2$/SiCl$_4$ ratio	Equilibrium mole %		Rate constant k$_1$
psig.	Atm.		SiHCl$_3$	SiCl$_4$	$\times 10^{-3}$ s^{-1}
25	2.70	2.0	18.4	81.6	14.5
73	5.97	2.8	23.4	76.6	13.0
100	7.80	2.0	22.9	77.1	12.2
200	14.6	2.0	31.0	69.0	8.33
500	35.0	2.8	37.5	62.5	7.06

Table 8.2 Table XXV [1]

TABLE XXV SUMMARY OF RATE CONSTANTS FOR THE HYDROCHLORINATION OF SiCl$_4$ AT 500 °C AS A FUNCTION OF REACTOR PRESSURE					
Reactor pressure		H$_2$/SiCl$_4$ ratio	Equilibrium mole %		Rate constant k$_1$
psig.	Atm.		SiHCl$_3$	SiCl$_4$	$\times 10^{-3}$ s^{-1}
25	2.70	2.0	18.4	81.6	14.5
73	5.97	2.8	23.4	76.6	13.0
100	7.80	2.0	22.9	77.1	12.2
200	14.6	2.0	31.0	69.0	8.33
500	35.0	2.8	37.5	62.5	7.06

18 in. Mixtures of hydrogen and liquid SiCl$_4$ were fed into the reactor. At the exit, the reaction mixture was condensed to produce liquid trichlorosilane.

Mui's [1] rate constants and equilibrium compositions are summarized in Tables 8.1 and 8.2 (Tables XXIV and XXV in [1]).

The rate is about 0.01 reciprocal seconds for all compositions and pressures. Hence, a quick design can be made to obtain conversion for a given velocity using a plug flow approximation discussed in the text. Figure 8.7 (Fig. IV in [1]) shows the conversions as a function of residence time for various pressures. Figure 8.8 (Fig. V in [1]) shows the effect of temperature on conversion.

Figure 8.9 (Fig. VI in [1]) gives the effect of hydrogen ratio. Figure 8.10 (Fig. VII in [1]) shows that an equilibrium of 0.2 is a satisfactory correction for an estimate of conversion

8.3.2 Ex. 2: Design of Deposition Reactors for Silicon Production

The thermal decomposition reaction for production of silicon based on the reaction studied by Mui [1] is given by Eq. (8.1). The rate of reaction, kmol/m^2 s, is given by Eq. (8.2) with the rate constant, k, equal to 0.0004T–0.2665, where T is in Kelvin.

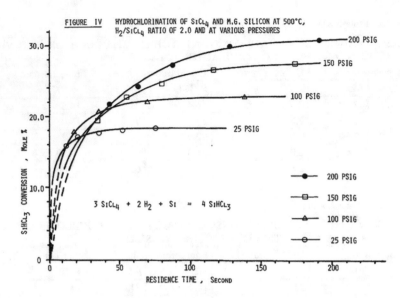

Fig. 8.7   Fig. IV [1]

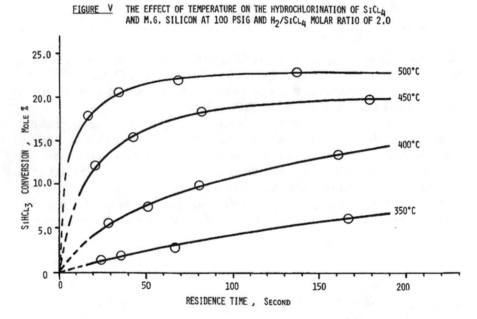

Fig. 8.8   Fig. V [1]

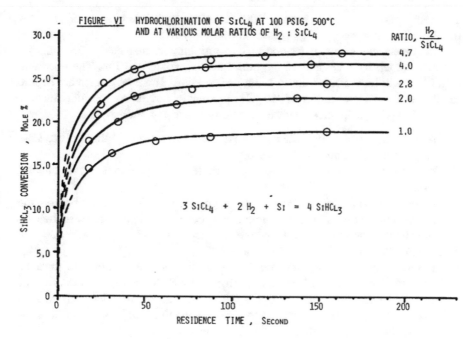

Fig. 8.9 Fig. VI [1]

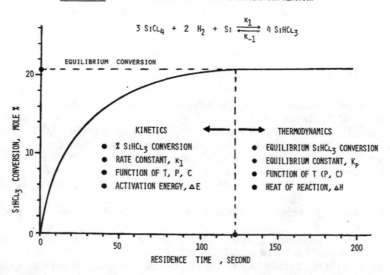

Fig. 8.10 Fig. VII [1]

The temperature should be above 400 °C. The equilibrium mole fraction of trichlorosilane is given by $Y_{SiHCl3,eq} = 0.0005T–0.1698$.

The dimensions of the deposition reactor are: internal diameter = 35 cm, with central inlet jet of 10 cm and 10 cm outlet at top, height = 10 m. The operating conditions will be at 600 °C and 4.1 atm (≈0.4 MPa). The inlet gas stream is pure trichlorosilane at a temperature of 350 °C to prevent its early decomposition. Its flow rate is 411 kg/h. The particle diameter is 850 μm as in Fig. 8.2b at 48 s.

1. Compare the simulation of the trichlorosilane concentration to that in Fig. 8.2b. Assume the temperature to be constant.
2. Compute the volume fraction of the particles. Are you in the core-annular regime?
3. Compute the axial particle velocities. How good will the heat transfer be from the heaters to the particles? Is the assumption of constant temperature reasonable?
4. Compute the rate of reaction. Is the present reactor too tall?
5. Compute the production rate of silicon.
6. In view of the very low production rate of silicon, the reactor size must be increased. But the rate of heat transfer from the wall limits the reactor size. The suggestion made in the text is to add the heat by means of silicon-coated rods placed into the reactor, connected to a high voltage supply, as in the Siemens reactor. Discuss the feasibility of this invention.
7. To produce silicon without contamination by chlorine, bubbles must be eliminated. The suggestion made in the text is to use mixtures of large particles. But then operation is restricted to be between the minimum fluidization and minimum bubbling velocities. Suggest a better alternative reactor design.

References

1. Mui JYP (1983) Investigation of the hydrochlorination of $SiCl_4$. Flat-plate solar array project, DOE/JPL/956061-7(DE83015173)
2. Gidaspow D, Jiradilok V (2009) Computational techniques: the multiphase CFD approach to fluidization and green energy technologies. Nova Science Publishers, New York
3. De Paola E, Duverneuil P (1998) Simulation of silicon deposition from $SiHCl_3$ in a CVD barrel reactor at atmospheric pressure. Comput Chem Eng, vol. 22: Suppl, S683–S686
4. Woditsch P, Koch W (2002) Solar grade silicon feedstock supply for PV industry. Sol Energy Mater Sol Cells 72:11–26
5. Hsu G, Rohatgi N, Houseman J (1987) Silicon particle growth in a fluidized bed reactor. AIChE J 33:784–791
6. Age S (1991) Manufacture of high purity/low chlorine content silicon by feeding chlorosilane into a fluidized bed of silicon particles. US Patent 5,077,028, Dec 31, 1991
7. Wikipedia (2018) Polycrystalline silicon. https://en.wikipedia.org/wiki/polycrystalline_silicon
8. Gidaspow D, Chaiwang P (2013) Bubble free fluidization of a binary mixture of large particles. Chem Eng Sci 97:152–161
9. Grace JR (1982) Chapter 8 Fluidization. In: Hetsroni G (ed) Handbook of multiphase systems. Hemisphere Publishing Corp

Chapter 9
Multiphase Hemodynamics Modeling (Blood Flow)

Hamid Arastoopour, Dimitri Gidaspow, and Robert W. Lyczkowski

9.1 Introduction

This chapter begins with a brief historical background that describes how the subject of multiphase hemodynamic modeling came about. Two of the authors of this book (DG and RWL) became interested in biomedical engineering, and hemodynamics (blood flow) modeling in particular, for quite different reasons. The first author (DG) learned of the efforts of Professor Richard Beissinger, then a faculty member in the Illinois Institute of Technology (IIT) Chemical Engineering department, and those of Vincent Turitto, then Pritzker Professor and Director of the Pritzker Institute of Biomedical Science and Engineering in the IIT Biomedical Engineering department, who was working on platelets. DG's interest in blood flow modeling increased when he experienced a mild heart attack requiring angioplasty and the insertion of a stent into his right coronary artery.

The third author (RWL) of this book and the author of this chapter contracted rheumatoid arthritis (RA), an autoimmune disease, and wanted to understand it and the concomitant medical physiology. When he was in the Energy Systems (ES) division at Argonne National Laboratory (ANL), his colleague, C. B. Panchal, had developed a model for fouling of hollow-fiber membranes undergoing pulsatile flows. The test results and analysis demonstrated that the shape and frequency of pulsatile flows have a direct impact on the fouling propensity of hollow-fiber membranes. Based on this investigation, he suggested an idea that modeling blood flow (hemodynamics) might constitute the basis of a novel and interesting project to pursue the pulsatile flow of blood (hemodynamics). Prior experience with modeling coal/water slurry flow led to the idea that blood flow resembles a slurry consisting of red blood cells. This subject of coal/water slurry flow will be discussed in Sect. 9.2. As time went on, the scope of the project was expanded to include vascular lesion formation, i.e., atherosclerosis. Discussions led to the preparation of a proposal together with Sanjeev Shroff, then Professor of Medicine at the University of Chicago and later Professor of Medicine and Gerald E. McGinnis Chair in

© Springer Nature Switzerland AG 2022
H. Arastoopour et al., *Transport Phenomena in Multiphase Systems*, Mechanical Engineering Series, https://doi.org/10.1007/978-3-030-68578-2_9

Bioengineering at the University of Pittsburgh, to procure funding from the National Institutes of Health (NIH). He visited ANL frequently to help and to offer advice. He became the driving force for the proposal because of his broad experience at the NIH reviewing proposals and obtaining research support. This began what would turn out to be a decade-long (2000–2010) research effort at ANL and resulted in collaboration with Professor Gidaspow at IIT to begin researching this area.

This chapter reports the results of this decade-long investigation of hemodynamics using multiphase flow. The objective of this research was to assess the probable causes and physiological sites for atherosclerosis initiation. The development of mechanistic models for the particulate-depleted layer in the vicinity of vessel walls (the Fahreaus-Lindqvist effect), the movement of platelets toward the walls of vessels with the simultaneous movement of red blood cells toward their center, and a monocyte adhesion model coupled to the multiphase computational fluid dynamics (CFD) model were achieved. These subjects as well as others will be described in subsequent sections.

About 610,000 people die of heart disease in the United States every year and it is the leading cause of death (Google: https://www.cdc.gov/heartdisease/facts.htm). Although atherosclerosis is generally considered to be a systemic disease, actual lesion formation is inherently local, being dependent on pathologic interaction of the vessel wall with blood and particulate elements circulating in a pulsatile hemodynamic field. Research using animal experiments and/or single phase CFD models has demonstrated that atherosclerotic lesions form preferentially in regions where there are: (1) high hemodynamic forces acting on the vessel walls as a result of high hemodynamic wall stress and pressure (which can produce high shear stresses within the walls) in regions of arterial bifurcations and curvatures; (2) repetitive periods of low shear, boundary layer separation, and flow recirculation within the pulsatile cardiac cycle, which tend to increase particle and cellular residence times; and (3) increased activation of blood elements, including inflammatory cells, cytokines, and adhesion molecules, especially monocytes. Incorporation of all of these characteristics into a single useful model has been difficult, given the inherent complexity of simultaneously modeling pulsatile flow fields in tubes with elastic walls interacting with particles in a multiphase blood stream.

The complex pulsatile three-dimensional (3D) flow patterns at vascular bifurcations and regions of high curvatures, shown in the left of Fig. 9.1, produce aberrant spatial and temporal wall stresses and distributions of blood-borne particulates in localized regions. These regions have high propensity for atherosclerotic lesion formation. The hemodynamic effects of branched and curved arteries on localized lesion formation are further illustrated in the enlarged view of coronary lesions in the right of Fig. 9.1 taken from an angiogram of an actual person who will remain anonymous. The atherosclerotic lesion formed in the left anterior descending (LAD) coronary artery is typical of those found in the vicinities of bifurcations. The atherosclerotic lesion formed in the circumflex (CX) to the posterior descending (PD) artery transition is typical of that found in highly curved arteries.

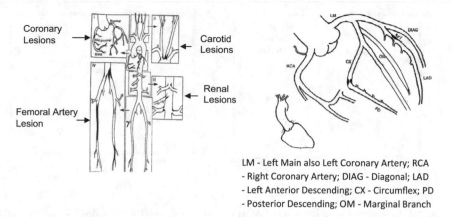

LM - Left Main also Left Coronary Artery; RCA - Right Coronary Artery; DIAG - Diagonal; LAD - Left Anterior Descending; CX - Circumflex; PD - Posterior Descending; OM - Marginal Branch

Fig. 9.1 Left: Distribution of atherosclerotic plaque formation. (Modified from [1]). Right: Enlarged view of coronary artery lesions

9.2 Origins in Non-Newtonian Coal/Water Slurry Modeling

The objective of this early work was to develop a non-equilibrium two-phase non-Newtonian hydrodynamic model of slurries in order to provide a better understanding of injector nozzle erosion and combustion performance of coal/water slurry (CWS) diesel engines. The most important results calculated from the model are the local velocities of liquid and solids, the local shear rate and viscosity of the mixture, the pressure distribution along the injection pipe, and the solids concentration at the nozzle outlet. The model was validated using Newtonian single-phase (pure water and No. 2 diesel fuel) and non-Newtonian two-phase CWS capillary tube viscometer shear rate, viscosity, mass flow rate, and pressure drop data measured at what was then ANL's Materials and Components Technology division. The range of Reynolds numbers was from about 500 to 2500.

The isothermal conservation equations of mass and momentum given in Chap. 4, Table 4.2 were written for the liquid and solids phases in two-dimensional cylindrical coordinates. Pure water, No. 2 diesel fuel, and the two-phase CWS carrier fluid (water) momentum equations are Newtonian. The solids (coal) stress terms in the CWS momentum equations are treated using the concept of the solids pressure for the normal components, and a non-Newtonian power-law model for the mixture instead of using the kinetic theory. The interfacial drag used is given in Chap. 2, Table 2.5 with a simple model for $\omega(\varepsilon)$, which accounts for the collective effect of the presence of particles in the fluid and acts as a correction to the usual Stokes law for free fall of a single particle. The simulations were performed with a computer code called SLUFIX, which was developed for liquid/solids slurry flow. This SLUFIX code extended the FLUFIX computer code [2], which had successfully been used to predict the hydrodynamic behavior of gas/solids flows. Results compared well with analytical solution, calculated and experimental non-Newtonian power-law single-phase flow friction factors, and with the experimental CWS

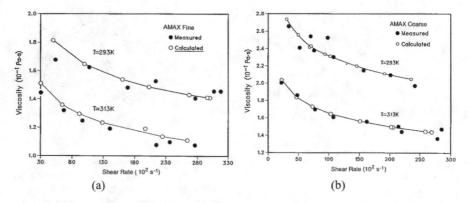

Fig. 9.2 Comparison of calculated (COMMIX) and experimental (NMR) AMAX fine (**a**) and coarse (**b**) viscosity data. (Modified from [3])

viscosity data for AMAX coarse and AMAX fine samples at temperatures of 293 and 313 K supplied by the AMAX Corporation, containing 52–52.5 weight percent of Kentucky #4 coal particles. Comparisons of experimental CWS viscosity data with the model predictions are shown in Fig. 9.2. More details of this study were published in Powder Technology journal [3]. This seminal work was to set the tenor and to serve as a model for subsequent work that will be described in Sect. 9.3 and a portion of Sect. 9.4.

9.3 Simulation of Concentrated Suspension Flows in Straight Pipe Geometries

As a result of Pittsburgh Energy Technology Center's (PETC) (now part of the National Energy Technology Laboratory, NETL) proposed Suspension Advanced Research Objective (SARO), ANL's Components Technology (CT) division, later changed to Energy Technology (ET), received funding for a project on concentrated suspensions, usually called slurries. A coordinated methodology was proposed involving theory (field and constitutive equation development), experimentation, and validation by computer modeling. ANL anticipated that synergism would result when a conscious effort was made to coordinate the execution of this research program involving complex and interdisciplinary phenomena requiring advances of both a theoretical and experimental nature.

9.3.1   Analysis of Lovelace Medical Foundation Experiments

William Peters, who was then in the Solids Transport Program (STP) at PETC, suggested analyzing some of the data for steady-state, fully-developed, and isothermal carrier-fluid velocity and solids concentration that was taken at the Lovelace Medical Foundation in Albuquerque, NM. Lovelace was part of the Granular Flow Advanced Research Objective (GFARO) established in 1990 under the auspices of the United States Department of Energy (USDOE) STP [4]. These data were obtained with three-dimensional, time-of-flight, nuclear magnetic resonance (NMR) imaging, also referred to as magnetic resonance imaging (MRI), to minimize the fear of any connection with things nuclear. NMR imaging is a powerful technique to non-intrusively determine three-dimensional velocity and concentration fields to assist development and validation of the constitutive models and the computer programs that describe concentrated suspensions. These experiments were carefully performed and probably represent the best available data of this kind in the open literature.

The Lovelace experiments [5] used a suspension of negatively buoyant, 1.03 g/cm^3, divinyl-benzene styrene copolymer (plastic) spheres having a mean diameter of 0.762 mm and a narrow size range flowing in a horizontal 2.54 cm inside diameter acrylic plastic tube. The carrier fluid was SAE 80 W gear oil with a viscosity of 3.84 P (0.384 Pa s) and a density of 0.875 g/cm^3. The flowing suspension can loosely be considered to represent idealized blood flow. NMR concentration and velocity data were taken over a range of 0–39 volume percent plastic spheres (1.0–0.61 fluid volume [void] fraction) and average fluid velocities of 1.7–22.3 cm/s using a 1.89 T superconducting solenoid.

The data were analyzed with the COMMIX-M computer program that was being developed for the USDOE to perform safety studies for nuclear power plants and the Nuclear Regulatory Commission (NRC) to analyze sodium-cooled fast nuclear reactors, then under consideration for development. The fluid viscosity was taken to be a constant for isothermal Newtonian flow. The solids viscosity was obtained from Krieger's empirical model for the reduced viscosity, η_r, of the fluid/solids mixture as,

$$\eta_r = \frac{\varepsilon_s \mu_s + \varepsilon_f \mu_f}{\mu_f} = (1 - \varepsilon_s/0.68) \tag{9.1}$$

where ε_s and ε_f are the volume fractions and μ_s and μ_f are the viscosities of solids and carrier fluid, respectively. Saffman's shear lift force expression for a single particle in a simple shear flow and the virtual mass force were extended to expressions for forces per unit volume for a continuum and programmed into COMMIX-M. Comparisons of the computed fluid velocities and volume fractions with the experimental data for Run Bl are shown in Fig. 9.3.

Run Bl had 9% average solids volume fraction and average fluid velocity of only 4.09 cm/s. COMMIX-M correctly predicts the trend of particle settling as shown in

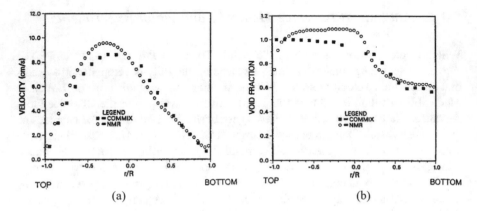

Fig. 9.3 Comparison of predicted (COMMIX) and experimental (NMR) data: (**a**) vertical fluid velocity, (**b**) vertical fluid volume fraction (void fraction); run Bl, 9% solids volume fraction, average fluid velocity, 4.09 cm/s. (Modified from [7])

the vertical fluid volume (void) fraction profile shown in Fig. 9.3b. It appears that the upper half of the tube is essentially pure fluid and the lower half is a loosely packed flowing dense suspension having about 60% fluid (40% solids). No attempt was made to correct the experimental data having void fractions more than 1.0. Particle settling distorts the vertical fluid velocity profiles shown in Fig. 9.3a with the position of the maximum velocity moving upwards as the average velocity is decreased. If there were no solids present, the maximum velocity would be 8.18 cm/s, twice the average. At this lowest fluid velocity, the vertical fluid velocity profile shown in Fig. 9.3a resembles that for stratified flow. For more details of the models and more comparisons of computations with experiment, see [5].

9.3.2 Sinton and Chow Experiments Analysis

The Sinton and Chow experiments [6] used a suspension of neutrally buoyant, 1.19 g/cm^3, polymethylmethacrylate spheres (Lucite 47G) with a median volume diameter of 0.131 mm, flowing in vertical pipes with diameters of 15.2, 25.4, and 50.8 mm, and a 500 mm entrance length. Intensity and velocity data were collected over a range of 21–52 volume percent plastic spheres and particle Reynolds numbers of 0.005–4.0. The carrier fluid having a viscosity of 4.8 Pa s was a mixture of polyether oil, water, and sodium iodide and was added to increase the fluid density to that of the solids. NMR data were taken with a vertically oriented 4.7-T superconducting solenoid. Three runs were analyzed: (1) 21 volume percent solids, an average fluid velocity of 22.7 cm/s, and pipe diameter of 25.4 mm; (2) 40 volume percent solids, an average velocity of 17.6 cm/s, and pipe diameter of 15.2 mm; and (3) 52 volume percent solids, an average velocity of 17.5 cm/s, and pipe diameter of 15.2 cm.

Ding et al. [5] analyzed the Sinton and Chow data [6] to study the shear-thinning phenomena of neutrally buoyant dense suspensions in vertical pipes using the Krieger model for solids viscosity given by Eq. (9.1). Results were in quite good agreement with the data. In later work, this data was reanalyzed by Ding et al. [7]. The suspension rheology data were fit to an expression for the relative viscosity, η, the ratio of mixture to fluid viscosity, $(\varepsilon_s \mu_s + \varepsilon_f \mu_f)/\mu_f$, using an extension of the Carreau-Yasuda viscosity model [8]. It serves to generalizes Krieger's model given by Eq. (9.1), where, as before, ε_s and ε_f are the volume fractions, and μ_s and μ_f are the viscosities of the solids and carrier fluid, respectively, as

$$\eta = (\varepsilon_s \mu_s + \varepsilon_f \mu_f)/\mu_f = m\left[1 + (\lambda \dot{\gamma})^2\right]^{(n-1)/2} \tag{9.2}$$

where $\dot{\gamma}$ (1/s) is the shear rate of the mixture. The parameters m (rest viscosity), λ, and n were fit to functions of the solids volume fraction, ε_s, generalizing the usual non-Newtonian power law model to account for concentration dependency because the solids concentration was found to affect the mixture viscosity. At the highest solids concentration, the mixture is shear thinning and at the lowest concentration it is essentially Newtonian with negligible shear dependency. Previous dense suspension models failed to explain these experimentally observed phenomena. The data as fit by Eq. (9.2) show that, at very low shear rates, the relative viscosity increases from 10 at 21% solids, to roughly 50 at 40% solids, and to roughly 100 at 52% solids.

The results computed by COMMIX-M using Krieger's model given by Eq. (9.1) were compared with those computed with the Carreau-Yasuda model given by Eq. (9.2) [5, 7]. For 21% solids concentration, both models agreed well with the data because the mixture is essentially Newtonian. As the solids concentration increases, results computed with the Carreau-Yasuda model agree better and were found to be in excellent agreement with the data. More details of the models and comparisons of computations with experiment are given in Ding et al. [7].

The hematocrit of human blood is in the range of 42–50%. Solids cannot pack much higher than about 60% (unless they deform). It is clear that even a modest change of solids volume fractions (40–52%) would change the mixture viscosity by a factor of 2 for a given shear rate if blood rheology behaved like the Sinton and Chow data [6]. With particle migration out of regions of high shear, the mixture viscosity could change tenfold [13]. For blood having a mixture viscosity of 3–4 cP, this range would be a factor of roughly 2 for a hematocrit of 50% for plasma having a viscosity of 1.8 cP. However, if the hematocrit goes higher, the relative viscosity would increase considerably as shown by the data of Goldsmith and Mason [9] contained in Lightfoot [10], which also shows that the viscosity of blood is strongly shear thinning at low shear rate.

In summary, this section shows that the CFD models developed provided a solid engineering basis for the development of multiphase pulsatile hemodynamics in the cardiovascular system that will be described in Sect. 9.4. The models account for fully three-dimensional two-phase non-Newtonian slurry flow at all concentration levels. The non-Newtonian power-law model was generalized to account for slurry rheology dependence upon particulate concentration and a rest (zero shear) viscosity by Ding et al. [7]. Previous single-phase slurry models failed to explain these experimentally observed phenomena. Also included is a generalized Saffman lift force model to predict particle migration in regions of high shear stresses. The models predicted particle sedimentation at the very low fluid flow rates that might exist in stagnation or recirculation regions near atherosclerotic lesions. Pulsatile flow will be produced by applying appropriate time-dependent boundary conditions obtained from experimental data. One set of boundary conditions to apply would be the time-dependent inlet flow rates and pressures; another set would be to apply the time-dependent inlet and outlet pressures.

9.4   Simulation of a Right Coronary Artery Using Basic Two-Phase Non-Newtonian and Kinetic Theory Models

As mentioned in the abstract, there were two key events; they are both related to money. Without it, no research could be initiated. Internal funding was applied for from ANL's Laboratory-Directed Research and Development (LDRD) program in 2002. The request was for $360,000 over a 2-year period ($180,000 per year). The proposal, approved in 2003, was funded in the amount of $145,000, just short of the amount requested for the first year of the project. This was the first key event that made it possible to initiate multiphase hemodynamics research. The LDRD was renewed in 2004 for $127,000. This section and Sects. 9.4.2 and 9.4.3 summarize the accomplishments achieved with this funding.

Fluent Inc. had a program in place on their website called FlowLab. It included a collection of problems with exercises submitted by users of their codes. One of these problems contained descriptions of models for a coronary artery. They were submitted by one of the authors of the paper by Berthier et al. [11] that reported modeling a realistic right coronary artery They used the single-phase Navier–Stokes equations solved by the finite-element method (FEM) implemented in Fluent, Inc.'s FIDAP code. The meshes files for a three-dimensional simplified and the realistic right coronary artery model geometry were used in this publication. They could be used in the FLUENT code and were available for downloading. They readily formed the basis of the subsequent modeling. Otherwise, much time would have been required to generate these models.

The commercial CFD code FLUENT 6.1, which contains the multiphase 3D Eulerian-Eulerian model, was used for both the simplified (idealized) and the realistic right coronary artery (RCA) models. In all of the literature studied, no one had ever used multiphase flow to model hemodynamics. The virtual mass and shear lift force contained in the code were included. The Carreau-Yasuda viscosity given by Eq. (9.2), which was used by Ding et al. [7] to model the Lovelace Medical Foundation concentrated suspension data as described in Sect. 9.3.1, was also used

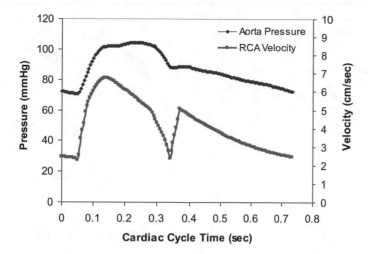

Fig. 9.4 Velocity waveform profile for RCA correlated with aorta pressure waveform

as the basis of the model for blood rheology. Constants in the model were refit to data for blood having a range of hematocrits from 30% to 75%. The fit for 45% hematocrit was used in subsequent studies since this is about the average value for humans. The Carreau-Yasuda viscosity model and the transient inlet blood velocity waveform for the cardiac cycle developed from several data sources were programmed into the FLUENT code as UDF and profile files. A pulsatile inlet velocity waveform was used as inlet boundary conditions for both the red blood cells (RBCs) and plasma. *In-vivo* pressure waveforms measured in the ascending aorta were used to adjust the timings of the systole, diastole, and total cardiac cycle period as shown in Fig. 9.4. The velocity profiles were maintained uniform across the inlet cross-section.

Zero slip velocity boundary condition was employed for both RBCs and plasma. RBCs were considered as spherical particles with an average diameter of 8 μm. The inlet volume fraction of the RBCs was maintained uniform and steady at 45%. For initial conditions, the RBC volume fraction was set to 45%, and both RBC and plasma velocities were set to zero. More details of the modeling are to be found in Jung et al. [12] for the idealized RCA and Jung et al. [13] for the realistic RCA.

The two-phase Eulerian-Eulerian model in the FLUENT code was first reanalyzed using the neutrally buoyant dense suspension NMR slurry data of Sinton and Chow for flow in a vertical tube [6] modeled by Ding et al. [7] as discussed in Sect. 9.3.2. The purpose was to perform preliminary validation of the FLUENT multiphase models before embarking on the hemodynamics simulations. Results are shown in Fig. 9.5 for a solids volume fraction equal to 0.52. As can be seen, the comparison is excellent. This is a particularly significant result because it shows that two completely different computer programs can produce essentially the same results, proving that the models are consistent.

Fig. 9.5 Comparison of
Sinton and Chow data [6]
with FLUENT code
computation for 52 volume
percent solids, an average
velocity of 17.5 cm/s, and
pipe diameter of 15.2 mm

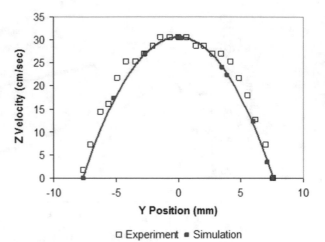

□ Experiment ■ Simulation

Summaries of the major results for both the idealized and realistic coronary artery simulations are contained in the next two sections. These resulted from the 2 years of funding for the ANL LDRD grants.

9.4.1   Idealized Model of a Right Coronary Artery

Figure 9.6 shows the distribution of RBC volume fractions along the length of the idealized curved artery model having a uniform diameter of 4.37 mm and a 41.51 mm radius of curvature. The objective was to determine the spatial and temporal distributions of RBCs during pulsatile flow. It is important to note that prediction of this RBC distribution is only possible using the multiphase Euler-Euler CFD model. Figure 9.6 shows that the recirculation pattern and higher residence times produce a buildup of RBCs on the inside curvature. The color bar shows the range of RBC volume fractions. This is an important result because, as discussed in the Sect. 9.1, atherosclerotic plaque occurs preferentially in such regions as shown in Fig. 9.1 [1]. The secondary flows shown in the vector plots are characterized by two vortices with a weak outward flow in the center and a stronger inward flow in the vicinity of the wall. The higher viscosity in the central portion of the curved vessel tends to block the RBC flow, causing the RBCs to migrate inward preferentially. The recirculation pattern weakens during the diastole cycle, and the ratio of the volume fraction of RBCs in the inside curvature to the outside curvature is about 1.09. The multiphase CFD model confirmed the well-known fact that RBCs determine the rheological behavior of blood, which is not possible with a single-phase CFD model. When the model includes monocytes and platelets, similar concentration buildups are expected to occur. The CFD model predicts that the wall shear stress (WSS) is lower on the inside curvature as compared to outside, thus correlating low WSS with particulate buildup.

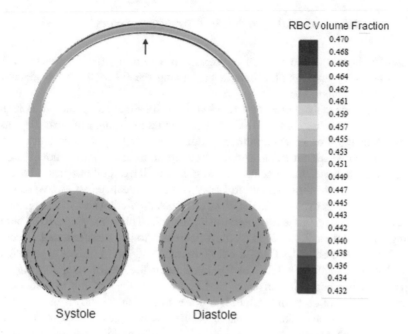

Fig. 9.6 RBC volume fraction and secondary flow (vector plots) in the third cardiac cycle. The cross-sections for the systole (1.61 s) and diastole (2.20 s) are at the center curvature indicated by the arrow. The inside curvature is at the bottom of each cross-section and the outside curvature is at the top. (This figure was originally published in [12] and has been reused with permission)

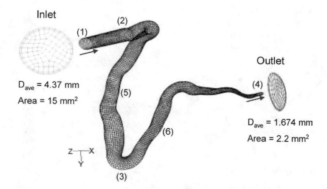

Fig. 9.7 Three-dimensional computational mesh for the realistic RCA generated using GAMBIT. Point (1): inlet; point (2): end of a straight entry tube; point (3): maximum curvature; point (4): outlet; and points (5) and (6): positions of side branches in the actual artery [13]. (This figure was originally published in [13] and has been reused with permission)

9.4.2 Realistic Model of a Right Coronary Artery

Figure 9.7 shows the geometry of the right coronary artery (RCA) generated from *in vivo* coronary angiographic images [11] using the GAMBIT mesh generation program from Fluent, Inc.

Figure 9.8 shows the predicted RBC volume fractions along the variable-area, realistic right coronary artery (RCA). RBC buildup is highest near the inside curvature in the area of maximum curvature, indicated by the curvature surface arrow, just downstream of a tortuous reduction in area. Also shown are the recirculation patterns that occur, resulting in the RBC buildup near the inside curvature, similar to those shown in Fig. 9.6 for the idealized RCA. A significant prediction shown is that, as is in the idealized RCA prediction shown in Fig. 9.6, the recirculation intensity varies during the cardiac cycle. This results in an RBC buildup during the diastole at 2.2 s (2) as the flow decreases to its minimum, and a decrease during the systole at 1.61 s (1) as the flow increases to its maximum, thus affecting the residence time of the RBCs. The RBC buildup in the area of maximum curvature increases at the beginning of each cardiac cycle. Note that the secondary flow patterns shown are quite unsymmetrical compared to those shown in Fig. 9.6 for the idealized coronary. This is because of the swirling flow that results from the complex three-dimensional geometry vs. the idealized two-dimensional geometry. It is important to note that the RBC buildup for the realistic RCA geometry shown in Fig. 9.8 is much higher than that for the idealized RCA geometry shown in Fig. 9.6,

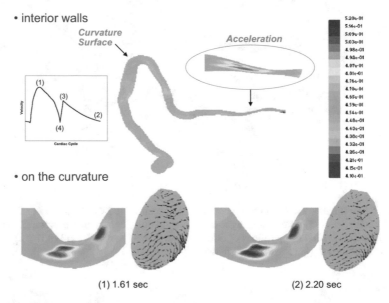

Fig. 9.8 RBC volume fraction (percent, color bar) and secondary flow (vector plots) in the third cycle. The cross-sections for the peaks of the systole (1) and diastole (2) cycles are in the area of the maximum curvature indicated by the arrow

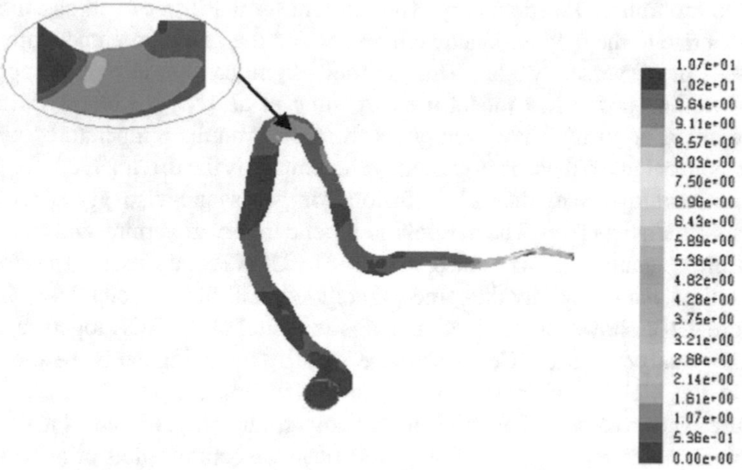
1.07e+01
1.02e+01
9.64e+00
9.11e+00
8.57e+00
8.03e+00
7.50e+00
6.96e+00
6.43e+00
5.89e+00
5.36e+00
4.82e+00
4.28e+00
3.75e+00
3.21e+00
2.68e+00
2.14e+00
1.61e+00
1.07e+00
5.36e-01
0.00e+00

Fig. 9.9 RBC wall shear stress (pascal, color bar)

with the ratio of the volume fraction of RBCs in the inside curvature to the outside curvature being about 1.27 for realistic geometry and 1.09 for idealized geometry. Figure 9.9 shows that the RBC WSS is lowest in the region of maximum curvature where the RBC buildup occurs, as shown in Fig. 9.8. More details may be found in Jung et al. [13].

9.4.3 *Realistic Model of a Right Coronary Artery Using Multiphase Kinetic Theory*

Jing Huang started her PhD thesis at IIT under the direction of DG early in 2007 to model various aspects of blood flow: the Fahreaus-Lindqvist effect, platelet deposition, and low-density lipoprotein (LDL) and high-density lipoprotein (HDL) transport using multiphase kinetic theory [14]. As part of her thesis in collaboration with ANL, she used the multiphase kinetic theory to analyze the realistic RCA that Jung had modeled as discussed in Sect. 9.3.2 [13]. She had already written a manuscript with her advisor, DG, on modeling the Fahreaus-Lindqvist effect of red cell depletion near the walls of capillaries (see Sect. 9.6.1). The reference report form for the IIT Fieldhouse Fellowship was prepared, highly recommending her. RWL was to be her co-advisor from ANL. Jing was awarded a Fieldhouse Fellowship by IIT in the amount of $15,000 for the period January 1 to December 31, 2008. Besides the ANL LDRD grants in 2004 and 2005 discussed in Sect. 9.4, which initiated the hemodynamics research, this was the second key event mentioned in the abstract which allowed progress on multiphase modeling of hemodynamics to continue.

Jing's work served to significantly improve the analysis of the pulsatile 3D hemodynamics in the same realistic RCA geometry reported by Jung et al. [13] by using the multiphase kinetic theory. The random oscillations of particles and shear rate gives rise to the random kinetic energy, called the granular temperature, which increases near vessel walls. This model significantly improves upon the non-Newtonian power-law model used by Jung et al. [13] for blood viscosity in that the blood viscosity is now computed from the granular temperature, instead of using empirical fits to data. This granular temperature is the driving force for particle migration, analogous to thermal diffusion for gases and also gives rise to the Fahraeus-Lindqvist effect. The multiphase kinetic theory governing and constitutive Euler-Euler equations programmed into the FLUENT code used to perform the computations are essentially the same as included in Tables 2.3 and 2.4 in Chap. 2.

Figure 9.10a shows the mesh of the realistic model of the RCA together with the 14 planes analyzed. The RCA is characterized by three segments: before (planes 1–5), in the vicinity of (planes 6–10), and after (planes 11–14) the maximum curvature. The cross-sectional area is increasing around planes 4 and 14 and decreasing around planes 6 through 13. Figure 9.10b shows contour plots of instantaneous RBC volume fraction, RBC granular temperature, and mixture viscosity at the peak of the systole, point (2) in the inset pressure waveform in panel a. Note the decrease in RBC volume fraction in Side A in the area of maximum curvature (planes 7 and 8) caused by increased migration due to increased granular temperature. The RBC volume fraction and granular temperature then increase to the end of the RCA because of decreased flow area. The mixture viscosity bears a strong resemblance to the RBC granular temperature. This is because the RBC viscosity depends much more strongly upon the granular temperature than the RBC volume fraction.

The WSS is higher on the inside curvature in the area of maximum curvature, as shown in plane 8. This is opposite to the idealized RCA WSS computed by Jung et al. [13]. However, lower WSS occurs on the inside curvatures before and after the area of maximum curvature. It is for reasons like this that researchers have found inconsistent correlations of atherosclerotic sites with WSS. This also points out the danger of utilizing highly idealized geometric models of arterial vessels. The multiphase kinetic theory model predicts somewhat lower WSS due to the presence of the Fahraeus-Lindqvist effect, which is absent in all other hemodynamic models. Note that the secondary flow patterns shown in Fig. 9.10c for the realistic coronary are quite asymmetric. This is indicative of complex swirling flow resulting from three-dimensional curvatures. There is a correlation between low WSS and RBC buildup on the outside curvature in the area of maximum curvature.

The inset table summarizes results for maximum recirculation (areas of increased residence time), temporal and spatial WSS gradients, WSS_t and WSS_s, and oscillating shear index (OSI) distributions computed from the multiphase kinetic theory model. These results are used, individually or in combination, to identify locations that are susceptible for developing atherosclerotic lesions. Based on these data, one can speculate that there are two areas with high likelihood of developing atherosclerotic lesions: (1) Planes 4 and 5—artery expansion with recirculation, high oscillatory shear index (OSI), and low wall shear stress (WSS), and (2) Side A of planes 7 and 8—area of maximum curvature with high WSS gradients. Such predictions provide a new level of understanding of how RBCs migrate in secondary flows.

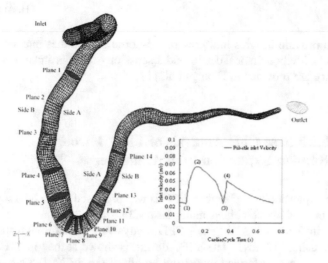

Atherosclerosis lesion sites	Characteristic planes	Factors affecting the occurrence of atherosclerosis lesion				
		Recirculation	WSS	WSSG$_t$	WSSG$_s$	OSI
Artery expansion	Plane 4 Plane 5	yes			Yes, but slightly	yes
Inside of maximum curvature	Side A of Plane 7 and Plane 8	yes but slightly	yes	yes	yes	
Artery contraction	Plane 13				yes	

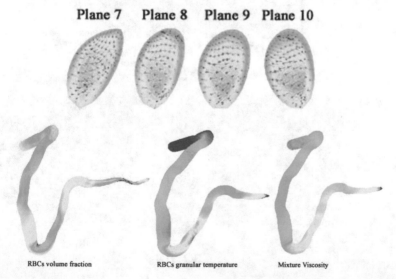

Fig. 9.10 (**a**) Computational mesh for realistic RCA model: Side A: inner wall, side B: outer wall. (**b**) Computed RBC volume fraction, granular temperature, and mixture viscosity contour plots at peak of the systole (2). Blue is lowest and red is highest. (**c**) Time-averaged RBC volume fractions contour plots showing the Fahraeus-Lindqvist effect and velocity magnitude vectors in the area of maximum curvature. Sites of likely atherosclerotic site initiations predicted from the computed recirculation, WSS, temporal and spatial WSS gradients, WSS$_t$ and WSS$_s$, and oscillation shear index (OSI)

They also provide insights into how blood-borne particulates interact with arterial vessel walls. Methodological details and discussion of results and their physiological significance are provided in Huang et al. [15].

9.5 Multiphase CFD Analysis of Flow Through a Sudden-Expansion Flow Chamber

A sudden-expansion flow chamber (SEFC) was proposed to the NIH for experimental validation of the combined multiphase CFD-cell adhesion model that will be discussed in Sect. 9.8. Multiphase CFD analysis of flow through the SEFC was performed using FLUENT. The SEFC design is shown at the top of Fig. 9.11. The CFD analysis was performed over 1 cm length of the SEFC, 0.5 cm upstream and downstream of the step. A uniform mesh was used having 31,500 mesh elements. Inlet flow was steady and set at 8 ml/min. Three conditions were simulated: (1) Newtonian single-phase flow having a density of 1050 kg/m^3 and a viscosity

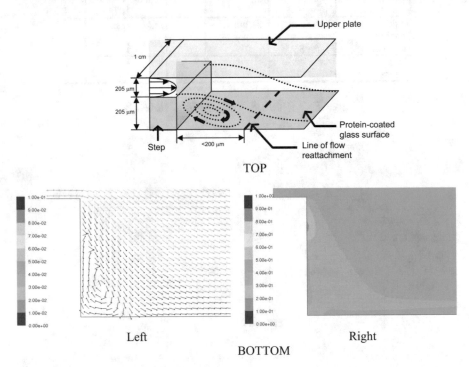

Fig. 9.11 Top: Schematic diagram of the SEFC. A flow channel is 2 cm long × 1 cm wide. The inflow channel is sufficiently long (1 cm) to ensure a parabolic velocity profile before the expansion point, and after <0.2 mm, flow assumes a parabolic profile. Bottom left: RBC velocity vector plot for SEFC. Scale: 0–10 cm/s. Bottom right: RBC volume fraction contour plot for SEFC. Scale: 0–100%

of 0.00427 kg/m s (4.27 cP), (2) non-Newtonian single-phase flow that mimics blood rheology at a hematocrit of 41.5%, and (3) non-Newtonian two-phase flow (plasma and 41.5% hematocrit at inlet). All three conditions predicted recirculation zones of about 0.1 mm or one half of the step height. As a point of reference, for Newtonian single-phase flow and fully developed steady laminar flow downstream of the step, wall shear rate, WSS, and Reynolds number were about 430 (s^{-1}), 20 dynes/cm^2 (2.0 Pa), and 3.0, respectively.

The non-Newtonian two-phase condition (3) is the most realistic and used the viscosity model given by Jung et al. [13]. The plasma density used was 1020 kg/m^3 and the RBC phase density was 1092 kg/m^3. The plasma viscosity used was 0.0012 kg/m s (1.2 cP). A no-slip boundary condition was applied to the plasma phase while a free-slip boundary condition was applied for RBCs. Gravity (pointing down) was also included in the simulation. The predicted RBC velocity vector and volume fraction contour plots are shown at the bottom of Fig. 9.11. The velocity vectors show a recirculation zone downstream of the step, with an attachment length (where the velocity arrows next to the lower wall change direction) of about 100 μm (about half the step height). The prediction also suggests that the RBCs are significantly depleted along the vertical wall of the step and in the corner of the flow cell.

The velocity vector plot clearly indicates a recirculation zone of sufficient length to allow for adhesion measurements to be made in this zone and also a reattachment zone downstream of the step expansion. The RBC volume fraction contour plot shows that there are significant spatial gradients of particle concentration. Thus, these data demonstrate the importance of the multiphase CFD analysis in predicting the streamline patterns, the position of the reattachment line, and the cell velocities near the wall.

9.6 Solutions to Two Important Phenomena

Computational fluid dynamics, combined with noninvasive medical imaging, has been used to better understand hemodynamic factors contributing to the initiation of atherosclerosis as shown in Sect. 9.4. However, there are shear-dependent mass transfer phenomena that are also important. It is therefore incumbent to explore multiphase models to get a better understanding of the transport of blood-borne particulates and their buildup. The multiphase kinetic theory introduces a new variable, granular temperature, the random fluctuation of particles. This granular temperature is the driving force for particulate migration, analogous to diffusion of gases.

This powerful theory has been applied to begin understanding two important phenomena known to exist in the cardiovascular system for a long time, but for which no rational solution existed. These are: (1) RBC depletion in the vicinity of vessel walls, the Fahreaus-Lindqvist effect, and (2) the motion of platelets toward vessel walls. This section serves to summarize two analyses addressing these phenomena. They are only the first steps, albeit important ones, and are the first and, thus far, unique.

9.6.1   Modeling of the Fahraeus-Lindqvist Effect

The existence of a region of partially or completely depleted particulates forming a clear layer next to the walls of a flowing suspension has been noted in several instances, for example, in the food industry during the flow of fruit juices and for coal/water slurries as discussed in Sect. 9.2. In the latter instance, the apparent viscosity becomes a function of the capillary viscometer tube diameter. This chapter has shown thus far that blood can no longer be treated as a Newtonian single-phase mixture. Red blood cells (diameter 8 μm) are the main constituents of blood and are responsible for many blood properties, including viscosity and non-Newtonian behavior, and diseases, and play an important role for blood flowing through small vessels. Experimental investigations reported by Cokelet [16] have shown that, for blood flowing through small vessels, there is a core region consisting of all the erythrocytes and a cell-free plasma (peripheral) layer.

The two-phase model consists of the continuous plasma phase and the dispersed red blood cell phase, which is treated as a continuum. In the kinetic theory model, the red blood cell viscosity is computed using an equation for the random kinetic energy of the red blood cells, called granular temperature. This kinetic theory based two-phase flow model for plasma and RBCs was developed to explain the Fahraeus-Lindqvist effect without which this effect cannot be predicted mechanistically. Figure 9.12 shows a comparison of the FLUENT calculation of the relative

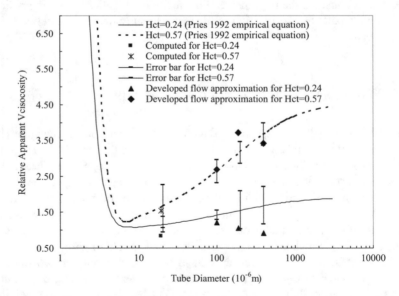

Fig. 9.12 Comparison of computed and measured blood viscosity. Relative apparent viscosity of RBC suspension is dependent on the hematocrit and tube diameter. The Fahraeus-Lindqvist effect correlated by Pries [17] was compared with the results assuming developed flow compared with FLUENT computational results [19]. (This figure was originally published in [19] and has been reused with permission)

Fig. 9.13 Migration of red blood cells away from the wall due to shear induced diffusion. Comparison of computational and experimental RBC volume fraction distribution profiles at a hematocrit (Hct) = 0.57 shows that the RBCs have the highest volume fraction at the center and are lower near the wall [20]

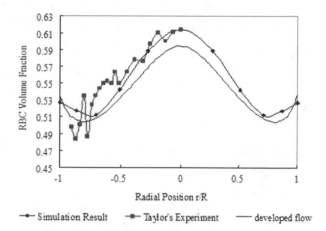

apparent viscosity dependence on the tube diameter and hematocrit, with the empirical equation given by Pries et al. [17]. The migration of RBCs from the wall to the center in narrow tubes is shown in Fig. 9.13. Both the computational results and Poiseuille flow approximation for developed flow show that the RBCs have the highest volume fraction at the center, but are lower at the wall, which are qualitatively in accordance with experimental observation [18]. There is a small increase of the RBC volume fraction near the wall, due to the use of the Johnson-Jackson boundary condition that accounts for the inelasticity of the wall.

More results and a description of the two-phase kinetic theory are contained in Gidaspow and Huang [19], who explained that the fluid/particle interaction term (drag) in the FLUENT code had to be set to zero in the granular temperature equation; otherwise, the RBC-depleted layer near the wall would not form.

9.6.2 Application to Platelet and RBC Transport

The previous section helped solve the riddle of why RBCs migrate toward the center of cardiovascular vessels, thus creating a depleted region in the vicinity of their walls. In addition to modeling this Fahraeus-Lindqvist effect, as well as the realistic RCA discussed in Sect. 9.4.3 for her PhD thesis (with DG and RWL as Adviser and Co-Adviser, respectively), Jing Huang performed a preliminary analysis to model the transport of platelets, a minor but important blood-borne particulate [20]. Jing left after graduating in 2009 to work at RTI International. In 2014, Vishak Chandra took up the subject for his master's thesis [21].

Many previous researches have concluded that the migration of platelets towards the wall of the vessels is through diffusion. Turitto et al. [22] explained the near-wall buildup as resulting from diffusion of platelets in blood flow, which is dominated by the movement of RBCs. They modeled the platelet transportation and deposition

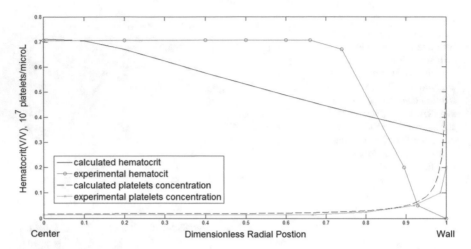

Fig. 9.14 A comparison of computed hematocrit and platelets concentration to the Aarts et al. [24] experiment. Inlet hematocrit = 0.6 [21]

using the convection-diffusion equation by relating the diffusivity to the local shear rate and RBC diameter. However, Goldsmith and Turitto [23] concluded that diffusion does not fully explain the migration of platelets. The biggest shortcoming for most convective diffusion-based models is the lack of explanation for the diffusion coefficient. As the diffusivity measured heavily depends on empirical coefficients and measurements, it does not fully explain the driving force behind the migration. The interaction between RBCs and platelets as observed by Turitto et al. [22] is the primary reason behind the motion of platelets, and the collision of platelets with RBCs and collision with themselves cause them to migrate towards the wall. To obtain the motion of red blood cells and platelets, it is necessary to use the unequal granular temperature kinetic theory reviewed by Gidaspow and Jiradilok [14].

The model of Gidaspow and Huang [19] was extended by Gidaspow and Chandra [25] to compute the concentration profiles of the red blood cells as well as platelets. The Aarts et al. [24] experiments show that the platelets move towards the wall, while the red blood cells move towards the center. For each phase, the driving force for migration is the granular pressure. For fully developed flow, the conventional boundary layer approximation shows that it is only possible to compute this driving force using the platelet and RBC granular temperatures through the use of their equations of state.

Figure 9.14 shows a comparison of computed red blood cells concentration, hematocrit, and platelets concentration, to the Aarts et al. [24] experiment, for an inlet hematocrit of 0.6 and a shear rate of $1200 \, s^{-1}$ [25]. The hematocrit at the center is much larger than the computed wall hematocrit in agreement with the experiment. However, with the prescribed dimensionless wall granular temperature of 0.5, the computed wall hematocrit is much larger than the experimental values. The most probable reason for the difference between the Aarts experiment and theory is that the equation of state for RBCs lacks cohesion. The agreement between the measured and the computed platelet concentrations is excellent as seen in Fig. 9.14. At the wall, the approximate analytical solution given by Eq. (2.78) in Chap. 2 gives an infinite concentration. More details may be found in Gidaspow and Chandra [25].

9.7 Analysis of LDL and HDL Transport

Numerous studies have shown that the development of atherosclerosis is characterized by the formation of lipid-filled foam cells and extracellular lipid. The formation was recognized to correlate with the concentration of low-density lipoprotein (LDL) and inversely related to the concentration of high-density lipoprotein (HDL). The atherosclerotic process begins when LDL becomes trapped in the vascular wall, the possibility of which arises with an elevated level of LDL. It has been demonstrated by experiments that the electrostatic charge, composition, and structure of lipoproteins are important determinants on their physical transport and biochemical reactions [26].

Study of the mechanisms resulting in atherosclerosis using computational techniques has been the subject of interest in the last decades. Simulations of the lipoproteins transport have been limited to modeling their mass transfer to the artery wall and the blood flow dynamics in the lumen neglecting the multi-component nature of blood. However, the concentration of lipoproteins is not uniform in the lumen and the transport of lipoproteins towards the wall in the blood flow plays an important role in the initiation of atherosclerosis and the further transfer in the arterial wall. Some researchers modeled the lipoprotein transport in the flowing of blood using the convection-diffusion equation for lipoproteins by assuming a constant diffusion coefficient [27]. Lipoproteins transport in flowing blood is a result of collisions with RBCs, leukocytes, and platelets, and by collisions with themselves.

The kinetic theory based-multiphase model was extended by Huang in her PhD thesis [20] to incorporate the influence of lipoprotein surface charge. The extended model assuming fully developed flow is used to explore the characteristics of LDL and HDL transport. The new model computes the migration of lipoproteins by the interactions with RBCs and the electric force induced by their surface charge. The transport of HDL to the vicinity of arterial vessel walls serves to protect them from LDL deposition. It should be emphasized that this model is of a preliminary nature subject to improvement and refinement in future studies.

It is well known that LDL and HDL have negative charges on their surfaces. The plasma carrier fluid becomes ionized as a result. The distribution of the ions is not uniform. In the vicinity of a negatively charged particle surface, positive charges develop and the numbers of positive and negative ions in a small element of volume become nearly equal. The negatively charged LDL particles and the non-uniformly distributed positive and negative ions in the plasma give rise to the electric field, which affects both the lipoprotein and plasma phases. Because of the small number of ions in the plasma, the magnitude of electric potential will be much lower than that

Table 9.1 Simulation conditions and system properties

Tube diameter	4.37 mm
Plasma density	1020 kg/m^3
RBC diameter	8 μm
RBC density	1092 kg/m^3
Central line velocity	0.1 m/s
Restitution coefficient of RBCs	0.999999
Wall restitution coefficient of RBCs	0.9999
Hematocrit	0.45
LDL diameter	23 nm
LDL density	1020 kg/m^3
LDL level	130 mg/dl
LDL ζ potential	−4.5 mV
Restitution coefficient of LDLs	0.9999
Restitution coefficient of LDLs with RBCs	0.99
Restitution coefficient of LDLs with HDLs	0.99
HDL diameter	10 nm
HDL density	1200 kg/m^3
HDL level	30 mg/dl
HDL ζ potential	−10.5 mV
Restitution coefficient of HDLs	0.9999
Restitution coefficient of HDLs with RBCs	0.99
Packing limit	0.7
Relative dielectric constant of plasma	80.0

produced by the negative charged LDL and HDL particles. The structure of the charge distribution in the ionized plasma near the negatively charged LDL and HDL particles can be obtained by the Gouy-Chapmann model for a diffusion layer by solving the electrostatic Poisson-Boltzmann equation [20].

The presence of the low concentration of platelets has little effect on the RBC granular temperature. Therefore, in calculating the granular temperature for the mixture of RBCs, LDL, and HDL, it is assumed that the fluctuating kinetic energy equation described in Gidaspow and Jiradilok [14] can be solved for the RBC phase, neglecting the presence of the low concentration LDL and HDL phases. The momentum equation for the fully developed flow, including the electric force, is solved using electric potential calculated from the Poisson equation. More details of the model can be found in [20].

The set of model equations listed in the previous section was finite differenced using central differences and programmed into a FORTRAN code. Table 9.1 gives the simulation conditions and the properties of RBCs, LDL, and HDL, which were assumed to be spherical rigid particles. The hematocrit was set to be 0.45. Their mixture flows through a channel having a height of 4.37 mm. In order to explore the electrostatic effect on the lipoprotein transport, the blood flow with neutral and negatively charged LDL is simulated. Flow of RBCs and negatively charged HDL and LDL is also simulated to investigate the effect of HDL on the transport of LDL.

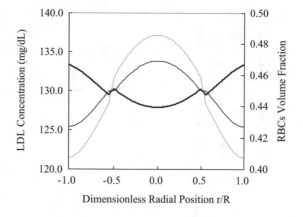

Fig. 9.15 Computed RBC volume fraction and LDL concentration with and without surface charge on the LDL phase [20]

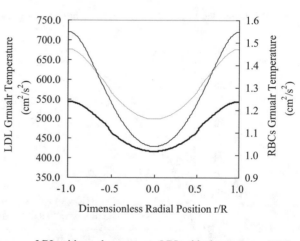

Fig. 9.16 Computed RBCs and LDL granular temperatures with and without surface charge on the LDL phase [20]

Gidaspow and Huang [19] have shown as summarized in Sect. 9.6.1 that RBCs migrate toward the center of narrow arteries due to shear, called the Fahraeus-Lindqvist effect. As expected, the nano-sized LDL particles without surface charge are also carried along with RBCs toward the center as shown in Fig. 9.15. Also shown is that, with negative surface charge included on the LDL particles, the force generated by the electric field transports the LDL particles move away from the center area towards the wall. Figure 9.16 shows that both RBC and LDL granular temperatures are higher near the wall with neutral LDL particles because the fluctuating kinetic energy production by shear is higher there. Also shown in Fig. 9.16 is that, when the electrostatic effect caused by the negative surface charge on the LDL phase is included, the LDL granular temperatures decreases because of the electric force acting on them. As shown in Fig. 9.17, when the negatively

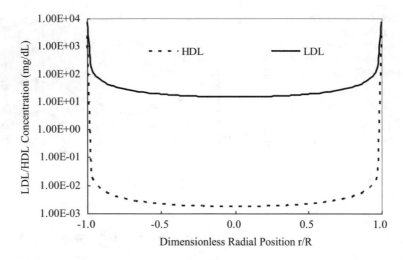

Fig. 9.17 Computed LDL and HDL concentrations for charged LDL and HDL [20]

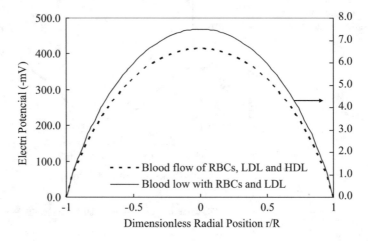

Fig. 9.18 Comparison of computed electric potential for the flow of RBCs, LDL, and HDL [20]

charged HDL phase is added to the flow of RBCs and LDL, the additional electric force acting on the charged LDL and HDL particles moves more LDL from the flow towards the wall. Both HDL and LDL concentrations are higher near the wall due to the electric force. The electric potential produced by the surface charges carried by LDL and HDL increases as shown in Fig. 9.18. The ratio of HDL to LDL near the wall is much larger than in the center as shown in Fig. 9.19.

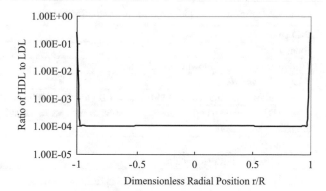

Fig. 9.19 Computed ratio of HDL and LDL concentrations [20]

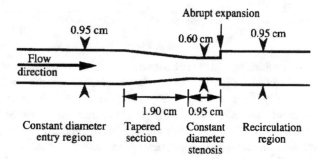

Fig. 9.20 Geometry of the flow cell used in the experiment by Pritchard et al. [30]. Horizontal orientation; 3D flow path; Steady flow ($Re = 200$); U937—monocyte-like cells (2×10^4 cells/cm^3 at the inlet); Surface coating: none

9.8 Application to Analysis of Monocyte Adhesion Data for Atherosclerosis

In the published version of the paper in which the combined CFD monocyte adhesion model was developed, the experimental adhesion data of Hinds et al. [28] for U937 monocyte-like human cells was analyzed in detail by Lyczkowski et al. [29] using the monocyte adhesion model developed there. This experiment was performed in a vertically oriented three-dimensional flow-cell model coated with E-selectin. Hinds et al. [28] examined the effects of local hemodynamics, computed using a single-phase CFD model. One of the main observations of their study was that, for steady flow, the degree of adhesion was generally shown to be inversely proportional to local wall shear stress (WSS) as shown in their Fig. 5A. Although this inverse relationship appears to hold in the taper and stenosis regions, this relationship is not unique, i.e., adhesion for a given shear stress varies depending on spatial location and Reynolds number. There were passing references to the experimental adhesion data of Pritchard et al. [30] for U937 monocytes, but no analysis. This earlier study used a *horizontally* oriented *uncoated* three-dimensional flow-cell model experiment shown schematically in Fig. 9.20. The geometry is quite similar to that used in the Hinds experiment. This earlier experiment was, thus, a check on the receptor-independent adhesion mechanism. They also computed the WSS with a single phase CFD model. These investigators arrived at similar

Table 9.2 Pritchard et al. [30] experiment simulation conditions

2-D geometry
Steady-state
Gravity points in $-y$ (downward) direction
Material properties
–   Culture medium (carrier fluid): $\rho_m = 1.0$ g/cm^3, $\mu_m = 0.008$ P
–   U937—monocyte like cells: $\rho_m = 1.065$ g/cm^3, $\mu_m = 0.008$ P, cell diameter, $d_m = 20$ μm
Boundary conditions
Inlet conditions:
– $v_z = 1.68$ cm/s (Re $= 200$)
– $v_r = v_\theta = 0$ for all runs
–   Cell number at inlet: 2×10^4 cells/cm^3 = monocyte volume fraction, $\varepsilon_m = 8.38 \times 10^{-5}$
On all walls:
No-slip for the culture medium and full-slip for the cells

conclusions, but there are contradictory results especially in the taper sections. In the Hinds et al. [28] experiment, the number of adhered cells *increased* in the taper section with increasing WSS, while in the Pritchard et al. [30] experiment, the number of adhered cells *decreased* with increasing WSS. These observations indicate that there are covariates that can independently affect the adhesion process. The resolution of these contradictory results for these two experiments had gone unanalyzed.

There was a very good reason for the fact that the Pritchard data were not included in the final manuscript submitted for review. The experiment was first modeled in three dimensions (3D) and difficulties were found introducing the right number of monocytes into the experiment through an inlet section. Monocytes were not building up to the correct amount reported for the experiment (see Table 9.2). This problem was resolved by switching to a two-dimensional model. However, even though the inlet problem turned out to be resolved, and comparison between the experimental adhesion data and computed results was excellent, as will be shown in this section, the authors felt that a two-dimensional model (with gravity facing downward) might be difficult to defend. It could be interpreted as a vertical slice through the round 3D geometry. Another, weaker, argument was that the manuscript was already overly long. Included in this section is the two-dimensional analysis of the Pritchard data. The explanation of the differences between the two- and three-dimensional models is a challenge to future investigators as an exercise. To fully appreciate what follows in this section, the reader is encouraged to consult the published paper by Lyczkowski et al. [29].

The commercial CFD code FLUENT 12.0 available from ANSYS Inc. [31] was used to simulate the experiment. This code contains the Euler-Euler multiphase implementation as described in Chap. 4, Table 4.2. The numerical solution method uses a finite-volume, unstructured-mesh, staggered-grid arrangement. The pressure-based solver was used. This solver uses a co-located scheme that stores the pressure and velocity at the cell centers. The momentum equations are solved using a staggered mesh, while the continuity equations are solved using a donor-cell method. The mesh for the flow chambers was created using the pre-processor GAMBIT v2.3. Table 9.2 contains the conditions used for the simulation.

The multiphase CFD model enables the calculation of the velocities, volume fractions, and WSS of the culture medium carrier fluid and monocytes phases. These results are then used as inputs to the monolayer population balance adhesion model to describe the effects of mechanical milieu upon adhesion as a function of time and space in the computational meshes which are next to the adhesive surface (wall surfaces of flow chambers or endothelial surfaces for physiological vessels). When monocytes (20 μm in diameter for U937 cells) used in the Hinds et al. [28] and Pritchard et al. [30] experiments are in close proximity to these surfaces, adhesive interactions can take place according to the scheme described by Lyczkowski et al. [29]. Given that a complete set of time-dependent adhesion data (i.e., rolling and firmly adherent cell densities and rolling velocities as a function of time) were not available in the Pritchard et al. [30] and Hinds et al. [28] experiments, an equilibrium steady-state analysis was performed [29]. The Pritchard et al. [30] data showed a linear increase in percent surface covering up to roughly 90 min, in agreement with the predictions of the adhesion model of Munn et al. [32], and then leveled off between 150 and 180 min, i.e., the number of adherent cells did not change with time beyond this point.

First, the multiphase CFD model was used to calculate the spatial distribution of near-wall, free cell surface density (n_f, cells/mm^2), and WSS (τ_w, dyne/cm^2) experiments as shown in Fig. 9.21 A. Several observations can immediately be made: (1) Even though the inlet concentration of U937 cells was an order of magnitude lower for the Pritchard et al. [30] study (2×10^4 cells/ml) as compared to that in the Hinds et al. [28] study (5×10^5 cells/ml) [29], n_f for the Pritchard et al. [30] study is significantly higher at all locations. (2) There is a significant spatial variation in n_f (about 2.5-fold) in both experiments; however, the patterns are quite different. While n_f increases in the tapered section for the Hinds experiment [29], it declines for the Pritchard et al. [30] experiment as shown in Fig. 9.21a. In addition, there is a rise in n_f in the recirculation zone (post-expansion region) for the Pritchard et al. [30] experiment only. We believe that both the differences in the magnitude of n_f and spatial distribution patterns are related, at least in part, to the gravity effect in the Pritchard et al. [30] experiment that carried cells to the bottom of the perfusion model (site of experimental measurements). (3) WSS patterns are comparable between the two experiments. Figure 9.22 shows the monocyte velocity magnitude and volume fraction contours computed by the CFD simulation in the vicinity of the rapid expansion. Clearly shown is the buildup of monocytes in the recirculation region after the rapid expansion.

Both studies provided experimentally determined spatial distribution of bound cell density (n_b, cells/mm^2). Together with n_f and WSS computed from the multiphase CFD model and the equilibrium adhesion relationship described in [29], a non-linear iterative search technique was used to determine optimal parameter values that best fit the experimentally observed adhesion data. The four parameters (K_1–K_4) were estimated for the Pritchard et al. [30] data. The estimated optimal

Fig. 9.21 Analysis of data from the Pritchard et al. [30] experiment. (**a**) Spatial distribution of free cell surface density (n_f) and wall shear stress (WSS) predicted by the multiphase CFD model for the Pritchard et al. [30] experiment. Horizontal orientation; Steady flow ($Re = 200$); U937—monocyte-like cells (2×10^4 cells/cm^3 at the inlet). (**b**) Comparison of experimentally measured adhesion (data) and combined multiphase CFD-adhesion model-predicted (model) bound cell density. (This figure was reused with permission from Sanjeev G. Shroff)

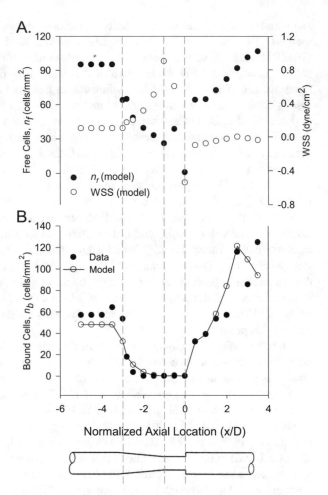

values of the model parameters which were obtained are: $K_1 = 0.1$, $K_2 = 10$, $K_3 = 0.05$, and $K_4 = 6$. The results from this comparative analysis are depicted in Fig. 9.21b, which will be discussed shortly. As can be seen, the agreement is excellent, even better than for the Hinds experiment [29]. This is thus considered to be a success for the combined multiphase CFD-adhesion model for the following reasons: (1) The agreement between model-predicted and experimentally measured bound cells is within the error bars reported in the experimental studies. (2) Although the WSS increases in the tapered region for both experiments, experimentally-measured bound cell density changes in opposite direction between the two datasets. This phenomenon is reproduced by the model and the primary reason for this is the differences in spatial variations in free cell density (n_f) between the two datasets, as calculated by our multiphase CFD model as shown in Fig. 9.21a *vs.* either Figs. 7 or 8 in [29]. (3) The convection term in the adhesion model was ignored due to insufficient experimental data and, therefore, our multiphase CFD and adhesion

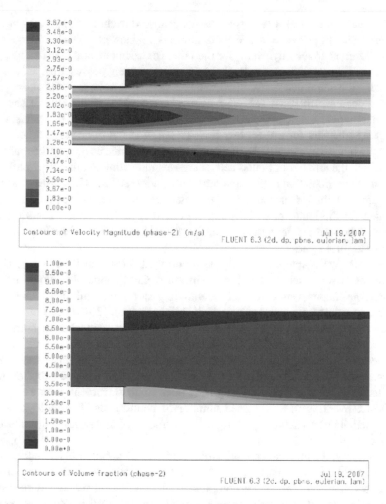

Fig. 9.22 Computed monocyte velocity magnitude (left) and volume fraction (right) for the Pritchard et al. [30] simulation in the vicinity of the rapid expansion

models should be considered as an important first step. (4) The estimated values of the estimated parameters in the population balance adhesion model all have physiological significance. Further analysis of these and other flow cells experiments should serve to refine and validate these parameters.

Next, the Pritchard et al. [30] adhesion data shown in their Fig. 6 were carefully analyzed. As mentioned above, no coating was applied to the plastic flow cell walls. The monocyte adhesion rate data were converted to cells/mm^2 over the length of the experiment (1 h) and plotted in Fig. 9.21b. Using Eq. (12) in [29], the estimated fully developed WSS at the end of the stenosis region is 0.4 dynes/cm^2 in reasonable agreement with about 0.5 dynes/cm^2 as shown in Pritchard et al. Fig. 4 and Fig. 9.21a. The effect of the WSS for bound cell detachment is more pronounced than for the Hinds et al. [28] experiment because detachment starts at about 0.2 dynes/cm^2 and drops to zero at a WSS of about 0.4 dynes/cm^2. The bound cells drop to almost zero at about 1 dyne/cm^2 for the Hinds experiment as shown in Figs. 7 and 8 in [29]. These trends are remarkably similar, which proves that, at least for essentially the same geometry, there exists a critical threshold WSS above which adhesion decreases because the WSS begins to detach adherent cells. It should be noted that there is a serious typographical error in Pritchard's Fig. 6. They state that there is no adhesion for a WSS *less than or equal* to 0.4 dynes/cm^2, whereas it is clearly the other way around. As in the Hinds experiments, as the rapid expansion is approached, the number of bound cells increases indicating another flow redevelopment. In the recirculation region, the number of adherent cells increases, reaching a maximum and then decreasing as the reattachment point is reached. The WSS computed for the Hinds et al. [28] and Pritchard et al. [30] experiments for U937 monocytes in buffer cover a range typically found in the sinus of the human carotid on the order of 4 dynes/cm^2.

There are two points that should be mentioned. First, the Pritchard et al. [30] experiment was modeled with the multiphase CFD model in two dimensions (2D) and the shapes were essentially identical. Second, the multiphase CFD model was able to resolve the paradox noted by Pritchard et al. [30]. They noted that the adhesion rate decreased with decreasing WSS near the reattachment point in the rapid expansion section where theoretically it is zero. The reason for this was hypothesized by them to be one of two reasons: decreased delivery of cells to the surface or decreased adhesion. The monolayer population balance adhesion model predicted a decrease in freely flowing cells near the reattachment point as shown in Fig. 9.21a, resulting in a decreased number of bound cells. The multiphase CFD model predicted an essentially straight-line buildup of freely flowing cells in the entrance region.

In summary, the proposed coupled multiphase CFD-population balance adhesion model was able, for the first time, to reconcile the seemingly contradictory effect of WSS upon adhesion. This was accomplished by simultaneously considering spatial variations in freely flowing cell concentration and the local WSS. Global correlations such as Fig. 5 in Hinds et al. [28] can be deceiving. In the same way, it may be possible to explain the findings in the literature that atherosclerotic sites cannot be uniquely correlated with WSS. In addition, the estimated adhesion model parameter values were consistent with *a priori* expectations of this study based on the physical design of the experiments, which provides mechanistic significance to the model parameters (as opposed to parameters used for "curve-fitting").

9.9   Analysis of an Actual Right Carotid Artery

A criticism surfaced about using coronary arteries in the strategy to use the combined CFD-monocyte adhesion model to investigate the potential sites for atherosclerotic lesion formation. The criticism was that it was just not possible to obtain good enough images of the coronary arteries lying on a beating heart. Therefore, it was decided to concentrate on modeling carotid arteries. The aim was changed to elucidate the **relative** role of local hemodynamics and blood-borne particulate concentration in determining the localization of early atherosclerotic lesions in carotid arteries of individuals at risk, and not the **absolute** probability of disease initiation or the rate of disease progression. Subha V. Raman and Orlando P. Simonetti at Ohio State University (OSU) had and still have the MRI equipment and expertise to perform imaging on actual patients. The following is a summary of their state-of-the-art accomplishments that can be used to validate the multiphase CFD model.

The close proximity of the carotid bifurcation to the surface of the neck lends itself to the use of small, multi-channel array surface coils to greatly improve the signal-to-noise ratio. As a result, high spatial resolution (0.5 mm × 0.5 mm × 0.5 mm) 3D static morphological images of the vessel wall were obtained using bright-blood or dark-blood MRI techniques. These in-vivo images provide the raw data required for the 3D finite-element mesh model of the carotid artery and obtaining boundary conditions for the CFD model of carotid artery blood-flow dynamics. Additionally, high spatial (0.5 mm × 0.5 mm × 4 mm) and temporal resolution (28 ms) phase velocity mapping were performed. Cross-sectional 2D images are obtained orthogonal to the artery at several slices above and below the bifurcation. From these images, velocity-time plots were generated at any location. Measurements taken above and below the carotid bifurcation serve as the inlet/outlet boundary conditions for the CFD model. Additionally, the spatial distribution of velocities within the vessel was mapped. The same technique was used to measure vessel wall motion. Tonometry allows noninvasive recording of carotid artery pressure waveform, which can serve as another input to the multiphase CFD model. Wall thickness and presence or absence of plaque were determined using a new high resolution (0.5 mm × 0.5 mm × 0.5 mm) 3D black-blood MRI technique. This 3D scan allows reformatting of the image data in any orientation to evaluate the total plaque burden along the entire length of the vessel.

The OSU group obtained segmented imaging data for the right carotid artery, along with MRI-based measurements of velocities at three different sites. With a great deal of help from Professor Shroff from the University of Pittsburgh, the 3D solid model and corresponding finite element mesh for these data were generated as shown in Fig. 9.23. Marc Horner from ANSYS, Inc. Evanston, IL, took these data and performed the multiphase CFD analysis. This was quite challenging as it took 8–9 h of computation time to run the analysis over one cardiac cycle. The results of these computations are shown in Fig. 9.24. This analysis considered two phases: plasma and RBCs (45% hematocrit at the inlet, evenly distributed). The multiphase kinetic theory model incorporated into the commercial code ANSYS FLUENT, as employed by Huang et al. [15] to model the realistic right coronary artery in Sect. 9.4.2, was used. There clearly is a spatial redistribution of RBCs, especially near the bifurcation region. There is RBC depletion in the vicinity of the wall clearly showing the Fahraeus-Lindqvist (see Sect. 9.6.1). These patterns of RBC redistribution are

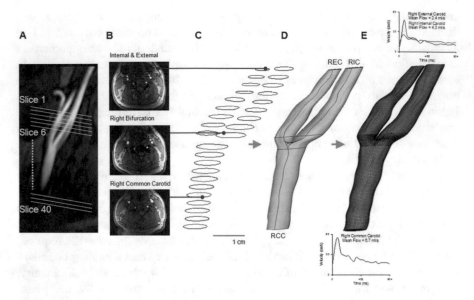

Fig. 9.23 3D MR imaging data for right carotid (RC) artery from a normal volunteer. (**a**) Sagittal view with dashed lines indicating 40 cross-sectional imaging planes (2 mm apart). (**b**) Cross-sectional MR images at three selected levels. (**c**) Stacked segmented images of all 40 slices. (**d**) Non-Uniform Rational B-Spline (NURBS)-based solid model generated by lofting the stacked segmented images showing right common carotid (RCC), right external carotid (REC), and right internal carotid (RIC) arteries. (**e**) Hex-dominant finite-element mesh for the multiphase CFD analysis generated from the solid model (204,837 elements). Measured velocity waveforms (MR-based measurements) at different sites are also shown. (Figures **a**, **b**, and **e** have been reused with permission from Subha V. Raman and Orlando P. Simonetti. Figures **c** and **d** have been reused with permission from Sanjeev G. Shroff)

quite complex, depending on the spatial location and the time during the cardiac cycle. They are the very first of their kind and remain unique.

9.10 Conclusion

The phenomena summarized in this chapter clearly demonstrate significant first steps in implementing a multiphase CFD model that accurately predicts hemodynamic parameters that are critical to evaluating the propensity for the local initiation of atherosclerosis. The commercial ANSYS FLUENT code is ideally suited for multiphase hemodynamics with complex geometries and boundary conditions. In principle, it can handle RBCs, monocytes, platelets, LDL, and HDL blood-borne particulates utilizing the models described herein to greatly facilitate coupling of the multiphase hemodynamic CFD model with a mechanistic adhesion model.

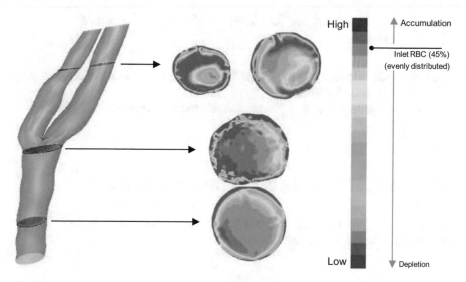

Fig. 9.24 Results of the multiphase CFD analysis for the right carotid geometry (Fig. 9.23e) showing the spatial redistribution of RBCs at three cross-sectional planes and at 0.4 s during the cardiac cycle. (This figure was reused with permission from Marc Horner ANSYS, Inc. Evanston, IL)

9.11   Exercises

9.11.1   Ex. 1: Model the Altobelli and Sinton and Chow Experiments

Model the Altobelli and Sinton and Chow concentrated suspension experiments of Sect. 9.3 using the multiphase kinetic theory model.

9.11.2   Ex. 2: Model the Realistic RCA Using the Carraeu-Yasuda Model

Using the models described in Sect. 9.4 for a realistic RCA, compare the shear-thinning model using the Carraeu-Yasuda model with multiphase kinetic theory model. The mesh and case files are available from the author (RWL).

9.11.3 Ex. 3: Model the Realistic RCA with Bifurcating Arteries

The realistic RCA model used in Sect. 9.4 does not have the addition of bifurcating branch arteries along its length.

Develop a model containing them and repeat the computations. See Borkin et al. Fig. 1C. [33].

9.11.4 Ex. 4: Extend the Fahraeus-Lidqvist Model

Extend the analysis for the Fahraeus-Lidqvist model developed in Sect. 9.6.1 to developing flow and to idealized bifurcations.

9.11.5 Ex. 5: Extend the LDL and HDL Model

Extend the LDL and HDL model developed in Sect. 9.6.2 to developing flow and idealized bifurcations.

9.11.6 Ex. 6: Model the Pritchard Experiment

Model the Pritchard experiment described in Sect. 9.8 in 3D and use the monocyte adhesion model to compare with the 2D analysis.

References

1. Thubrikar M, Robicsek F (1995) Pressure-induced arterial wall stress and atherosclerosis. Annu Thorac Surg 59:1594–1603
2. Lyczkowski RW, Bouillard JX, Folga SM (1992) Users manual for FLUFIX/MOD2: a computer program for fluid-solids hydrodynamics. Argonne National Laboratory Sponsor Report, Argonne, IL, April 1992. Reprinted by USDOE METC as DOE/MC/24193-3491/NTIS no. DE94000033, available from NTIS, Springfield, VA, 1994
3. Lyczkowski RW, Wang CS (1992) Hydrodynamic modeling and analysis of two-phase non-Newtonian coal/water slurries. Powder Technol 69:285–294
4. Passman SL, Peters WC (1992) United States Department of Energy granular flow advanced research objective, SAND92-0069, DE93 006447
5. Ding J, Lyczkowski RW, Sha WT, Altobelli SA, Fukushima E (1993) Analysis of liquid-solids suspension velocities and concentrations obtained by NMR imaging. Fourth NSF-DOE workshop on flow of particulates and fluids proceedings, pp 281–297. U.S. Government Printing

Office: 1993-341-832/83044,1993. Published as: Numerical analysis of liquid-solid suspension velocities and concentrations obtained by NMR imaging. Powder Technol 77:301–312

6. Sinton SW, Chow AW (1991) NMR flow imaging of fluids and solid suspensions in Poiseuille flow. J Rheol 35:735–772

7. Ding J, Lyczkowski RW, Sha WT (1995) Modeling of concentrated liquid-solids flow in pipes displaying shear-thinning phenomena. Chem Eng Comm 138:145–155

8. Bird RB, Armstrong O, Hassager RC (1987). Dynamics of polymeric liquids, 2nd edn. Wiley, New York

9. Goldsmith, HL, Mason SG (1967) The micro-rheology of dispersions. In: Eirich FR (ed) Rheology, vol 4. Academic Press

10. Lightfoot EN (1974) Transport phenomena and living systems. John Wiley and Sons, New York

11. Berthier B, Bouzerar R, Legallais C (2002) Blood flow patterns in an anatomically realistic coronary vessel: influence of three different reconstruction methods. J Biomech 35, 1347–1356

12. Jung J, Lyczkowski RW, Panchal CB, Hassanein A (2006) Multiphase hemodynamic simulation of pulsatile flow in a coronary artery. J Biomech 39:2064–2073

13. Jung J, Hassanein A, Lyczkowski RW (2006) Hemodynamic computation using multiphase flow dynamics in a right coronary artery. Ann Biomed Eng 34(3):393–407

14. Gidaspow D, Jiradilok V (2009) Computational techniques: the multiphase CFD approach to fluidization and green energy technologies. Nova Science Publishers, New York

15. Huang J, Lyczkowski RW, Gidaspow D (2009) Pulsatile flow in a coronary artery using multiphase kinetic theory. J Biomech 42:743–754. Supplemental data available at https://doi.org/10.1016/j.jbiomech.2009.01.038

16. Cokelet GR (1972) The rheology of human blood. In: Fung YC (ed) Biomechanics. Prentice-Hall, Englewood Cliffs, NJ, pp. 63–103

17. Pries, AR, Neuhaus D, Gaehtgens P (1992) Blood viscosity in tube flow: dependence on diameter and hematocrit. Am J Physiol Heart Circ Physiol 263: H1770–H1778

18. Taylor M (1955) The flow of blood in narrow tubes II. The axial stream and its formation, as determined by changes in optical density Austral J Exp Biol 33:1–16

19. Gidaspow D, Huang J (2009) Kinetic theory based model for blood flow and its viscosity. Ann Biomed Eng 37(8):1534–1545

20. Huang J (2009) Simulation of blood flow using kinetic based multiphase models. PhD thesis, Illinois Institute of Technology, Chicago

21. Chandra V (2015) Unequal granular temperature theory for motion of red blood cells and platelets. MS thesis, Illinois Institute of Technology, Chicago

22. Turitto V, Benis A, Leonard E (1972) Platelet diffusion in flowing blood. Ind Eng Chem Fundam 11(2):216–223

23. Goldsmith HL, Turitto VT (1986) Rheological aspects of thrombosis and haemostasis; basic principles and Applications. Thrombosis and Haemostasis 55(3):415–435

24. Aarts PA, van den Broek S, Prins G, Kuiken G, Sixma J, Heethar R (1988) Blood platelets are concentrated near the wall and red blood cells in the center in flowing blood. Arteriosclerosis 8 (6):819–824

25. Gidaspow D, Chandra V (2014) Unequal granular temperature model for motion of platelets to the wall and red blood cells to the center. Chem Eng Sci 117:107–113

26. Dobiasova M, Urbanov Z, Samanek M (2005) Relations between particle size of HDL and LDL lipoproteins and cholesterol esterification rate. Physiol Res 54:159–165

27. Wada S, Koujiya M, Karino T (2002) Theoretical study of the effect of local flow disturbances on the concentration of low-density lipoproteins at the luminal surface of end-to-end anastomosed vessels. Med Biol Eng Comput 40:576–587

28. Hinds MT, Park YJ, Jones SA, Giddens DP, Alevriadou BR (2001) Local hemodynamics affect monocytic cell adhesion to a three-dimensional flow model coated with E-selectin. J Biomech 34:95–103

29. Lyczkowski, RW, Alevriadou BR, Horner M, Panchal CB, and Shroff SG (2009) Application of multiphase computational fluid dynamics to analyze monocyte adhesion. Annals of Biomedical Engineering 37(8):1516–1533
30. Pritchard WF, Davies PF, Derafshi Z, Polacek DC, Tsao R, Dull RO, Jones SA, Giddens DP (1995) Effects of wall shear stress and fluid recirculation on the localization of circulating monocytes in a three-dimensional flow model. J Biomech 28:1459–1469
31. ANSYS (2009). ANSYS Fluent user's guide, release 12.0. ANSYS, Inc.
32. Munn LL, Melder RJ, Jain RK (1996) Role of erythrocytes in leukocyte-endothelial interactions: mathematical model and experimental validation. Biophys J 71:466–478
33. Borkin MA, Gajos KZ, Peters AE, Mitsouras D, Melchionna S, Rybicki FJ, Feldman CL, Pfister H (2011) Evaluation of artery visualizations for heart disease diagnosis. IEEE Trans on Visualization and Computer Graphics 17(12):2479–2488

Chapter 10
Multiphase Flow Modeling of Explosive Volcanic Eruptions

Augusto Neri, Tomaso Esposti Ongaro, Mattia de' Michieli Vitturi, and Matteo Cerminara

10.1 Introduction

Among all natural risks, explosive volcanic eruptions are one of the most complex and difficult to observe and understand. The first observation of a Plinian column and associated pyroclastic density currents (PDCs) reported in historical chronicles is the one by Pliny the Younger about the famous eruption of Mount Vesuvius in AD 79 [1, 2] and the consequent destruction of Pompeii and Herculaneum. More direct observations and descriptions of a lethal explosive eruption occurred with the eruption of Montagne Pelée (Martinique) in 1902, when the town of St. Pierre was completely destroyed causing about 28,000 fatalities [3, 4].

However, an increased interest in explosive eruptions followed the large explosive eruption at Mount St. Helens, Washington USA, in 1980. This event led to extensive new data as well as a new understanding of volcanic column development, highly mobile PDCs produced by column collapse, and lateral volcanic blast as shown in Fig. 10.1 [5, 6].

Even closer and more detailed observations of explosive eruptions occurred since 1995 at the Soufrière Hills Volcano on the small island of Montserrat in the Caribbean Sea [7] as shown in Fig. 10.2 [8], at Unzen Volcano (Japan) in 1900–1995 [9], Mt. Pinatubo in 1991 [10], as well as more recently at Eyjafjallajokull Volcano (Iceland) in 2010 [11], Merapi (Indonesia) in 2010 and 2016 [12], and Fuego (Guatemala) on June 3, 2018 [13].

These observations always increased our knowledge of volcanism but, at the same time, gave evidence of complex and often non-intuitive processes occurring during explosive eruptions and, most importantly, posed new scientific questions and challenges.

A first basic understanding of explosive eruption dynamics has been largely derived from studies of the field deposit sequences [14, 15]. These studies, together with the direct observation of phenomena, allowed inferences on the main mechanisms of generation, transport, and deposition of pyroclastic particles as well as the

© Springer Nature Switzerland AG 2022
H. Arastoopour et al., *Transport Phenomena in Multiphase Systems*, Mechanical
Engineering Series, https://doi.org/10.1007/978-3-030-68578-2_10

Fig. 10.1 Picture sequence of the initial stage of the Mount St. Helens May 18 volcanic blast. (Copyright photo courtesy of Gary Rosenquist)

estimation of some of the key eruption variables, such as the erupted volume, column height, deposit area, PDC runout, inundation area, and characteristic grain-size distribution of the gas/particle mixture [16–19].

However, volcanic processes involve the movement of diverse phases (fluids and solids) and their interaction with the surrounding environment. For explosive eruptions, the ascent of magma and its interactions with the surrounding rocks and the atmospheric and superficial environments determine the type and scale of processes

Fig. 10.2 (**a**) Pyroclastic density currents (PDC) produced by the dome collapse of the June 25, 1997, Soufriere Hills volcano, Montserrat, West Indies. (**b**) Deposit left by the PDC shown in (**a**) as appeared the day after. (Pictures of P. Cole [8])

occurring during the eruption. Such processes are often extremely complex because they involve multiphase mixtures (whose properties are not well-known) and an extremely wide range of eruptive conditions and spatial and temporal scales. Moreover, such processes are largely governed by the dynamics of the involved fluids, which are often strongly non-linear and, therefore, difficult to predict in a sufficiently accurate way. Finally, and this is probably the most difficult source of uncertainty to remove, explosive eruption processes can be observed and characterized to a quite limited extent, due to the remote, often inaccessible, and hazardous nature of the phenomena involved.

For all of the above reasons, in the last four decades, mathematical models of volcanic processes represented a valuable tool to better understand the key controls on explosive eruption dynamics. Early works in the 1970s provided first-order descriptions of magma ascent in conduit and volcanic plumes [20–22]. Since then, mathematical models of volcanic phenomena have been widely developed for both explosive and effusive eruptions, as well as for the variety of phenomena associated with them. Moreover, the development of numerical models able to describe some of the key complexities of volcanic phenomena has represented a significant advancement and opportunity for hazard assessment studies [23, 24].

Regarding explosive volcanism, since the late 1980s, 2D Eulerian-Eulerian multiphase flow models of volcanic plumes and collapsing columns generating PDCs allowed a first investigation of the transient and turbulent dynamics of these phenomena [25–29]. Additional non-equilibrium processes associated with the multi-particle nature of the eruptive mixture, the role of overpressure in determining the structure of volcanic jets and that of large vortices and umbrella cloud in controlling the large-scale dynamics of volcanic plumes, and the complex influence of volcano topography on PDCs are just some of the key processes identified and described by further developments of these Eulerian-Eulerian multiphase flow models [25, 30–41] as well as by new multiphase flow techniques and approaches such as the method of moments [42], the Equilibrium-Eulerian method [43–45], and the Lagrangian particle method [46]. All these models have also allowed the representation of eruptive scenarios of specific interest for hazard assessment.

Finally, a further use of such multiphase and multidimensional models has been the calibration of simpler integral models or emulators used for carrying out sensitivity analyses and Monte Carlo simulations aimed at the production of probability hazard maps. For instance, a recent inter-comparison study of volcanic plume models has allowed the evaluation of the accuracy of classic plume theory models, based on the self-similarity assumption of the flow field, for the cases of weak and strong plumes [47]. Such comparison has allowed the calibration of key unknown model parameters, such as the entrainment coefficient used by integral plume models in the description of the Eruption Source Parameters [48]. Similar approaches, including calibration of simple models through comparison of their outcomes with real events, laboratory experiments, or more complex 4D models, have been followed in the development of simplified models of PDCs emplacement [49, 50].

10.2   Scaling Properties and Regimes of Volcanic Gas/Particle Flows

10.2.1   Scaling Properties

In their atmospheric and sub-aerial development, explosive eruptions display several different regimes, from compressible supersonic to diffusive turbulent flows, from dilute to concentrated, encompassing a wide range of characteristic temporal and spatial scales, and can always be characterized by means of dimensionless numbers. While a full treatment of the scaling properties of volcanic multiphase flow is beyond the scope of this chapter [51], some key parameters are discussed here because they are relevant to illustrate the applications of the multiphase flow models presented in Chaps. 1 and 2, as well as some modifications of them presented later in this chapter, to the case of explosive eruptions. In all applications described in this chapter, heavy particles are considered (volcanic pyroclasts in air), i.e., $\frac{\rho_s}{\rho_f} \gg 1$.

The first two parameters characterize the interaction regime between phases: the particle volumetric concentration, ε_s, and the particle mass fraction, Y_s. Three other parameters characterize the response of the individual particles and particle assemblages to the flow dynamics: the kinematic, thermal, and collisional Stokes numbers.

The kinematic Stokes number measures the coupling between fluid and individual particles and is defined as the ratio between the gas/particle relaxation time and the characteristic fluid time scale, i.e.,

$$St = \frac{\tau_s}{\tau_F}. \tag{10.1}$$

The gas/particle drag relaxation time, τ_s, is the time needed for a particle to equilibrate to a change of gas velocity by the effect of gas/particle drag. It can be evaluated from the semi-empirical drag models reported in Table 2.5 of Chap. 2 as,

$$\tau_s = \frac{\varepsilon_s \rho_s}{\beta(Re_s)}, \tag{10.2}$$

where ρ_s is the density of the solid particles and $\beta(Re_s)$ is a function of the particle Reynolds number, which is defined by the gas volume fraction ε_g, gas density ρ_g, gas viscosity μ_g, particle size d_p, and gas and particle velocities $\vec{v}_g$ and $\vec{v}_s$, respectively,

$$Re_s = \frac{\varepsilon_g \rho_g d_p |\vec{v}_g - \vec{v}_s|}{\mu_g}. \tag{10.3}$$

For non-interacting particles, at $Re_s < 10^3$, its expression can be simplified as [43, 44],

$$\tau_s \cong \frac{\rho_s d_p^2}{18 \, \mu_g \varphi(Re_s)},$$ (10.4)

where μ_g is the gas dynamic viscosity and $\varphi = 1 + 0.15Re^{0.687}$ is a correction factor for finite particle Reynolds number.

Analogously to the drag relaxation time, we can define a thermal relaxation time τ_T, which is the time needed for a solid particle to equilibrate to a change of the surrounding fluid temperature. This can be approximated,

$$\tau_T \cong \frac{2}{Nu_k} \frac{\rho_s C_s}{k_s} \frac{d_s^2}{12}$$ (10.5)

where ρ_s is the particle density, C_s is the specific heat capacity, Nu_k is the Nusselt number, and k_s is the conductivity of the solid phase. This defines the thermal Stokes number as $St_T = \frac{\tau_T}{\tau_F}$. It is easily shown that $\frac{\tau_T}{\tau_s} \sim 1$ [43, 44].

The collisional Stokes number is the ratio between the particle/particle collision time and the characteristic fluid time scale,

$$St_c = \frac{\tau_c}{\tau_F}.$$ (10.6)

The time scale between collisions, τ_c, is evaluated as the product of the mean free-path of particles and the variance of the velocity distribution of the particles. By definition, the latter is proportional to the square root of the granular temperature (see Chap. 2). Following Gidaspow [52] (see also Eq. (2.29), Chap. 2), for elastic collisions,

$$\tau_c = \frac{\lambda}{<v>} \cong \frac{\sqrt{\pi}}{12} \frac{\frac{d_p}{\varepsilon_s}}{\sqrt{\Theta}}.$$ (10.7)

In dilute regimes (i.e., when $\varepsilon_s < 10^{-3}$), the collisional Stokes number is generally much larger than one, so that particle/particle momentum exchange and the granular stress can be neglected, compared to the viscous stress and gas/particle drag.

The typical flow time scale, τ_F, can be estimated as the inverse of the shear rate, i.e., $\frac{\tau_F}{\gamma} \sim 1$. This estimate can be problematic, however, for compressible and for high-Reynolds number flows (typical of explosive eruptions), which are characterized by a wide range of interacting length and time scales (a distinctive feature of the turbulent regime). In Sect. 10.3, the relevant flow time scales are introduced for the different applications of the multiphase flow theory to volcanic flows.

10.2.2   Regimes in Multiphase Flows

Depending on the value of the previously identified parameters, some multiphase flow regimes can be identified, along with the adoption of the most appropriate and efficient mathematical description of the multiphase flow [53].

Four main regimes of gas/particle coupling can be described:

- $St < 0.001$ (strong coupling). The *dusty gas* approach is the most effective approach in the presence of very fine particles (<50 μm, in the volcanic case).
- $0.001 < St < 0.2$ (weak coupling). The *Equilibrium-Eulerian* approach is the most suited. This is used for particles up to about 1 mm in the volcanic case (volcanic ash).
- $0.2 < St < 1$ (particle decoupling). The *Eulerian-Eulerian* approach is the most effective approach for modeling both kinematic and thermal decoupling of particles up to coarse ash (up to about 10 mm).
- The *Lagrangian* point-particle approach, when the effect of the flow at the scale of the particle is not crucial (e.g., volcanic lapilli and bombs, diameter larger than a few cm).

As previously mentioned, the thermal Stokes time, St_T, has the same order of magnitude as St. Thermal decoupling becomes relevant only when $St_T \approx St > 0.2$ [43, 44].

For the interaction regime between phases, three conditions can be identified:

- $\varepsilon_s \lesssim 0.001$ and $Y_s < 0.01$ (one-way coupling). Particles do not significantly influence the gas motion (e.g., in the atmospheric dispersal of volcanic ash).
- $\varepsilon_s \lesssim 0.001$ (two-way coupling). Particles affect the gas flow but particle/particle interactions are negligible (e.g., in the dilute portions of volcanic jets and plumes).
- $\varepsilon_s \gtrsim 0.01$ (four-way coupling). In such a case, $St_c < 1$ and particle/particle interactions control the dynamics (e.g., in concentrated pyroclastic density currents).

In the following sections, different multiphase flow models and their application to volcanic flows will be illustrated adopting some of these formulations.

10.2.3   Grain-Size Distribution

For applying continuous (fluid) models, there are several ways to describe multiphase granular flows. The most common approach, in volcanology, is the discretization of the *grain-size distribution* into a finite number of *bins*, representing particle classes belonging to different size intervals. More precisely, in volcanological studies, it is common practice to adopt a uniform discretization in the logarithmic Krumbein φ scale, where $\varphi = -log_2\left(\frac{d_p}{d_0}\right)$, d_p is the particle diameter expressed

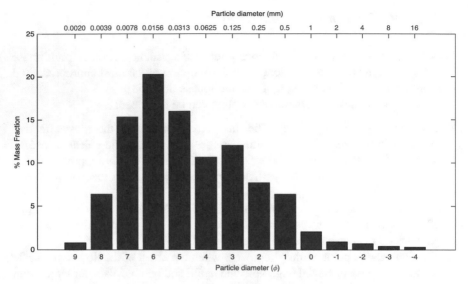

Fig. 10.3 Typical grain-size distribution of pyroclastic material. The distribution shown in the figure is representative of the May 18, 1980, sub-Plinian eruption of Mount St. Helens and indicates that particles are distributed over a significantly large range of dimensions. (Redrawn from [55])

in millimeters, and $d_0 = 1$ mm [14, 54]. The Eulerian fluid transport equations for mass, momentum, and energy can then be written for the interstitial gas and for every particle class, treated as interpenetrating continua (see Chap. 2). A typical grain-size distribution of a pyroclastic mixture is reported in Fig. 10.3 and Table 10.1 [55]. Depending on the specific application, particles can be characterized by their Stokes number instead of diameter and density [40].

10.3 Eulerian-Eulerian Multiphase Flow Modeling

In the most general multiphase flow regime and, in particular, for volcanic flows displaying the coexistence of regions of dilute and concentrated flow, compressible and turbulent regimes, all terms of the multiphase flow equations (Table 2.3, Chap. 2 and [79]) for a polydisperse mixture must be considered and retained. In the following subsections, some applications of the multiphase flow theory to volcanic jets, volcanic blasts, and pyroclastic density currents are presented.

Table 10.1 Total grain-size distribution considered as representative of the pyroclastic mixture of the May 18, 1980, sub-Plinian eruption of Mount St. Helens (redrawn from [55]). Particle diameter (in φ scale and microns) density, volume and mass fractions are reported for each of the 14 grain sizes

Grain size d_p [φ]	Particle diameter d_p (μm)	Density ρ_p (kg/m^3)	Volume fraction ε_p	Mass fraction Y_p
9	2	2500.0	0.0005	0.00799
8	4	2500.0	0.004	0.06389
7	8	2500.0	0.0096	0.15334
6	16	2500.0	0.0127	0.20285
5	31	2500.0	0.01	0.15973
4	63	2224.8	0.0075	0.10661
3	125	1939.8	0.0097	0.12022
2	250	1654.9	0.0073	0.07718
1	500	1369.9	0.0073	0.06389
0	1000	1084.9	0.003	0.02079
−1	2000	800.0	0.0018	0.00920
−2	4000	800.0	0.0014	0.00716
−3	8000	800.0	0.0008	0.00409
−4	16,000	800.0	0.0006	0.00307

10.3.1 Compressible Multiphase Flow Regime: Volcanic Jets and Blasts

Explosive eruptions are characterized by the rapid ascent and decompression of volatiles-rich, viscous magmas [54]. There are several mechanisms in nature controlling the excess of pressure of magma when it reaches the surface and its potential enrichment in volatiles (e.g., primary exsolution of dissolved volatiles, interaction with external ground waters, and fluids). In any case, the initial stage of explosive eruptions is always characterized by a decompression/expansion of the eruptive gas/particle mixture, whose relevance in the eruption dynamics depends on the particular conditions leading to fragmentation of magma. In this section, we summarize the analysis of two volcanic phenomena in which compressible phenomena strongly control eruption dynamics and, consequently, also their associated hazards: volcanic jets and blasts.

10.3.1.1 Volcanic Jets

During Plinian and sub-Plinian eruptions (e.g., the May 18, 1980, eruption of Mount St. Helens, USA, Fig. 10.4), a polydisperse mixture of gases, magma, and rock fragments is steadily injected into the atmosphere at high velocity, pressure and temperature.

Fig. 10.4 Volcanic jet generated by Mount St. Helens during the paroxysmal stage of the sub-Plinian eruption of May 18, 1980. (Photo by Robert Krimmel, public image of USGS)

The gas/particle mixture is generated by the fragmentation of magma at depth, during its rise in the volcanic conduit. Above fragmentation depth, the flow is accelerated by the rapid decompression of the pyroclastic mixture to reach choked flow conditions at the conduit exit [22, 56, 57]. It is well known that eruptive vent conditions characterized by sonic velocity and gas pressure larger than atmospheric result in a rapid expansion and acceleration of the fluid to a high Mach number. A series of expansion waves (Prandtl-Meyer expansion) form at the vent exit, which are reflected as compression waves at the jet flow boundary. The compression waves coalesce to form a barrel shock and a standing normal shock wave (Mach disk), across which the vertical velocity is reduced to subsonic regime and the pressure in the core of the jet increases. Above the Mach disk, the fluid moves slowly in the core of the jet and is surrounded by a supersonic moving shell, with a slip line or a shear layer dividing these regions [58]. Figure 10.5 shows such flow structures for typical volcanic conditions [40].

Analogously, explosive dynamics in Plinian eruptions are controlled by the rapid decompression in the proximity of the vent and the formation of an underexpanded multiphase volcanic jet [37, 59–61], which, in turn, directly affect the volcanic plume behavior. To better understand the influence of grain-size distribution on the eruption dynamics, non-equilibrium polydisperse volcanic jets in transonic regimes have been studied [40, 62] by using the Eulerian-Eulerian PDAC multiphase flow model [30, 36]. Because of the dominantly dilute flow regime in volcanic jets, the solid stress tensor and particle/particle drag terms were neglected in the

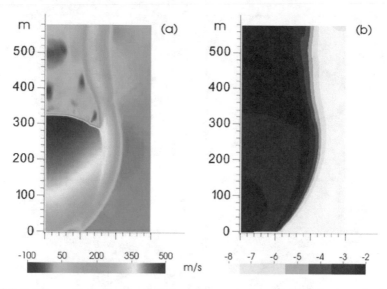

Fig. 10.5 (**a**) Gas vertical velocity and (**b**) Logarithm to the base 10 of solid volume fraction of a sonic volcanic jet with velocity of 150 m/s, temperature of 1200 K, and overpressure of 20 bar leaving a vent of 80 m in diameter. Particles are monodisperse with a diameter of 10 μm. The snapshots refer to 20 s from the beginning of the injection. (Modified from [40])

simulations; the Gidaspow-Wen-Yu drag and the Ergun models (see Chap. 2 for details) were adopted to describe interphase gas/particle momentum and energy exchange.

As a flow characteristic time scale, the Mach disk formation time τ_{Ma} [63] was considered, estimated as $\tau_{Ma} = R_v/c_m$ where R_v is the vent radius and c_m is the mixture speed of sound. This allowed quantitative discernment of *fine* ($St < 1$) from *coarse* ($St > 1$) particles and identification of two different decompression regimes associated to St. For monodisperse jets with $St \ll 1$, the gas/particle mixture behaves as an isothermal dusty-gas [59, 64, 65] and the volcanic jet presents the typical Mach disk structure of under expanded gas jets (Figs. 10.5 and 10.6, curves a, b, c). On the contrary, jets laden with coarse particles ($St \gg 1$) display non-equilibrium phenomena associated with the inertia of particles, eventually leading to the obliteration of the Mach disk structure (Fig. 10.6, curves d, e).

For bidisperse and polydisperse mixtures, this behavior is quantitatively confirmed also for "dominantly fine" and "dominantly coarse" grain-size distributions, as suggested by scaling analysis and demonstrated by numerical simulations performed with the multiphase flow model and with a hybrid dusty-gas-multiphase flow model, in which the fine components are coupled to the gas phase whereas coarse particles are treated as non-equilibrium interpenetrating continua [40, 62]. In this situation, it is shown that even a relatively small amount of fine particles (typically less than 10% in weight) is able to significantly lower the effective Stokes number of the coarsest particles due to particle/particle interaction, and the Mach disk structure can be retrieved.

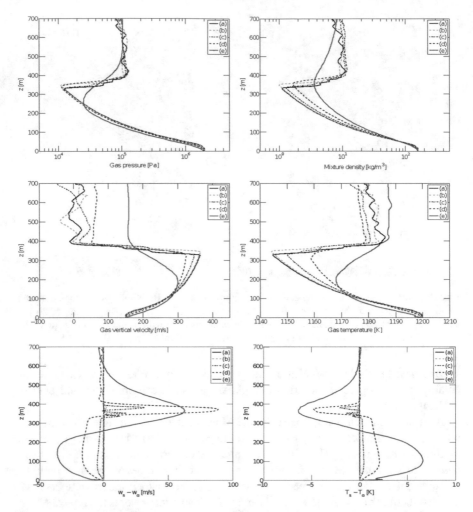

Fig. 10.6 Time-averaged axial profiles computed over the period [16–20 s] since the beginning of the injection of gas pressure (top left), mixture density (top right), gas vertical velocity (medium left), gas temperature (medium right), difference between particle and gas vertical velocity (bottom left) and difference between particle and gas temperature (bottom right). Curves refer to (a) $d_s = 8\,\mu m$, (b) $d_s = 27\,\mu m$, (c) $d_s = 95\,\mu m$, (d) $d_s = 346\,\mu m$, and (e) $d_s = 1300\,\mu m$. The volcanic jet is characterized by a vent diameter of 80 m and an overpressure of 20 bar. (Modified from [62])

Because the presence of the Mach disk strongly affects the stability of the jet [37, 60, 61] and the potential generation of lethal pyroclastic gravity currents, this study has relevant implications for the assessment of volcanic hazards and emphasizes the importance of properly accounting for non-equilibrium effects in the modeling of explosive eruptions.

10.3.1.2   Volcanic Blasts

Lateral blasts represent another peculiar eruptive category where compressible multiphase flow effects are relevant. Volcanic blasts are characterized by the violent release of a relatively small amount of magma producing a remarkably broad area of significant damage. In most cases, the occurrence of volcanic blasts is associated with the explosive destruction of a partly crystallized magma body (dome or cryptodome) situated in the upper (or shallow) part of the volcanic edifice, as a result of major edifice or lava dome failure. Explosive fragmentation occurs when the magma body is (asymmetrically) exposed to atmospheric pressure and the pressurized magma very rapidly decompresses causing a directed explosion or series of explosions (see Fig. 10.1 for the Mount St. Helens May 18, 1980, lateral blast).

The non-equilibrium multiphase flow PDAC model has been used to simulate the dynamics of two recent volcanic blast events, namely the Boxing Day, 1997, blast at Soufrière Hills volcano (Montserrat, Lesser Antilles, UK) [66], and the May 18, 1980, blast of Mount St. Helens, USA [38, 39]. Three-dimensional numerical simulations have described the effects of the dome and proximal volcano morphologies, which focus the expansion of the fragmented mixture into a laterally directed flow and further drive the propagation of pyroclastic density currents generated by the collapse of the blast cloud. Numerical simulations were able to reproduce, with minimal constraints on the initial conditions, most of the area invaded by the blast (Fig. 10.7) and some of the main features of the depositional system.

The application of a non-equilibrium multiphase flow model, in this case, has been crucial to elucidate some of the main but still puzzling features of the blast dynamics. The eruption was initiated by the explosive decompression of the

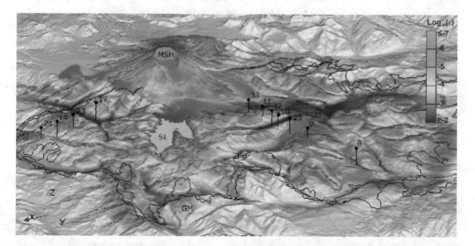

Fig. 10.7 Deposit boundary, topography, simulated particle concentration in the basal cell, and section locations, in perspective view from North (direction of the Y axis). Red line: tree blow-down limit; brown line, seared zone boundary. *SL* spirit lake, *MSH* Mount St. Helens, *GM* goat mountain. (Modified from [39])

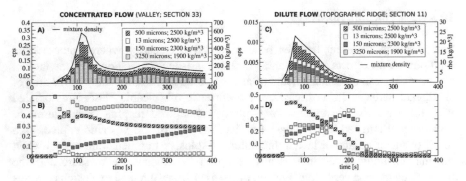

Fig. 10.8 Simulation of the segregation process during the 1980 Mount St Helens blast. Sampling flow properties in the basal computational cell. (**a, b**) In a valley (Section 11, concentrated flow). (**c, d**) On a topographic ridge (Section 33, dilute current). (**a–c**) Solid line: variation in time of mixture density (ρ-rho; right-hand side scale). Histogram: cumulative volumetric fraction of particles (ε-eps; left-hand side scale) in the first computational cell above the topography. (**b–d**) Mass fraction (m) of each particle class (identified by different textures). (Modified from [39])

multiphase magmatic mixture (with sound speed as low as ~80 m/s), which determined the ejection speed of pyroclasts (at Mach number M > 2) and the proximal dynamics. Nonetheless, the flow rapidly evolved into a gravity-driven, inertial current, with flow front advancement velocity, runout and morphology controlled by buoyancy. In the case of the blast of Mount St. Helens, it was speculated that particle/particle dissipative processes occurring in the most concentrated part of the current did not play a controlling role in the large-scale flow dynamics. Instead, non-equilibrium gas/particle processes controlled particle segregation and sedimentation, flow dilution, and transition to positive buoyancy in response to topographic blocking effects of the basal concentrated layer (Fig. 10.8 [38, 39]).

10.3.2   Stratified Flow Regime: Pyroclastic Density Currents

Pyroclastic density currents (PDC) are hazardous volcanic flows characterized by the propagation, along the ground surface, of a mixture of hot gases and particles (pyroclastic material and eroded substrate). They can be generated by a variety of explosive volcanic phenomena, including Plinian column instability and collapse, lateral blast, and gravitational dome collapse [67].

After an initial proximal generation phase, when compressible phenomena can be relevant [38, 68], PDCs propagate mostly as subsonic gravity-driven flows. Gravity controls their longitudinal acceleration along volcanic slopes, the sedimentation of particles, the formation of deposits and, ultimately, their buoyancy, lift-off, and, eventually, runout. Sedimentation provides the main mechanism for transfer of horizontal momentum into the basal layer of the current, where it is mostly dissipated by viscous and frictional forces. This produces a much faster deceleration of the flow front with respect to other particle-laden gravity driven currents [69–71]. Such a mechanism is of fundamental importance in PDCs, where particle load is much higher than in common idealized models and experiments, and momentum transfer to the basal flow largely controls flow front velocity and runout [72].

Modeling the multiphase flow dynamics of PDCs is an extremely challenging task and still an open problem. Thus, a simplified situation is discussed here where PDCs propagate along a flat surface, as relevant for assessing volcanic hazard in caldera settings. Calderas are large volcanic systems characterized by a negligible topographic mean slope (due to the ancient cave-in of the whole volcanic edifice occurred after some major eruptions) and by the presence of a complex morphology characterized by nested craters, cones and rings, remnants of old volcanic edifices, and rims. Examples include the Campi Flegrei caldera in Italy, one of the regions at the highest volcanic risk in the world. To evaluate PDC hazards in a caldera, it is necessary to understand the mechanism and the controlling factors of flow sedimentation and deposition during its propagation because these control PDC stratification and their capability to overcome topographic obstacles and eventually inundate the region outside the caldera margins [49, 73].

The application of the PDAC multiphase flow model has made it possible to better describe the complex dynamics of PDCs and, at the same time, to calibrate a fast numerical model to emulate PDC kinematics for hazard assessment studies [49, 50]. Multiphase flow simulations suggest that, on a flat or rugged surface, PDCs propagate as inertial currents with Froude number $0.8 < Fr < 1.6$, and that the granular stress developed in the concentrated basal layer does not significantly affect the overall PDC runout. For instance, Fig. 10.9 shows the difference in the PDC propagation using two different formulations for the collisional stress tensor: the first one (shown by the top plots) is the semi-empirical viscosity model adopted by Neri et al. [30], while the second one (shown by the bottom plots) is based on the kinetic theory constitutive models described in Chap. 2. The plots represent the particle volume concentration for a simulation characterized by an initial volume concentration of 0.1 of particles of 100 μm. Despite the initial high concentration, from the plot it is evident that the difference of runout is less than 10% and perceivable only at the latest propagation stages.

Nevertheless, during their propagation, PDCs develop strong vertical stratification as a consequence of particle non-equilibrium and sedimentation. In this stage of propagation, a concentrated basal flow and a turbulent dilute upper layer are formed.

As shown in Fig. 10.10, a vertical quasi-equilibrium profile develops under the concurrent effects of sedimentation, turbulence diffusion, and mixing with atmospheric air. Particle concentration in the basal flow can reach values of 10–20%, up to the maximum packing of particles. However, on a flat topography, the basal flow has a negligible influence on the dynamics of the overlying dilute ash cloud and on the propagation of the flow front. Modeling deposition through the removal of particles from the basal layer does not affect the structure and dynamics of the dilute upper layer.

The high-resolution multiphase flow simulations also show that the propagation of the PDC (see Fig. 10.11) is characterized by the formation of a current head (the

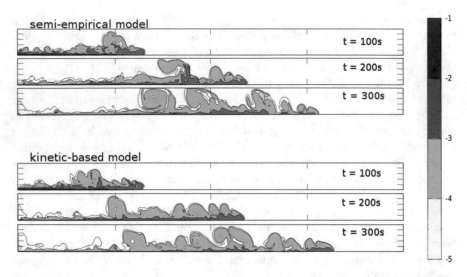

Fig. 10.9 PDCs traveling along a flat topography. The two plots represent the particle volume concentration at 100, 200, and 300 s from the start of the flow initially characterized by a volume concentration of 0.1 of particles of 100 mm. The upper plot shows the evolution of the PDC by adopting a semi-empirical viscosity model whereas the lower plot refers to a PDC with a rheology described by a kinetic theory-based model. The horizontal and vertical dimensions of the computational domain are 4 and 0.5 km, respectively. The grey colors of the bar refer to the logarithm to the base 10 of the particle volume concentration of the PDC

PDC nose), the development of a Kelvin-Helmoltz instability, which generates transversal eddies at the upper interface between the current and the atmosphere, and the Lobe-and-Cleft (LC) instability, which is associated to the engulfment of air by the flow front that generates positive buoyancy at the current head [74].

Finally, it is worth mentioning that the PDAC model has been extensively used to simulate 3D column collapse scenarios at explosive volcanoes [31, 37–39, 66] in order to investigate the complex effect of topography on the propagation of PDCs and provide useful information in terms of hazard zonation. Figure 10.12 illustrates one of these 3D simulations representing a Plinian-scale eruption scenario at Campi Flegrei.

10.4 The Method of Moments

As discussed in Sect. 10.2.3, volcanic gas/particle flows are composed of pyroclastic particles of different sizes, ranging from a few microns up to several decimeters. A proper description of the multi-particle nature of the flow is therefore crucial to properly model the evolution of the grain-size distribution in the different portions of the phenomenon. As shown above, in the past decades, numerical simulation of volcanic eruptions has greatly advanced and models are now often able to deal with

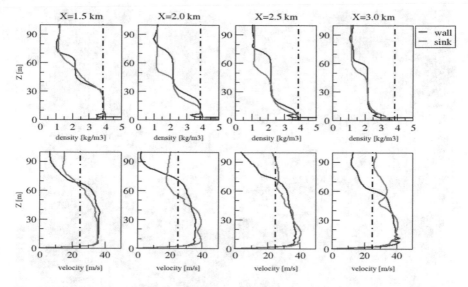

Fig. 10.10 Vertical profiles of the PDC density (top row) and velocity (bottom row) at different distances from the flow origin. The red curves refer to a wall condition of no-outflow of particles that therefore accumulate in the bottom cells of the domain. Vice versa, the green curve assumes the removal of particles from the ground. As shown by the profiles, the different boundary conditions do not significantly affect the large-scale features of the PDC

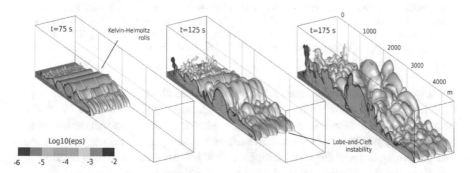

Fig. 10.11 Structure of the PDC represented by the isosurface of the volume particle concentration ($\varepsilon_s = 10^{-6}$) at 75 s (left), 125 s (center) and 175 s (right) from the flow injection. The isolines of the volume concentration, every unit of the log to the base 10, are also plotted on the front (y = 0) plane (see color bar). The high-resolution simulation clearly captures the generation of the Kelvin-Helmoltz rolls and the Lobe-and-Cleft instability. (Modified from [74])

the multi-phase nature of volcanic flows, but polydispersity in size is often largely simplified and a proper treatment is still a challenging issue. This is particularly true when physical processes able to change particle size distribution, such as aggregation and breakage in volcanic plumes and PDCs, are important. In such cases, an effective and viable alternative to Eulerian-Eulerian (see Sect. 10.3) and Eulerian-Lagrangian (see Sect. 10.6) models is the Method of Moments (MoM), a theoretical

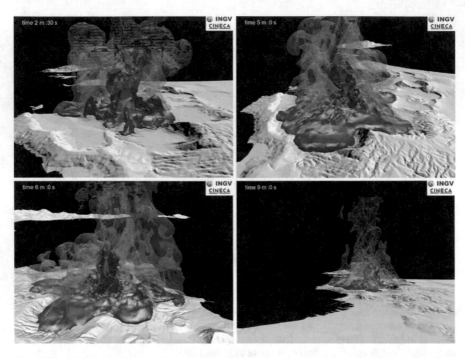

Fig. 10.12 Bird's-eye view of a hypothetical explosive eruption scenario at Campi Flegrei caldera. The eruption parameters are representative of a Plinian-scale eruption with vent location in the Agnano plain similar to the Agnano Monte Spina Eruption of 4100 BP. The plots show the isosurface of the flow temperature with the pink and orange colors representing the 100 °C and 350 °C isosurfaces, respectively, and refer to 2′30″, 5′, 6′, and 9′ from the injection of the mixture in the atmosphere

framework with a corresponding computational method, developed in the past decades mostly in the chemical engineering community [75–79]. The MoM for a polydisperse particle mixture can be seen as a mesoscale[1] model where first a transport equation depending on mesoscale variables (called *internal coordinates*) and describing the polydisperse multiphase flow is introduced, and then the transport equations for the integral moments of the mesoscale variables are derived.

This approach is briefly described here focusing on a particular application, i.e., the modeling of the continuous variability of particle size distribution in a rising volcanic plume. In the past decades, several studies used and adapted the Buoyant Plume Theory (BPT) of Morton et al. [80] for turbulent buoyant plumes to describe the multiphase character of volcanic eruptions. In such models, a common velocity for a homogeneous mixture of pyroclasts, water (in any state), and entrained air, is

---

[1]Here the term mesoscale is not used in the meteorological sense, but as referred to a modeling description in between a microscale model (resolving all relevant length and time scales at the level of single disperse particles) and a macroscale model (describing only a limited set of disperse-phase variables needed to represent "average" properties of the multiphase flow).

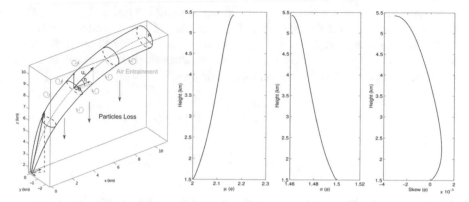

Fig. 10.13 Left panel: sketch of the plume rise model adopted (see text for explanation). Right panel: distribution of the mean, variance, and skewness of the particle-size distribution with height for a low-flux (i.e., 1.5×10^6 kg/s) plume with wind, showing the continuous change with height. (Modified from [42])

assumed. Pyroclasts of different sizes are transported within the plume until the vertical velocity of the mixture falls below the particle settling velocity w_s [81–83], resulting in a plume particle size distribution becoming poorer in coarse particles with increasing height. On the other hand, favored by the presence of liquid water or ice, aggregation between particles may occur in the plume, forming ash aggregates with typical sizes of few hundred μm. Both phenomena, i.e., settling and aggregation, contribute to change significantly the particle size distribution of the plume.

The equations set for the plume rise model are derived and solved by considering the bulk properties of the eruptive mixture (see left panel of Fig. 10.13).

The plume is assumed having, in the plane normal to the centerline trajectory with curvilinear coordinates, a circular section with radius R, and a *top-hat* profile of the velocity component along the centerline $U = U(s,t)$. As discussed in Sect. 10.3, most of the volcanic plume models discretize the continuous particle grain-size distribution by grouping particles in different bins and writing transport equations for the mass of each bin. Here, instead, as mesoscale transport equation, the so-called Population Balance Equation (PBE) is considered, describing the evolution in time and space of the particle-size distribution, $f(V)$, where V is the internal variable representing particle volume. Particle-size distribution represents the number concentration of particles (particles per unit volume) with diameter between V and $V + dV$ which, building upon the work of Bursik et al. [84], satisfies the following equation,

$$\frac{\partial}{\partial t}\left[\pi R^2 f(V,t,s)\right] + \frac{\partial}{\partial s}\left[\pi R^2 f(V,t,s)U(s,t)\right] + \frac{\partial}{\partial L}\left[\pi R^2 f(V,t,s)G(V)\right]$$

$$= -2\pi R\rho w_s(V)f(V,t,s) - \pi R^2[D(V,t,s) + B(V,t,s)], \qquad (10.8)$$

where p is the probability that an individual particle will fall out of the plume, G is the particle growth, D is the rate of death of particles of size V due to aggregation with other particles, and B is the rate of birth of particles due to aggregation of smaller particles. These last two terms are defined, accordingly to the Smoluchowski coagulation equation [85], by the following integrals,

$$D(V,t,s) = -\int_0^{+\infty} K(V,\lambda)f(V,t,s)f(\lambda,t,s)d\lambda, \tag{10.9}$$

$$B(V,t,s) = \int_0^{+\infty} K(V-\lambda,\lambda)f(V,t,s)f(\lambda,t,s)d\lambda, \tag{10.10}$$

where $K(V,\lambda)$ is the aggregation kernel describing the rate at which particles of volume V aggregate with particles of volume λ. Only the dependence of the kernel on particle sizes is expressed here implicitly but, generally, it is also a function of other variables, such as the local amount of water in the plume, ambient fluid shear, and differential sedimentation.

The PBE represents an integral-differential equation of the particle-size distribution, and a direct resolution of such equation at every time step, for every particle size, and at every spatial location, is prohibitive. For this reason, several numerical methods have been proposed to solve the PBE, whose advantages and limitations have been compared in several review articles [86]. Here, it is shown how the MoM can be used to describe the evolution of the moments of the PBE, reducing considerably the dimensionality of the problem. Given a particle size distribution f (V), its "shape" can be quantified through the moments $M^{(i)}$ [87], defined,

$$M^{(i)} = \int_0^{+\infty} V^i f(V)dV. \tag{10.11}$$

With the choice made for the internal variable of the particle-size distribution, it can be observed that the first moments have the following physical interpretation: $M^{(0)}$ is the total number of particles per unit volume of the multiphase mixture in the plume, while $M^{(1)}$ is the particle volume fraction.

Now, if a steady condition is assumed and continuous particle growth ($G = 0$) is neglected, multiplying both sides of the PBE for V^i and integrating over the size spectrum $[0, +\infty]$, the following conservation equations for the moments $M^{(i)}$ are obtained:

$$\frac{\partial}{\partial s}\left[\pi R^2 M^{(i)} U(s,t)\right] = -2\pi R p w_s^{(i)} M^{(i)} - \pi R^2 M^{(i)}\left[D^{(i)} + B^{(i)}\right], \tag{10.12}$$

where the terms $w_s^{(i)}$, $D^{(i)}$, and $B^{(i)}$ are called the moments of the respective quantities and are defined,

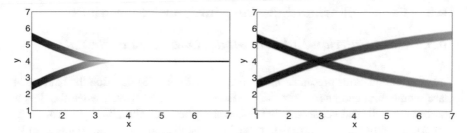

Fig. 10.14 Modeling of very dilute jets in a crossflow. Left: result from the Eulerian approach, right: result from the mesoscale approach based on the MoM. (Redrawn from [90])

$$(\cdot)^{(i)} = \frac{1}{M^{(i)}} \int_0^{+\infty} (\cdot) V^i f(V) dV \tag{10.13}$$

These terms represent averaged values of the original quantity, where the averaging is different accordingly to the order of the moment. For example, $w_s^{(0)}$ represents the mean settling velocity, averaged with respect to the number of particles, while $w_s^{(1)}$ is the volume-weighted average velocity. The calculation of these integral terms represents one of the key points of every MoM, because, in principle, a proper computation would require the knowledge of the particle size distribution $f(V)$. In the quadrature method of moments (QMOM, [88]), the integral in the definition of these moments is replaced by a quadrature formula, where the weights ω_l and nodes V_l are obtained from a Gaussian quadrature rule applied to the known moments. For example, for the i-th moments of the settling velocity,

$$w_s^{(i)} = \frac{1}{M^{(i)}} \int_0^{+\infty} w_s(V) V^i f(V) dV \approx \sum_l w_s(V_l) V_l^i \omega_l. \tag{10.14}$$

This approach has been successfully used in de' Michieli Vitturi et al. [42] and the results of an application for a low-flux plume (i.e., 1.5×10^6 kg/s) with wind are reported in the right panels of Fig. 10.13, showing the continuous change with height of the mean, variance, and skewness of the particle-size distribution.

It is worth noting that, in gas/particle flows, the particle velocity also can be used as an internal variable and a formulation based on the mesoscale description of particle velocity distribution (leading to the so-called Generalized Population Balance Equation-GPBE) and transport equation for its moments can be derived [89]. In very dilute regimes, where particles' collisions are negligible and a Eulerian description of the solid phase as a continuous is not valid, the MoM can represent a valid alternative, as shown by the application to very dilute jets in a crossflow presented in Fig. 10.14 [90]. The volume fraction field highlights that the Eulerian formulation (left panel) cannot predict particle crossing, resulting in zero y-momentum, whereas the mesoscale approach (right panel) allows, in the absence of collisions, particle clouds to cross.

10.5 The Equilibrium-Eulerian Multiphase Flow Model

10.5.1 The Eulerian-Eulerian Model in Mixture Formulation

The general equations presented in Chaps. 1 and 2 and [78] describe the balance of mass momentum and energy for each phase singularly. However, when the Stokes number is small, and consequently the coupling between the phases is strong, the numerical solution of this formulation can be demanding because the coupling terms become very important. As discussed in Cerminara et al. [43, 44], an alternative is to reformulate the problem in terms of the mixture fields (ϱ_m, u_m, h_m), using the explicit form of the coupling terms only to determine the velocity and temperature difference between the phases. This is possible by summing the equations of all the phases and by defining a field ψ in mixture formulation,

$$\psi_m = \sum_{j \in J} Y_j \psi_j \tag{10.15}$$

where J is the set of all the phases and components. In this way, thanks to the Newton's third law and to the energy conservation equation, the momentum and energy coupling terms cancel out. The resulting equations in mixture formulations read,

$$\partial_t \varrho_m + \nabla \cdot (\varrho_m \, u_m) = 0 \tag{10.16}$$

$$\partial_t(\varrho_m u_m) + \nabla \cdot (\varrho_m \, u_m \otimes u_m + \varrho_m T_m) + \nabla P = \nabla \cdot T + \varrho_m g \tag{10.17}$$

$$\begin{aligned}\partial_t(\varrho_m h_m) + \nabla \cdot [\varrho_m \, (u_m + v_h) h_m] - \nabla \cdot (T \cdot u_f - q) = \\ = \partial_t P - \partial_t(\varrho_m k_m) - \nabla \cdot [\varrho_m \, (u_m + v_K) k_m] + \varrho_m g \cdot u_m\end{aligned} \tag{10.18}$$

where,

$$T_m = \sum_{j \in J} Y_j (v_j \otimes v_j) - (v_m \otimes v_m) \tag{10.19}$$

$$v_j = u_j - u_f \tag{10.20}$$

$$v_m = \sum_{j \in J} Y_j v_j \tag{10.21}$$

$$\psi_m v_\psi = \sum_{j \in J} Y_j v_j (\psi_j - \psi_m), \quad \psi = h, K \tag{10.22}$$

Thus, five equations were obtained for a system with $5 * J$ fields plus the pressure that should be determined by the equation of state. The remaining equations needed to close the system are the continuity equations for all the phases excluding that for the carrier fluid which is replaced by the continuity equation of the mixture, and the momentum and energy equations for the decoupling fields v_j and h_j. In this way, the balance equations of the mixture can be solved using the standard methods for the compressible Navier-Stokes equations, once the decoupling equations containing the drag and the heat exchange between the phases are solved. When $v_j \to 0$ and the temperature $T_j \to T_f$ for each $j \in J$, the previous equations reduce to the dusty-gas model.

10.5.2   The Equilibrium-Eulerian Model

This sub-section is focused on the presentation of the equilibrium-Eulerian approach [91, 92] in the dilute regime (two-way coupling) and for heavy particles. This is the typical regime of volcanic plumes and dilute type of PDCs (so called *pyroclastic surges*). In this case, the equation describing the kinematic decoupling reduces to,

$$\partial_t u_j + u_j \cdot \nabla u_j = \frac{1}{\tau_j}(u_f - u_j) + g \tag{10.23}$$

The equilibrium-Eulerian model represents the first-order asymptotic approximation for small Stokes times $\tau_{sj} \ll 1$. As derived in Maxey [91], Ferry and Ferry and Balachandar [92], and Cerminara et al. [43], the decoupling velocity can be approximated with the following analytical solution depending on the settling velocity w_j and on the fluid acceleration $a_f = d_t u_f$,

$$v_j = \mathbb{G}_j^{-1} \cdot (w_j - \tau_j a_f) + O(\tau_j^2) \tag{10.24}$$

$$\mathbb{G}_j = I + \tau_j(\nabla u_f)^T \tag{10.25}$$

Regarding the thermal decoupling, it has been shown by Ferry and Balachandar [93] that it is second order with respect to the kinematic decoupling, even when the kinematic and thermal decoupling times are comparable. This model has been tested for various applications, showing a very good behavior up to $St \approx 0.2$. For example, the preferential concentration induced by the kinematic decoupling in isotropic turbulence modeled with the equilibrium-Eulerian model has been compared with Lagrangian direct numerical simulations obtaining good results [43].

10.5.3   Application of the Equilibrium-Eulerian Model to Volcanic Plumes

The equations presented in the previous section have been implemented in the ASHEE code [43] using the C++ libraries of OpenFOAM. Because the Reynolds number of explosive volcanic phenomena is very high, the resulting multiphase flow is turbulent and subgrid-scale models are needed to correctly deal with the eddy viscosity and diffusivity. In the ASHEE code, the equations are filtered in order to solve them using the Large Eddy Simulation (LES) approach. In particular, dynamic LES models have been implemented to avoid the use of any turbulence empirical parameter.

The ASHEE code has been used to investigate the complex dynamics of volcanic plumes and also to calibrate steady-state one-dimensional (1D) models by using the outcomes of transient three-dimensional (3D) models [41, 45, 47, 94]. Two volcanic scenarios representative of *weak* and *strong plumes* have been studied (Fig. 10.15).

The main difference between the two scenarios is the Mass Eruption Rate (MER), chosen to be similar to the January 26, 2011, Shinmoe-dake eruption (MER $= 1.5 \times 10^6$ kg/s) for the weak plume, and to the climactic phase of the Pinatubo eruption, Philippines, on June 15, 1991 (MER $= 1.5 \times 10^9$ kg/s), for the strong plume.

The 3D results were also time-averaged and spatially filtered coherently with 1D model formulation, allowing for the comparison of profiles between different models and calibration of the entrainment coefficient. The comparison between 1D and 3D models highlighted the capability of the former to capture the maximum plume height with an error of about 20%, even if the vertical profiles of the relevant fields presented important differences with respect to the 3D models. Figure 10.16 shows a comparison of column height as computed by a variety of 1D and 3D models and semi-empirical correlations [47].

In the sensitivity studies carried out in this study, the entrainment coefficient appeared to be the most critical parameters for 1D models, especially above and near the neutral buoyancy level and around the collapsing region, where the self-similarity assumption of 1D models is not accurate, particularly for strong plumes (Fig. 10.17).

On the other hand, 3D simulations highlighted that turbulence should be carefully modeled in order to reproduce the correct entrainment rate. In Cerminara et al. [45], the effect of kinematic decoupling, of mesh resolution, and of different turbulent sub-grid scale models was studied, allowing the quantification of the variability of vertical profiles and entrainment rate due to the formulation of the numerical solver.

10.6   The Lagrangian Particle Approach

The Eulerian-Eulerian and the Equilibrium-Eulerian approaches adopted by the previously described multiphase flow models, in which particles are treated as a continuous phase to model non-equilibrium dynamics between a gas phase and solid particles, are not always the best choice to catch the dynamics of particles moving inside a fluid. The dense or dilute character of the particle phase and some characteristic time scales are fundamental to determine the best approach to be used for modeling the dispersed phase. In fact, when particle concentration in the fluid is low and the particle relaxation time (see Sect. 10.2), which quantifies the inertial effect of particles, is much smaller than the characteristic time of collisions between particles,

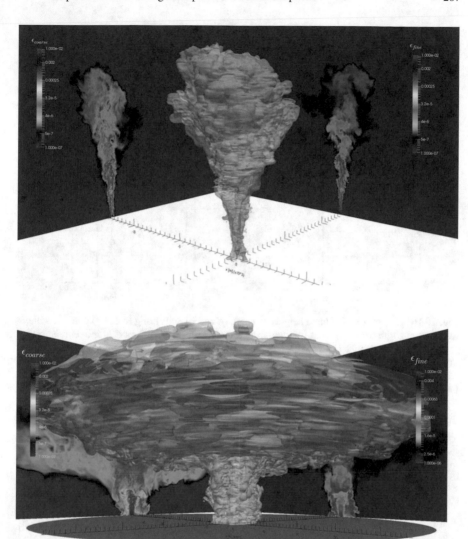

Fig. 10.15 Numerical simulation of volcanic plumes by using the Equilibrium-Eulerian model ASHEE. The top plot refers to a *weak plume*, 400 s after the beginning of the eruption, whereas the bottom plot to a *strong plume*, 1000 s after the beginning of the eruption. The two isosurfaces in 3D rendering plots correspond to the threshold $\epsilon_s = 10^{-7}$ of the fine (white) and coarse (sand colored) ash volume fractions. The two-dimensional sections on the sides of both plots represent the distribution of the volume concentration of coarse (left) and fine (right) particles across vertical orthogonal slices crossing the plume axis. See color palette for volume concentration values. (Modified from [45])

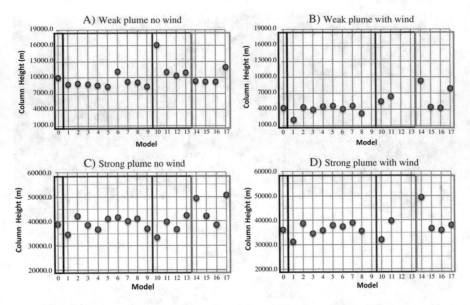

Fig. 10.16 Predictions of column heights of weak (**a** and **b**) and strong (**c** and **d**) plumes, without (**a** and **c**) and with (**b** and **d**) wind, returned from a variety of 1D and 3D models and semi-empirical correlations (all denoted by numerical labels on the x axis) for fixed MER (modified from [47]). Red group indicates 1D models, blue 3D models, green empirical relationships, and black the average of 1D and 3D models (for the description of the model features, the reader is referred to Costa et al. [47])

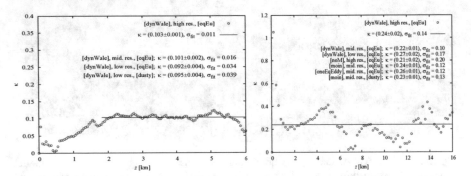

Fig. 10.17 (left) Entrainment coefficient κ for the *weak plume* case in the plume region, i.e., $2 \div 6$ km above the crater. The points represent the results from the high-resolution 3D simulation carried out with the [dynWale] sub-grid model, whereas the line represents the best fit to the data. Fit results for the other simulations performed are also shown. (right) Entrainment coefficient κ for the *strong plume* below the Neutral Buoyancy Level, i.e., $0 \div 16$ km. We show the fit graph obtained at high resolution with the [dynWale] sub-grid model and report in the legend the fit result for the other simulations performed. (Modified from [45]).

Fig. 10.18 Sketch of a particle motion within a fluid to illustrate the Lagrangian modeling approach

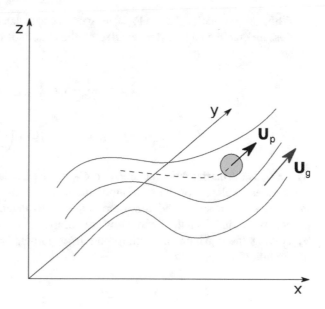

a one-way coupling between gas and particles can be assumed and a Eulerian-Lagrangian approach is better suited. In this approach, the Eulerian conservation equations are solved for the gas phase whereas, by the Lagrangian approach, equations of motion are solved for the particulate phase in order to determine the trajectories of single particles. This approach is particularly well suited for the simulation of the dynamics of large volcanic clasts (particles of any origin with an average diameter of a few centimeters or larger) ejected in volcanic explosions and initially engulfed in a rapidly expanding mixture of gas and fine particles.

The equation for the motion of a particle in an unsteady flow field was first derived by Basset, Boussinesq, and Oseen, and is commonly known as the BBO equation [95–97]. For each particle with mass m_p, diameter d_p, and velocity U_p (see Fig. 10.18), the BBO equation of motion expresses the acceleration of a particle as a function of the sum of the forces acting on the particle, i.e., steady-state drag, virtual mass force, Basset force, pressure gradient force, and gravitational force,

$$m_p \frac{dU_p}{dt} = \sum F_p = F_D + F_{VM} + F_B + F_P + F_G. \tag{10.26}$$

For the large particles considered in the volcanological applications of interest, virtual mass and Basset forces' magnitude are usually much smaller than the other terms and the sum of forces reduces to,

$$\sum F_p = F_D + F_P + F_G. \tag{10.27}$$

While the gravitational force $\mathbf{F}_G = m_p\,\mathbf{g}$ can be calculated directly, the calculation of pressure gradient and the drag exerted by the continuous fluid phase require some care,

$$\mathbf{F}_P = -\frac{m_p}{\rho_p}\nabla P, \qquad (10.28)$$

$$\mathbf{F}_D = \frac{1}{2}\rho_c\left|\mathbf{U}_c - \mathbf{U}_p\right|\left(\mathbf{U}_c - \mathbf{U}_p\right)C_D\pi\left(\frac{d_p}{2}\right)^2, \qquad (10.29)$$

where $\mathbf{U}_c$ is the carrier fluid velocity, and C_D is the particle drag coefficient. In fact, pressure and velocity of the continuous phase are defined on a discrete grid, and interpolations (in space and time) are required to determine their values at the particle position. When the force terms are computed, particle velocity is computed by solving the equation of motion, and new particle positions are obtained by integration of the velocity,

$$\frac{d\mathbf{x}_p}{dt} = \mathbf{U}_p \qquad (10.30)$$

The advantages of the Lagrangian approach are clear: the equations to integrate are general and direct and the implementation in a numerical code, with respect to the implementation of a Eulerian method, is relatively simple. The main drawback is that, for a very large number of Lagrangian particles, the computational cost can become very large.

As discussed in Sect. 10.2, a key parameter to characterize the dynamics of particles transported by a gas is the Stokes number, St, defined as the ratio of the particle drag relaxation time to the characteristic fluid time scale. The Stokes number gives first-order information on particle dynamics: when it is significantly larger than one, the particle continues to move along its trajectory although the fluid turns, whereas when St is significantly smaller than one, its trajectory closely follows the fluid streamlines.

This is clearly shown in Fig. 10.19, where the Lagrangian model LPAC [46] has been applied to the analysis of ballistic particles produced by the Vulcanian explosions occurring in August 1997 at Soufrière Hills Volcano, Montserrat (West Indies, UK). The model has been coupled with the Eulerian code PDAC [30, 36], defining the background flow field affecting Lagrangian particles, and a one-way coupling between the Eulerian and the Lagrangian codes was assumed. The investigated scenario is that named Sim3 in Clarke et al. [98], with an initially over pressurized mixture of gas and fine particles confined in the volcanic conduit exposed to atmospheric pressure, expanding, accelerating, and then evolving with the separation between a collapsing portion forming a PDC and a buoyant plume. In particular, Fig. 10.19 shows the trajectories of three different sizes of particles, with a Stokes number (in the plume region) smaller than, of the same order of, and larger than one, respectively. The paths clearly illustrate the different behavior of the three particles:

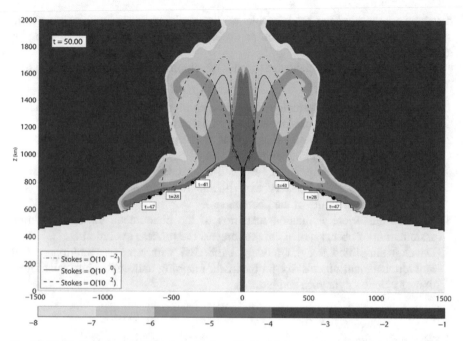

Fig. 10.19 Trajectories and flight times (reported in the labels near the particle landing positions) of three paired particles with different diameters and Stokes numbers. In particular, the trajectories represent the condition of a strong coupling (St = O(10^{-2}), dp = 10^{-3} m), an intermediate coupling (St = O(1), dp = 10^{-2} m), and a weak coupling (St = O(10^2), dp = 10^{-1} m) with respect to the plume. The color contours represent the exponent to the base 10 of the total particle volume concentration of the eruptive mixture as computed by the Eulerian PDAC model at 50 s from the onset of the explosion. (Modified from [46])

the smallest ones behave like fluid tracers, the intermediate-sized particles exhibit a more complex trajectory, and the largest particles have a parabolic-like trajectory.

From the modeling and hazard perspective, it is important to note that, as shown in de' Michieli Vitturi et al. [46], in the initial stages of a Vulcanian eruption, the pressure gradient term is of the same order of magnitude as the drag term and the neglecting pressure gradient entails a decrease of the maximum distance reached by decametric particles by about 25%.

10.7   Conclusion

As briefly shown in this chapter, the development of multiphase flow models has illuminated several aspects of the dynamics of explosive eruptions in the last few decades. This progress was also reflected in the quantification of volcanic hazards that became more quantitative through the application of such models to represent plausible future scenarios or probabilistic hazard maps. Nevertheless, the development of physical models able to accurately replicate, within acceptable statistical uncertainty, the evolution of explosive eruptions remains a challenging goal still to be achieved.

In particular, the next challenges in the development and application of future multiphase flow models are:

1. Improve the physical formulation and constitutive equations of the pyroclastic mixture (e.g., sub-grid and kinetic theory-based models).
2. Improve the accuracy and efficiency of numerical codes to fully exploit the capabilities of modern computing resources.
3. Model testing against quantitatively well-described volcanic events (*case-studies*) and large-scale laboratory experiments (e.g., to test the multiscale and multidimensional nature of the phenomena).
4. Develop numerical simulation scenarios of the phenomena through real-time assimilation of observational data to forecast the process evolution.
5. Develop simplified integral models, calibrated with accurate multiphase and multidimensional models, for probabilistic mapping and uncertainty quantification of volcano dynamics and hazards.

10.8   Exercises

10.8.1   Ex. 1: Independent Eruption Source Parameters of Volcanic Plumes

There are four independent variables defining the fluid and thermodynamic properties of the fragmented magma at the eruptive vent. A possible choice is to provide the mass eruption rate Q, the mixture temperature T, the erupted gas mass fraction Y_g, and the magma overpressure at the vent p_v. Then, by assuming an adiabatic transformation, it is possible to estimate the velocity, U_0, after the jet decompression with this equation,

$$U_0 = U_v + \frac{p_v - p_0}{\rho_v U_v},$$

where U_v is the velocity at the vent, equal to the speed of sound of the mixture, i.e.,

$$U_v = \sqrt{\gamma_m p_v / \rho_v},$$

while γ_m is the ratio of specific heats of the mixture and ρ_v is the density of the mixture before decompression (see [44, 64, 99]).

1. Given $Q = 10^7$ kg/s, $T = 1200$ K, $Y_g = 0.04$, and $p_v = 10^6$ Pa, calculate: U_0 and U_v; the jet radius before and after the decompression R_v and R_0; the Mach number after decompression; the density of the mixture before ρ_v and after

ρ_0 decompression; the Richardson and the Reynolds number after decompression; and the volumetric concentration ε_s and the bulk density ϱ_s of solids.

In the calculations, assume the International Standard Atmosphere at the sea level conditions; use constant values for the gas and air gas constants $R_g = 461$ J/kg/K and $R_{atm} = 287$ J/kg/K; for the specific heats at constant pressure $C_g = 1952$ J/kg/K, $C_{atm} = 998$ J/kg/K, $C_s = 1250$ J/kg/K; for the density of the solids $\rho_s = 2350$ kg/m^3.

10.8.2   Ex. 2: Calculation of Stokes Time

1. Using the expression for the gas/particle drag coefficient of Table 2.5 of Chap. 2, demonstrate that, in the dilute regime and for coarse particles, with $Re > 1000$, the Stokes time τ_s can be computed analytically as,

$$\tau_s = \sqrt{\frac{4}{3} \frac{\rho_s d_p}{C_D \rho_g g}}$$

2. Assuming that particles have perfectly spherical shape, compute the Stokes time τ_s for every particle class of Table 10.1.
3. Assuming, for a Plinian eruption plume, velocity U_v and radius R_v, and Strouhal number Str $= 0.3$, the characteristic large-eddy turnover time scale can be estimated as [44, 50],

$$\tau_f = \text{Str} \frac{R_v}{U_v}$$

Compute the particle Stokes number $St = \frac{\tau_s}{\tau_f}$ for every particle class of Table 10.1, with velocity U_v and radius R_v calculated in Exercise 1.

10.8.3   Ex. 3: Sauter Diameter of a Grain-Size Distribution

Many volcanic processes involve pyroclastic mixtures of particles of different sizes (polydisperse mixtures, see Sect. 10.2.3). However, for modeling and computational requirements, it is often necessary or convenient to describe a polydisperse mixture with a single average grain-size class. Given a particle size distribution, the average particle diameter can be calculated in different ways.

The Sauter mean diameter, s, is defined as the particle diameter whose surface-to-volume ratio is equivalent to the (arithmetic) mean of the distribution. Analytically, it is defined by the following equation,

$$s = \frac{\overline{d^3}}{\overline{d^2}}$$

where $\overline{d^n}$ represents the n-th moment of the particle size distribution, i.e.,

$$\overline{d^n} = \int_0^\infty x^n n(x) dx$$

and $n(x)$ = number of particles with a diameter between x and $(x + dx)$.

For a discrete distribution, it can thus be written,

$$s = \frac{\sum_{i=1}^\infty n_i d_i^3}{\sum_{i=1}^\infty n_i d_i^2}$$

with n_i defined as the number of particles in the range between x_{i-1} and x_i.

1. Grain size distributions in volcanology are usually expressed as a function of the mass fraction of the single granulometric class $w_i(x_i)$. Assuming that the density of the particles is the same for all granulometric classes, show that Sauter's diameter for a discrete mass distribution of particles is expressed by the following relation,

$$\frac{1}{s} = \sum_{i=1}^\infty \frac{w_i}{d_i}$$

2. Compute the Sauter mean diameter of the grain size distribution of Table 10.1.
3. Compute the weighted mean diameter $d_w = \sum_{i=1}^\infty w_i d_i$ and compare it with the Sauter diameter computed in part (2) above.
4. Compute the Stokes time, τ_s, for the Sauter mean particle and compare it with that of the single particle classes computed in Exercise 2.

10.8.4   Ex. 4: Numerical Simulation of Phreatic Explosions

Phreatic explosions are typically described as a sudden explosion of a confined and high-pressure geothermal system that abruptly becomes mechanically unstable with respect to the overlying cover due to the occurrence of an external trigger such as an earthquake, slope instability, self-sealing processes, or temperature increase.

A simplified description of such dynamics in ideal conditions was obtained by using a multiphase flow model. Neri et al. [100] adopted a Eulerian-Eulerian three-phase formulation (two particulate phases of different sizes and one carrier gaseous two-component phase) to investigate, in axisymmetric conditions, the first tens of seconds of the explosion and its hazard. In particular, they investigated the effect of a limited number of parameters characterizing the explosions such as its magnitude (in terms of mass involved) and energy.

1. By adopting the same geometrical conditions and physical properties of the gas and particulate phases of Neri et al. [100], perform numerical simulations to:

   (a) Reproduce the results of Neri et al. [100].
   (b) Investigate the effects of different cylindrical geometries of the exploding pocket of steam and particles at constant magnitude.
   (c) Investigate the effects of different particle porosities of the exploding pocket of ground at constant specific energy.
   (d) Investigate the effects of different two particle size distributions at constant Sauter diameter.

Nomenclature

a	Fluid acceleration
B	Rate of birth of particles
C	Specific heat
C_D	Drag coefficient
c	Speed of sound
D	Rate of death of particles
d	Particle diameter
$\overline{d^n}$	n-th moment of the particle size distribution
F	Body force
F_B	Basset force
F_D	Drag force
F_G	Gravitational force
F_P	Pressure gradient force
F_{VM}	Virtual mass force
Fr	Froude number
$f(V)$	Particle-size distribution
G	Particle growth
g	Gravity acceleration
K	Particle aggregation kernel
k	Thermal conductivity
h	Enthalpy
I	Identity matrix
J	Number of phases and components
j	Index of phases and components
$M^{(i)}$	i-th moment
m	Mass
$n(x)$	Number of particles with a diameter between x and $(x + dx)$

Nu	Nusselt number
P	Pressure
p	Probability of fall-out of individual particles
Q	Mass eruption rate
$\boldsymbol{q}$	Heat flux
R	Radius of vent or plume
Re	Reynolds number
s	Curvilinear coordinate or Sauter diameter
St	Kinematic Stokes number
St_c	Collisional Stokes number
St_T	Thermal Stokes number
Str	Strouhal number
$\boldsymbol{T}$	Stress tensor
T	Temperature
t	Time
$\boldsymbol{U}$	Velocity vector
U	Velocity component along the centerline of the plume
$\vec{\boldsymbol{u}}$	Mean velocity vector
V	Particle volume
$\vec{\boldsymbol{v}}$	Mean velocity vector
$\vec{\boldsymbol{w}}$	Mean particle settling vector
$w_i(x_i)$	Mass fraction of the single granulometric class
w_S	Particle settling velocity
$\boldsymbol{x}$	Position vector
Y_{ik}	Mass fraction of component k in phase i

Subscripts

atm	Atmospheric
c	Continuous or carrier phase
f	Fluid phase
g	Gas phase
m	Mixture
p	Particle
s	Solid phase
v	Vent
0	Jet conditions after flow decompression

Greek Symbols

ε_i	Volume fraction of phase i
γ	Shear rate or ratio of specific heats
μ	Viscosity
λ	Particle volume
ρ	Microscopic density
ϱ	Bulk density
θ	Particle granular temperature

τ_C	Collision characteristic time
τ_F	Fluid characteristic time
τ_{Ma}	Mach disk formation time
τ_S	Gas/particle relaxation time
τ_T	Thermal relaxation time
φ	Krumbein scale
Φ	Field
ω	Weight

References

1. Sigurdsson H, Cashdollar S, Sparks RSJ (1982) The eruption of Vesuvius in A.D. 79: reconstruction from historical and volcanological evidence. Amer J Archaeol 86:39–51
2. Cioni R, Marianelli P, Sbrana A (1992) Dynamics of the A.D. 79 eruption: stratigraphic, sedimentological and geochemical data on the successions from the Somma-Vesuvius southern and eastern sectors. Acta Vulcanologica 2:109–123
3. Lacroix A (1904) La Montagne Pelée et ses eruptions. Paris, Masson, pp. 1–650
4. Tanguy JC (1994) The 1902–1905 eruptions of Montagne Pelée, Martinique: anatomy and retrospection. J Volcanol Geotherm Res 60(2): 87–107
5. Lipman PW, Mullineaux DR (eds) (1981) The 1980 eruptions of Mount St. Helens, Washington. US Geological Survey, p 1250
6. Newhall CG (2000) Mount St Helens, master teacher. Science 288:1181–1183
7. Wadge G, Robertson RE, Voight B (eds) (2014) The eruption of Soufriere Hill Volcano, Montserrat from 2000 to 2010. Memoir 39, Geological Society, London, 2014
8. Cole P, Neri A, Baxter PJ (2015) Pyroclastic density current hazard. In: Sigurdsson H 9 (chief ed) Encyclopedia of volcanoes. Academic Press Publisher, Chapter 54
9. Nakada S, Eichelberger JC, Shimizu H (1999) Special Issue Unzen Volcano. J Volcanol Geoth Res 89(1)
10. Newhall CG, Punongbayan R (eds) (1996) Fire and mud: eruptions and lahars of Mount Pinatubo, Philippines. Philippine Institute of Volcanology and Seismology, Quezon City, p 1126
11. Gudmundsson M et al. (2012) Ash generation and distribution from the April-May 2010 eruption of Eyjafjallajokull, Iceland, Scientific Reports 2(572)
12. Komorowski JC, Jenkins S, Baxter PJ, Picquout A, Lavigne F, Charbonnier S, Gertisser R, Preece K, Cholik N, Budi-Santoso A, Surono (2013) Paroxysmal dome explosion during the Merapi 2010 eruption: processes and facies relationships of associated high-energy pyroclastic density currents. J Volcanol Geotherm Res doi:https://doi.org/10.1016/j.jvolgeores.2013.01.007, 2013
13. CONRED, National Coordinator for Disaster Reduction. Situation Report No. 1 (2018). Internal Report, available online: https://reliefweb.int/sites/reliefweb.int/files/resources/2018-06-04%20GT%20SITREP%20-%20Volcanic%20Activity%20%28ENG%29.pdf
14. Cas RAF, Wright JV (1987) Volcanic successions, modern and ancient: a geological approach to processes, products, and successions. Allen & Unwin
15. Druitt TH (1998) Pyroclastic density currents. In: Gilbert JS, Sparks RSJ (eds) The physics of explosive volcanic eruptions. Geological Society Special Publications no. 145. Geological Society, London, 21, 145–182
16. Hayashi JN, Self S (1992) A comparison of pyroclastic flow and debris avalanche mobility. J Geophys Res 97(B6): 9063–9071
17. Yamamoto T, Takarada S, Suto S (1993) Pyroclastic flows from the eruption of Unzen Volcano, Jpn, Bull Volcanol 55:166–175

18. Calder ES, Cole PD, Dade WB, Druitt TH, Hoblitt RP, Huppert HE, Ritchie L, Sparks RSJ, Young SR (1999) Mobility of pyroclastic flows and surges at the Soufriere Hills volcano, Montserrat. Geophys Res Lett 26(5):537–540
19. Widiwijayanti C, Voight B, Hidayat D, Schilling SP (2008) Objective rapid delineation of areas at risk from block-and-ash pyroclastic flows and surges. Bull Volcanol. doi: https://doi.org/10.1007/s00445-008-0254-6
20. Wilson L (1976) Explosive volcanic eruptions III: Plinian eruption columns. Geophys J Roy Astron Soc 45:543–556
21. Sparks RSJ, Wilson L, Hulme G (1978) Theoretical modeling of the generation, movement, and emplacement of pyroclastic flows by column collapse, J Geophys Res 83(B4):1727–1739
22. Wilson L, Sparks RSJ, Walker, GPL (1980) Explosive volcanic eruptions. IV. The control of magma properties and conduit geometry on eruption column behaviour. Geophys J R Astron Soc 63:117–148
23. Textor C, Graf H-F, Longo A, Neri A, Esposti Ongaro T, Papale P, Timmreck C, Ernst GGJ (2005) Numerical simulation of explosive volcanic eruptions from the conduit flow to global atmospheric scales. Annals of Geophysics 48(4/5):817–842
24. Sigurdsson H, Houghton B, Rymer H, Stix J, McNutt S (eds) (2015) The encyclopedia of volcanoes. ISBN: 9780123859389, Elsevier Inc., Academic Press
25. Wohletz KH, McGetchin TR, Sandford II MT, Jones EM (1984) Hydrodynamic aspects of caldera-forming eruptions: Numerical models. J Geophys Res 89:8269–8285
26. Valentine GA, Wohletz KH (1989) Numerical models of Plinian eruption columns and pyroclastic flows. J Geophys Res 94(B2):1867–1887
27. Dobran F, Neri A, Macedonio G (1993) Numerical simulation of collapsing volcanic columns. J Geophys Res 98:4231–4259
28. Neri A, Dobran F (1994) Influence of eruption parameters on the thermofluid-dynamics of collapsing volcanic columns. J Geophys Res 99:11833–11857
29. Neri A, Macedonio G (1996) Numerical simulation of collapsing volcanic columns with particles of two sizes. J Geophys Res 101:8153–8174
30. Neri A, Esposti Ongaro T, Macedonio G, Gidaspow D (2003) Multiparticle simulation of collapsing volcanic columns and pyroclastic flows. J Geophys Res 108(B4):2202
31. Neri A, Esposti Ongaro T, Menconi G, de' Michieli Vitturi M, Cavazzoni C, Erbacci G, Baxter PJ (2007) 4D Simulation of explosive eruption dynamics at Vesuvius. Geophys Res Lett 34. doi: https://doi.org/10.1029/2006GL028597
32. Dartevelle S (2004) Numerical modeling of geophysical granular flows: 1. A comprehensive approach to granular rheologies and geophysical multiphase flows. Geochem Geophy Geosy 5 Q08003. doi:https://doi.org/10.1029/2003GC000636
33. Di Muro A, Neri A, Rosi M (2004) Contemporaneous convective and collapsing eruptive dynamics: the transitional regime of explosive eruptions, Geophys Res Lett 31. doi:https://doi.org/10.1029/2004GRL019709
34. Dufek J, Bergantz G (2007) Suspended load and bed-load transport of particle-laden gravity currents: the role of particle–bed interaction. Theor Comp Fluid Dyn. 21:119–145
35. Dufek J, Bergantz GW (2007) Dynamics and deposits generated by the Kos Plateau Tuff eruption: controls of basal particle loss on pyroclastic flow transport, Geochem Geophys Geosyst 8, Q12007. doi:https://doi.org/10.1029/2007GC001741
36. Esposti Ongaro T, Cavazzoni C, Erbacci G, Neri A, Salvetti MV (2007) A parallel multiphase flow code for the 3-D simulation of explosive volcanic eruptions. Parallel Comput 33:541–560
37. Esposti Ongaro T, Neri A, Menconi G, de' Michieli Vitturi M, Marianelli P, Cavazzoni C, Erbacci G, Baxter PJ (2008) Transient 3D numerical simulations of column collapse and pyroclastic density current scenarios at Vesuvius. J Volcanol Geotherm Res 178(3):378–396. doi:https://doi.org/10.1016/j.jvolgeores.2008.06.036
38. Esposti Ongaro T, Widiwijayanti C, Clarke AB, Voight B, Neri A (2011) Multiphase-flow numerical modelling of the May 18, 1980 lateral blast at Mount St. Helens (USA), Geology 39 (6):535–538. doi:https://doi.org/10.1130/G31865.1

39. Esposti Ongaro T, Clarke AB, Voight B, Neri A, Widiwijayanti C (2012) Multiphase-flow dynamics of pyroclastic density currents during the May 18, 1980 lateral blast at Mount St. Helens (USA), J Geophys Res 117. doi:https://doi.org/10.1029/2011JB009081
40. Carcano S, Bonaventura L, Esposti Ongaro T, Neri A (2013) A semi-implicit, second-order accurate numerical model for multiphase underexpanded volcanic jets. Geosci Model Dev 6:1905–1924
41. Suzuki YJ, Costa A, Cerminara M, Esposti Ongaro T, Herzog M, Van Eaton A, Denby LC (2016) Inter-comparison of three-dimensional models of volcanic plumes. J Volcanol Geotherm Res
42. de' Michieli Vitturi M, Neri A, Barsotti S (2015) PLUME-MoM 1.0: A new integral model of volcanic plumes based on the method of moments. Geosci Model Dev 8.8, 2447–2463
43. Cerminara M, Esposti Ongaro T, Berselli LC (2016) ASHEE-1.0: a compressible, equilibrium–Eulerian model for volcanic ash plumes. Geosci Model Dev 9(2):697–730
44. Cerminara M (2016) Modeling dispersed gas-particle turbulence in volcanic ash plumes. PhD thesis, Scuola Normale Superiore
45. Cerminara M, Esposti Ongaro T, Neri A (2016) Large eddy simulation of gas–particle kinematic decoupling and turbulent entrainment in volcanic plumes. J Volcanol Geotherm Res 326:143–171
46. de' Michieli Vitturi M, Neri A, Esposti Ongaro T, Lo Savio S, Boschi E (2010) Lagrangian modelling of large volcanic particles: application to Vulcanian explosions. J Geophys Res 115. doi:https://doi.org/10.1029/2009JB007111
47. Costa A, Suzuki YJ, Cerminara M, Devenish BJ, Esposti Ongaro T, Herzog M, Van Eaton AR, Denby LC, Bursik M, de' Michieli Vitturi M, Engwell S, Neri A, Barsotti S, Folch A, Macedonio G, Girault F, Carazzo G, Tait S, Kaminski E, Mastin LG, Woodhouse MJ, Phillips JC, Hogg AJ, Degruyter W, Bonadonna C (2016) Results of the eruption column model inter-comparison study. J Volcanol Geotherm Res. doi:https://doi.org/10.1016/j.jvolgeores.2016.01.017
48. Mastin LG, Guffanti M, Servranckx R, Webley PW, Barsotti S, Dean KG, Durant AK, Ewert JW, Neri A, Rose WI, Schneider DJ, Siebert L, Stunder BJ, Swanson G, Tupper A, Volentik A, Waythomas CF (2009) A multidisciplinary effort to assign realistic source parameters to models of volcanic ash-cloud transport and dispersion during eruptions, J Volcanol Geotherm Res 186(1-2):10-21. doi: https://doi.org/10.1016/jvolgeores.2009.01.008
49. Neri A, Bevilacqua A, Esposti Ongaro T, Isaia R, Aspinall WP, Bisson M, Flandoli F, Baxter PJ, Bertagnini A, Iannuzzi E, Orsucci S, Pistolesi M, Rosi M, Vitale S (2015) Quantifying volcanic hazard at Campi Flegrei caldera (Italy) with uncertainty assessment: II. Pyroclastic density current invasion maps. J Geophys Res 120. doi:https://doi.org/10.1002/2014JB011776
50. Esposti Ongaro T, Orsucci S, Cornolti F (2016) A fast, calibrated model for pyroclastic density currents kinematics and hazard. J Volcanol Geotherm Res 327:257–272. doi: https://doi.org/10.1016/j.jvolgeores.2016.08.002
51. Burgisser A, Bergantz GW (2002) Reconciling pyroclastic flow and surge: the multiphase physics of pyroclastic density currents. Earth Planet Sc Lett 202:405–418
52. Gidaspow D (1994) Multiphase flow and fluidization, continuum and kinetic theory description. Academic Press Inc., San Diego, California
53. Balachandar S, Eaton JK (2010) Turbulent dispersed multiphase flow. Annu Rev Fluid Mech 42(1):111–133
54. Sparks RSJ, Bursik MI, Carey SN, Gilbert JS, Glaze LS, Sigurdsson H, Woods AW (1997) Volcanic plumes. John Wiley, New York
55. Carey S, Sigurdsson H (1982) Influence of particle aggregation on deposition of distal tephra from the May 18, 1980, eruption of Mount St. Helens volcano. J Geophys Res 87, 7061–7072, https://doi.org/10.1029/JB087iB08p07061
56. Dobran F (1992) Nonequilibrium flow in volcanic conduits and application to the eruptions of Mt. St. Helens on May 18, 1980, and Vesuvius in AD 79. J Volcanol Geotherm Res 49:285–311

57. Papale P, Neri A, Macedonio G (1998) The role of magma composition and water content in explosive eruptions. I. Magma ascent dynamics. J Volcanol Geotherm Res 87:75–93
58. Lewis C, Carlson D (1964) Normal shock location in underexpanded gas and gas particle jets. AIAA J 2:776–777 https://doi.org/10.2514/3.2409
59. Kieffer SW (1984) Factors governing the structure of volcanic jets. In: Boyd FR (ed) Explosive volcanism: inception, evolution, and hazards. National Academy Press, Washington, pp 143–157
60. Woods AW, Bower SM (1995) The decompression of volcanic jets in a crater during explosive volcanic eruptions. Earth Planet Sc Lett 131:189–205
61. Ogden DE, Wohletz KH, Glatzmaier GA, Brodsky EE (2008) Numerical simulations of volcanic jets: importance of vent overpressure, J Geophys Res 113(2):B02204 doi:https://doi.org/10.1029/2007JB005133
62. Carcano S, Esposti Ongaro T, Bonaventura L, Neri A (2014) Influence of grain-size distribution on the dynamics of under-expanded volcanic jets. J Volcanol Geotherm Res 285:60–80
63. Orescanin MM, Austin JM, and Kieffer SW (2010) Unsteady high-pressure flow experiments with applications to explosive volcanic eruptions. J Geophys Res 115(6): B06206. doi:https://doi.org/10.1029/2009JB006985
64. Marble FE (1970) Dynamics of dusty gases. Annu Rev Fluid Mech 2:397–446
65. Sommerfeld M (1994) The structure of particle-laden, underexpanded free jets. Shock Waves 3:299–311
66. Esposti Ongaro T, Clarke AB, Neri A, Voight B, Widiwijayanti C (2008) Fluid dynamics of the 1997 Boxing Day volcanic blast on Montserrat, West Indies. J Geophys Res 113. doi: https://doi.org/10.1029/2006JB004898
67. Branney MJ, Kokelaar BP (2002) Pyroclastic density currents and the sedimentation of ignimbrites. Mem Geol Soc 27(8)
68. Sweeney MR, Valentine GA (2017) Impact zone dynamics of dilute mono-and polydisperse jets and their implications for the initial conditions of pyroclastic density currents. Phys Fluids 29(9):093304
69. Gladstone C, Phillips JC, Sparks RSJ (1998) Experiments on bidisperse, constant-volume gravity currents: propagation and sediment deposition. Sedimentology 45(5):833–843
70. Necker F, Hartel C, Kleiser L, Meiburg E (2002) High-resolution simulations of particle-driven gravity currents. Int J Multiph Flow 28
71. Ishimine Y (2005) A numerical study of pyroclastic surges. J Volcanol Geotherm Res 139:33–57
72. Bursik MI, Woods AW (1996) The dynamics and thermodynamics of large ash flows. Bull Volcanol 58(2):175–193
73. Todesco M, Neri A, Esposti Ongaro T, Papale P, Rosi M (2006) Pyroclastic flow hazard in a caldera setting: application to Phlegrean Fields, G 3 - Geochemistry, Geophysics, Geosystems, 7, Q11003. doi:https://doi.org/10.1029/2006GC001314
74. Esposti Ongaro T, Barsotti S, Neri A, Salvetti MV (2011) Large-eddy simulation of pyroclastic density currents. In: Salvetti MV, Geurts BJ, Meyers J, Sagaut P (eds) Quality and reliability of large-eddy simulations II in series ERCOFTAC. Springer, 16, ISSN 1382-4309, pp 161–170
75. Hulburt HM, Katz S (1964) Some problems in particle technology: a statistical mechanical formulation. Chem Eng Sci 19:555–574
76. Marchisio DL, Vigil RD, Fox RO (2003) Quadrature method of moments for aggregation–breakage processes. J Colloid Interf Sci 258:322–334
77. Strumendo M, Arastoopour H (2009) Solution of bivariate population balance equations using the FCMOM. Ind Eng Chem Res 48(1):262–273
78. Strumendo M, Arastoopour H (2008) Solution of PBE by non-infinite size domain. Chem Eng Sci 63:2624–2640
79. Arastoopour H, Gidaspow D, Abbasi E (2017) Computational transport phenomena of fluid-particle systems, mechanical engineering series. ISBN 978-3-319-45488-7, Springer Press

80. Morton B, Taylor GI, Turner JS (1956) Turbulent gravitational convection from maintained and instantaneous source. Phil Trans Roy Soc Lond A 234:1–23
81. Wilson L, Walker GPL (1987) Explosive volcanic eruptions-VI. Ejecta dispersal in Plinian eruptions: the control of eruption conditions and atmospheric properties. Geophys J Int 89 (2):657–679
82. Woods AW, Bursik MI (1991) Particle fallout, thermal disequilibrium, and volcanic plumes. Bull Volcanol 53.7:559–570
83. Ernst GG, Sparks RSJ, Carey SN, Bursi MI (1996) Sedimentation from turbulent jets and plumes. J Geophys Res 101:5575–5589. doi:https://doi.org/10.1029/95JB01900
84. Bursik M, Sparks R, Gilbert J, Carey S (1992) Sedimentation of tephra by volcanic plumes: I. Theory and its comparison with a study of the Fogo A plinian deposit, Sao Miguel (Azores). Bull Volcanol 54:329–344
85. Von Smoluchowski M (1916) Three lectures on diffusion, Brownian motion, and coagulation of colloidal particles. Z Phys 17:557–585
86. Kraft M (2005) Modelling of particulate processes. Kona 23(23):18–35
87. Hazewinkel M (2001) Moment. Encyclopedia of Mathematics, Springer, the Netherlands
88. McGraw R (1997) Description of aerosol dynamics by the quadrature method of moments. Aerosol Sci Technol 27:255–265. doi:https://doi.org/10.1080/02786829708965471
89. Fox RO (2014) Quadrature-based moment methods for polydisperse multiphase flows. In: Stochastic methods in fluid mechanics. Springer, Vienna, pp 87–136
90. Fréret L, Laurent F, de Chaisemartin S, Kah D, Fox RO, Vedula P, Reveillon J, Thomine O, Massot M (2008) Turbulent combustion of polydisperse evaporating sprays with droplet crossing: Eulerian modeling of collisions at finite Knudsen and validation. In: Proceedings of the summer program, pp 277–288
91. Maxey MR (1987) The gravitational settling of aerosol particles in homogeneous turbulence and random flow fields. J Fluid Mech 174:441
92. Ferry J, Balachandar S (2001) A fast Eulerian method for disperse two-phase flow. Int J Multiphase Flow 27:1199–1226
93. Ferry J, Balachandar S (2005) Equilibrium Eulerian approach for predicting the thermal field of a dispersion of small particles. Int J Heat Mass Transf 48(3):681–689
94. Devenish BJ, Cerminara M (2018) The transition from eruption column to umbrella cloud. J Geophys Res Solid Earth 123(12):10418–10430
95. Maxey MR, Riley JJ (1983) Equation of motion for a small rigid sphere in a nonuniform flow. Phys Fluids 26(4):883–889
96. Crowe C, Sommerfeld M, Tsuji Y (1998) Multiphase flows with droplets and particles. CRC Press LLC, Boca Raton, Florida
97. Unluturk SK, Arastoopour H (2003) Steady-state simulations of liquid-particle food flow in a vertical pipe. Ind Eng Chem Res 42:3845–3850
98. Clarke A, Voight B, Neri A, Macedonio G (2002) Transient dynamics of Vulcanian explosions and column collapse. Nature 415:897–901
99. Trolese M, Cerminara M, Esposti Ongaro T, Giordano G (2019) The footprint of column collapse regimes on pyroclastic flow temperatures and plume heights. Nat Commun 10(2476). doi:https://doi.org/10.1038/s41467-019-10337-3
100. Neri A, Macedonio G, Gidaspow D (1999) Phreatic explosion hazard assessment by numerical simulation. Phys Chem Earth (A) 24:989–995

Chapter 11
Multiphase Flow Modeling of Wind Turbine Performance Under Rainy Conditions

Hamid Arastoopour, Dimitri Gidaspow, and Robert W. Lyczkowski

11.1 Introduction

Climate change due to utilization of carbon-based fuels and continuous depletion of fossil fuel reserves has been the main motivation behind a significant demand for renewable energy. Wind energy is one of the most promising sources of renewable energy because of its low negative environmental impacts, relatively low cost, and no water consumption requirements. The performance of a typical wind turbine is continuously influenced by the surrounding environmental conditions. Performance losses due to the sand-laden winds and dust accumulation in arid regions, or insect debris build-up on the blades of the wind turbine in tropical regions have been reported by many researchers. Power losses of about 25% on wind farms in California due to accumulated insect debris on the leading edge of the wind turbine blades and performance losses of 50% due to dust accumulation and increased roughness of the blade surface have been reported [1–3].

Moreover, other meteorological phenomena such as air moisture, rain, snow, and ice formation on the blades also significantly influence the performance of the wind turbine due to the impact on the geometry of the blades and constant or periodic load increase on the blades [4–6]. Investigating the effect of these phenomena is necessary to improve the design and performance of the wind turbines. There is therefore a need for comprehensive study of the performance of wind turbines under rainy conditions and reduction of the turbine efficiency under such conditions. Computational fluid dynamics (CFD) is capable of providing a reliable tool to simulate and analyze the physical phenomena involving fluids without the need for costly experimentation [7–9]. In the literature, the CFD approach to simulate the effect of rain in the form of an increase in air density and momentum of the droplets upon their collision with the surface of the blades has been studied [10, 11]. The formation of the water layer on the surface of the airfoils was not considered in these studies, while experimental studies showed formation of a water layer after hitting the surface [12, 13]. Cai et al. [14] were the first to consider using CFD models to

© Springer Nature Switzerland AG 2022
H. Arastoopour et al., *Transport Phenomena in Multiphase Systems*, Mechanical Engineering Series, https://doi.org/10.1007/978-3-030-68578-2_11

simulate the formation of a water layer due to rain droplets on the surface of a two-dimensional airfoil. They included in their model a continuous addition of rain droplet mass on the water layer on the surface of their two-dimensional airfoil by coupling the Lagrangian discrete phase model (DPM) for rain droplets with the Eulerian volume of fluid (VOF) model for the water layer flow on the surface of an airfoil. Cohan and Arastoopour [15] further improved the Cai et al. [14] model by adding the surface tension and the rain droplet total momentum in addition to droplet mass to simulate the effect of lift and drag forces exerted on an airfoil due to rain. Later, Arastoopour and Cohan [16] simulated the performance of a three-dimensional National Renewable Energy Laboratory (NREL) phase VI horizontal-axis wind turbine [17] for single-phase cases at various wind speeds, and the multiphase case (under rainy conditions) using the multiphase flow approach. The focus in this chapter is on the CFD simulation of an airfoil and a wind turbine at different wind speeds and directions, and the impact of rain on the performance of the wind turbines and airfoils [14–16].

11.2 Numerical Modeling

The governing equations of two multiphase models coupled to simulate the formation of a water layer from rain droplets, as well as the details of the implementation of the coupling are presented in this section. The computational domain was therefore divided into two subdomains. The first subdomain was the Lagrangian rain droplets and the second subdomain was the Eulerian free-stream air and water layer formed by droplets on the airfoil or turbine blades surfaces (see Fig. 11.1).

Fig. 11.1 Demonstration of rain droplets (DPM modeling) and continuous water layer on the surface of turbine blades (VOF modeling) [18]

For the first subdomain, DPM was used to simulate the motion of rain droplets before they enter the water layer. For the second subdomain, the Eulerian VOF model was used to simulate the air that carries the rain droplets and to capture the water layer formation and the interface between the two phases. The two subdomains were then coupled to carry out the simulation.

The Lagrangian DPM model was used to simulate the motion of rain droplets before they enter the water film; the fluid phase was treated as a continuum, while the dispersed phase was solved by tracking the droplets through the calculated flow field. The flow field was obtained using the Eulerian model. In these simulations, the VOF model was used to calculate the flow field. Additionally, to render the simulation computationally affordable, the parcel approach [19] was used to calculate droplet trajectories. In this approach, each parcel represents several droplets, all having the same diameter, density, and velocity. If the mass of the parcel is needed in any multiphase calculation, however, it should be manually set equal to the mass of an individual droplet times the number of droplets that the parcel is representing. The momentum equation for the droplets in the DPM written in a Lagrangian reference frame may be written,

$$\frac{du_{Pi}}{dt} = F_D(u_i - u_{Pi}) + g_i \left(\frac{\rho_D - \rho}{\rho_D} \right) \tag{11.1}$$

where u_{Pi} and u_i are parcel and fluid phase velocity in the i-direction, respectively. F_D is the drag coefficient for which the spherical drag expression was used, and g_i is the gravitational acceleration. ρ_D and ρ are densities of the droplets and the fluid phase, respectively.

The VOF model is a surface-tracking technique where water and air share a single set of momentum equations, and the volume fraction of air and water in each computational cell is tracked throughout the domain. The fields for all variables and properties are shared by the air and water phases. Properties appearing in the transport equations of the VOF model are determined therefore by the presence of the component phases in each control volume and represent volume-averaged values. The tracking of the interface between air and water is accomplished by the solution of a continuity equation for the volume fraction of the phases using the geometric reconstruction scheme. In general, for the *qth* phase, this continuity equation has the following form,

$$\frac{\partial}{\partial t} (\alpha_q \rho_q) + \nabla \cdot (\alpha_q \rho_q \boldsymbol{u}_q) = S_{\alpha_q} \tag{11.2}$$

where α_q, ρ_q and $\boldsymbol{u}_q$ are volume fraction, density, and velocity of the *qth* phase, respectively. S_{α_q} is the mass source term for the *qth* phase. The volume fraction equation will not be solved for the primary phase; the primary-phase volume fraction will be computed based on the constraint that, in each control volume, the volume fractions of all phases sum to unity.

A single momentum equation is solved throughout the domain for all phases, and the resulting velocity field is shared among the phases. The Reynolds averaged momentum equation of the VOF model is,

$$\frac{\partial \rho u_i}{\partial t} + \frac{\partial}{\partial x_j}\left(\rho u_j u_i\right) = -\frac{\partial p}{\partial x_i} + \frac{\partial \tau_{ji}}{\partial x_j} + F_i \tag{11.3}$$

$$\tau_{ji} = \mu\left(\frac{\partial u_i}{\partial x_j} + \frac{\partial u_j}{\partial x_i}\right) - \overline{\rho u'_j u'_i} \tag{11.4}$$

where F_i is the momentum source term and $-\overline{\rho u'_j u'_i}$ is the Reynolds stress term for which the shear-stress transport (SST) $k - \omega$ turbulent model was used. This model is proven to work well for wind turbine applications [20]. The turbulent kinetic energy, k, and the specific dissipation rate, ω, are obtained from the following transport equation [21],

$$\frac{\partial}{\partial t}(\rho k) + \frac{\partial}{\partial x_i}(\rho k u_i) = \frac{\partial}{\partial x_j}\left(\Gamma_k \frac{\partial k}{\partial x_j}\right) + \widetilde{G_k} - Y_k \tag{11.5}$$

$$\frac{\partial}{\partial t}(\rho \omega) + \frac{\partial}{\partial x_i}(\rho \omega u_i) = \frac{\partial}{\partial x_j}\left(\Gamma_\omega \frac{\partial \omega}{\partial x_j}\right) + G_\omega - Y_\omega + D_\omega \tag{11.6}$$

In these equations, $\widetilde{G_k}$ represents the generation of turbulence kinetic energy due to mean velocity gradients. G_ω represents the generation of ω, while Γ_k and Γ_ω are the effective diffusivity of k and ω, respectively. Y_k and Y_ω denote the dissipation of k and ω due to turbulence. D_ω is the cross-diffusion term.

The coupling between the DPM and VOF models was implemented by adding two source terms to the VOF model's continuity and momentum equations using ANSYS Fluent's User Define Functions (UDFs) [19], as each rain droplet hit the blade's surface or entered the water layer around it. At this moment, the droplet was removed from the DPM computational domain and its effect was accounted for by adding an instantaneous source term to both the VOF volume fraction equation of the liquid phase and the momentum equation of the mixture at the beginning of that time step. These source terms were then set to zero at the end of the time step. The magnitudes of these source terms were decided by the impacting particle properties from the DPM model.

The mass source term for the liquid water phase is,

$$S_\alpha = \frac{m_P}{V\Delta t} \tag{11.7}$$

The momentum source term of the mixture is,

$$F_i = \frac{m_P u_{ri}}{V \Delta t} \tag{11.8}$$

where

$$u_{ri} = u_{Pi} - u_{fi} \tag{11.9}$$

m_P is the mass of the parcel, which is equal to the sum of the masses of all of the particles it represents. u_{ri} is the droplet's relative velocity to the continuous fluid phase in the i-direction, V is the volume of the computational cell that the air/water interface lays in, and Δt is the time-step size of the continuous phase solver. The surface tension model used in this research is the continuum surface force (CSF) model proposed by Brackbill et al. [22]. In this model, the addition of surface tension to the VOF calculation results in additional tangential stresses implemented as source term in the VOF momentum equation. This surface tension force has the following form for the two-phase problem in this study,

$$F_{st} = \sigma \frac{\rho(\nabla \cdot \hat{n}_2)\nabla \alpha_2}{\frac{1}{2}(\rho_1 + \rho_2)} \tag{11.10}$$

The unit normal, $\hat{n}_2$, is defined,

$$\hat{n}_2 = \frac{\nabla \alpha_2}{|\nabla \alpha_2|} \tag{11.11}$$

where subscripts 1 and 2 refer to air and water, respectively. σ is the surface tension, and ρ is the volume-averaged density. For cells adjacent to a boundary cell, the contact angle that the fluid is assumed to make with the wall at the boundary cell is used to calculate $\nabla \alpha_2$.

11.3   Numerical Simulation

The objective is to simulate the performance of a three-dimensional horizontal-axis wind turbine under dry (single-phase flow) and rainy conditions (two-phase flow). To achieve this, the NREL phase VI horizontal-axis turbine (whose blade profile is an S809 airfoil) was used in the simulation. To obtain a better understanding of the effect of rain and its accumulation on the turbine blades, two-dimensional simulation of air and water flow on the performance of an S809 airfoil was first considered; the model was then applied to a three-dimensional horizontal axis wind turbine.

For simulation of an S809 airfoil, the computational domain was extended 10 and 13 chord lengths from the leading and trailing edges of the 1 m long airfoil with 220,000 computational cells with very small cells on the surface of the airfoil. This choice of the grid size guaranteed grid-independent simulations. The incoming air has a velocity of 29.2 m/s (Re $= 2.0 \times 10^6$). The number of rain droplets in each parcel was chosen such that the desired rainfall rate was achieved [14]. A uniform droplet of 2 mm with a terminal velocity of 6.01 m/s using 350 parcels was injected at the time step of 10^{-4} s.

For simulation of the NREL phase VI horizontal-axis two-bladed 10.058 m long (with an S809 airfoil profile) turbine [23], a cylindrical computational domain with radius of 25 m was used. The blade was located 15 m from the inlet and 60 m from the outlet. A cylinder with 1.2 m radius was also placed around the blades to be used in mesh generation for creating a finer mesh close to the blade surface. The experiments were conducted in a wind tunnel located at the NASA Ames Research Center at Moffett Field, California, at air velocity ranging from 5.00 to 25.00 m/s and rotor rotational speed of about 72 rpm.

A mesh system was then created using the Gambit program incorporated in the ANSYS Fluent computer software [19] on the computational domain having a total of 12 million cells. The first layer thickness was chosen such that it meets the SST k-ω turbulent model requirements for using wall function. To reduce the three-dimensional simulation computational time to a level that is reasonable and affordable, two additional assumptions were made: one-way coupling between the rain droplets and free-stream air. This means that air exerts force on the rain droplets and thereby affects the motion of the rain droplets. The rain droplets do not however have any impact on the air and do not affect its motion; second, only the mass of the rain droplets was added to the water layer formed around the wind turbine blades and neglected the momentum transfer from the rain droplets to the water layer.

11.4   Results and Discussion

11.4.1   S809 Airfoil

To obtain a better understanding of the water flow patterns and accumulation on the surface of the airfoils of the horizontal turbine, the airfoil was kept stationary while the angle of attack of air and rain on the S809 airfoil was changed. The two-dimensional, two-phase (air and water) airfoil simulations were run for five angles of attack at two different rainfall rates of 40 and 400 mm/h. The surface tension coefficient between water and air was assumed to be 0.072 N/m.

The water injection started at t $= 2.00$ s to make sure the single-phase simulation had reached steady state before rain injection. The simulations were run for 20 s for reaching a quasi-steady state under rainy conditions. Figure 11.2 shows the contours of the volume fraction of water for five angles of attack at two different rainfall rates of 40 and 400 mm/h.

The lift and drag coefficients of the airfoil are, respectively,

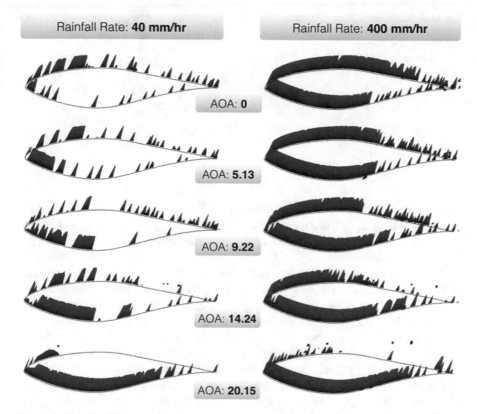

Fig. 11.2 Contours of the volume fraction of water on an S809 airfoil at two rainfall rates of 40 and 400 mm/h at five different angles of attack [18]

$$C_L = \frac{F_L}{\frac{1}{2}\rho U^2 A} \tag{11.12}$$

and

$$C_D = \frac{F_D}{\frac{1}{2}\rho U^2 A} \tag{11.13}$$

where F_L and F_D are the lift and drag forces. U is the magnitude of far field velocity, ρ is the density of air, and A is area. It is desirable to have higher lift force while keeping the drag force as low as possible for horizontal-axis turbine airfoils. The schematic diagram demonstrating angle of attack, and lift and drag forces on an airfoil is shown in Fig. 11.3.

The single-phase lift and drag coefficient calculated results were compared with experimental data in the literature [24]; however, no experimental data are available for rainy conditions. Figure 11.4 shows the time-averaged variation of lift and drag

Fig. 11.3 Schematic
diagram of angle of attack,
and lift and drag forces
exerted on an airfoil [18]

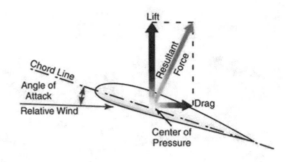

coefficients with respect to angle of attack at different rainfall rates. Figure 11.4 shows that the lift coefficient increases with increasing angle of attack up to a certain angle called the stall angle. As the angle of attack increases past the stall angle, the lift coefficient decreases.

This is due to the fact that, at stall angle, flow starts to separate from the airfoil as a result of the adverse pressure gradient on the upper surface of the airfoil at high angles of attack. It can be seen from Fig. 11.4 that, for both rainfall rates, the lift coefficient increases with angle of attack even past the single-phase stall angle [16]. This is in contrast with the single-phase case in which the lift coefficient decreases with increasing angle of attack past the stall angle. This probably could be due to a very complicated two-phase turbulent flow at the tail of the airfoil, caused by the pressure difference between the lower and top surfaces of the airfoil as a result of different water distribution on these surfaces. This favorable increase in lift coefficient past the stall angle however is accompanied by a significant increase in the drag coefficient, and this increase in drag coefficient is greater at larger angles of attack. This means that rainfall has a considerable effect on the performance of airfoils and, in turn, affects the power generation by the horizontal turbines.

Figure 11.4 also shows that lift and drag coefficient values for the case with a 400 mm/h rainfall rate are closer to the values of single-phase simulations compared to the case with a 40 mm/h rainfall rate. The reason for this is the fact that, at a rainfall rate of 40 mm/h, only one of the upper or lower surfaces of the airfoil gets wet with water and therefore the pressure on that surface increases. At the higher rainfall rate of 400 mm/h, however, both the upper and lower surfaces of the airfoil get wet with water and the increases in pressure at both surfaces offset each other (see Fig. 11.2). That is why, at the higher rainfall rate, the numerical values of the lift and drag coefficients are closer to those of the single-phase simulation.

At low angle of attack (less than 15°), water can reach the upper surface of the airfoil. At higher angles of attack (greater than 15°), the top surface does not get covered with water, even at the highest rainfall rate of 400 mm/h. The reasons for this are the sharp turn in the geometry of the airfoil at the leading edge and the gravitational force that prevents the water layer formed on the surface from reaching to the top surface, even though the airfoil has a wetting surface. By increasing the angle of attack from 0° to 20°, the surface that becomes wet with water gradually

Fig. 11.4 Time-averaged lift and drag coefficients for two rainfall rates of 40 and 400 mm/h at 0° to 20° angles of attack. (This figure was originally published in [16] and has been reused with permission)

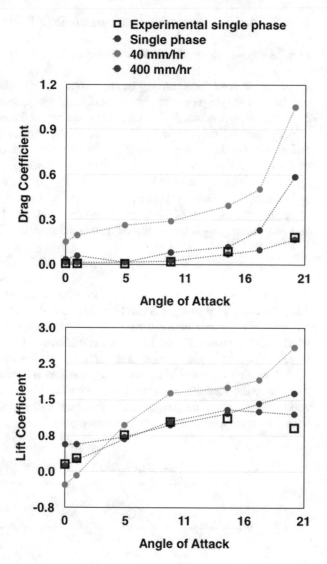

shifts from the upper surface of the airfoil to the lower surface. The numerical simulation of air and rain flow over the S809 airfoil provided us a clear understanding of the water layer formation, accumulation, and flow on the blades/airfoil of the horizontal wind turbines when blades are rotating and wind and rain attack the turbine at a constant angle.

11.4.2   NREL Phase VI Horizontal-Axis Wind Turbine

11.4.2.1   Single-Phase (Air) Flow Simulation

To avoid the need for a moving mesh, a rotating reference frame was used in the simulations. Single-phase simulations at different wind speeds ranging from 5.00 to 25.00 m/s were conducted. The boundary condition at the inlet was set to be uniform velocity, and the pressure outlet was used as the boundary condition at the outlet. The no-slip condition was enforced on the surface of the blades. A transient solver was used with a time step of 0.001 s. The wind turbine power output was calculated for different wind speeds and is shown in Fig. 11.5, along with the experimental data from the NREL report [17]. The power output of the wind turbine is calculated by calculating the moment of the forces acting on the blades, M, about the wind turbine axis and multiplying it by the blades' rotational velocity, Ω,

$$\text{power (W)} = \Omega \text{ (rad/s)} \times M \text{ (N.m)} \tag{11.14}$$

The numerical simulation showed very good agreement with the experimental data, especially at wind speeds less than 10 m/s and greater than 20 m/s. The reason for some discrepancy between the CFD calculations and the experimental data at wind speeds between 10 and 15 m/s is the difficulty of Reynolds-Averaged Navier-Stokes (RANS) turbulence models in capturing the initiation of the stall [25].

At low wind speeds of less than 10 m/s, the flow is attached. The CFD predictions therefore were accurate and agreed well with experimental data. At 20 and 25 m/s wind speeds, which are categorized as the deep stall region (flow is fully separated over the entire span and the blade is completely stalled), the CFD prediction also

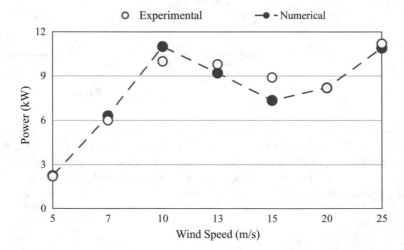

Fig. 11.5 Wind turbine power output versus wind speed. (This figure was originally published in [16] and has been reused with permission)

agreed well with the experimental data. At the onset of stall (10 m/s), however, separation occurs and RANS models have difficulty in capturing highly transient effects at the onset of stall. CFD predictions, therefore, deviated slightly from the experimental data. Various researchers [17, 25, 26] also have conducted CFD simulation of wind turbines using RANS turbulence models and observed similar discrepancy of their results with the experimental data at the onset of stall, even when the mesh was fine enough to resolve the near-wall regions.

11.4.2.2   Two-Phase (Air and Rain) Flow Simulation

Three-dimensional simulation of the NREL horizontal turbine at the rainfall rate of 40 mm/h and inlet air velocity of 7 m/s and blades rotational speed of 72 rpm was performed. The initial condition was such that the velocity and pressure in the entire domain were the same as those of the inlet boundary. A time step of 10^{-4} s was used for the continuous phase (VOF model) solver and droplets were also tracked with the fluid-flow time step.

Figure 11.6 shows the power output of the NREL two-blade wind turbine as a function of time. After 0.5 s of simulation with no rain injection, the steady state condition was reached. Rain injection was then started at time t = 0.6 s from the injection plane surface located 5 m from the blades. The droplets were injected with a velocity of 7 m/s and it took them about 0.7 s to reach the wind turbine blades at around time t = 1.3 s. The simulation accurately simulates the motion of rain

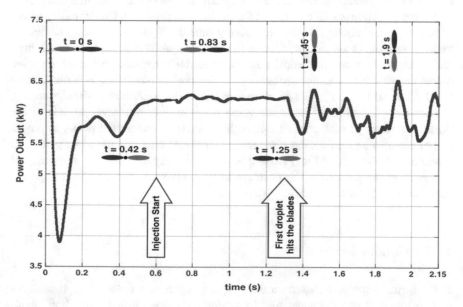

Fig. 11.6 Wind turbine power output at 40 mm/h rainfall rate and blades rotational speed of 72 rpm; blades in the figure are rotating counter clockwise. (This figure was originally published in [16] and has been reused with permission)

droplets as the effect of rain on the power output of the wind turbine also starts to show up at around time t = 1.3 s, when the rain droplets reach the blade for the first time. As soon as the rain droplets reached the wind turbine blades, the power output drops and becomes quasi-periodic with a period of about 0.42 s, which is the time needed for the blades to rotate 180° (half a rotation). The time-averaged wind turbine power output under rainy conditions is 5.95 kW compared to 6.25 kW for the no-rain conditions, which suggests about a 5% reduction in power output as a result of rain. Figure 11.5 also shows the blade orientation at several instances of time to explain the quasi-periodic behavior of wind turbine power output observed once the rain droplets reach the wind turbine blades. Each blade is shown with a different color to assist the visualization of blade rotation over time. At time t = 0 s, the blades are in a horizontal position as depicted in Fig. 11.6.

In general, the rain reduces the power output of the wind turbine. The reason is that, as the blades are rotating, one of them is always going up and the other one is going down while the downward moving rain droplets are hitting both blades. Therefore, the relative velocity of the rain droplets to the blade going upward is greater than the relative velocity of the rain droplets to the blade going downward. As a result, the rain droplets oppose the upward motion of one blade more than they contribute to the downward motion of the other blade, and, consequently, reduce the power output of the wind turbine. This is why the power output declines as the rain reaches the blades at time t = 1.30 s and the time-averaged power output under rainy conditions is almost 5% less that the no-rain condition.

The reason for the peaks in wind turbine power output at times t = 1.45 s and t = 1.9 s is that, when the blades are in the vertical position, rain droplets can barely hit the top or bottom surface of any of the blades and, therefore, the degrading effect of rain explained above is eliminated. Every time the blades are moving towards a vertical position, therefore, an increase in power output is observed; and each time the blades are leaving the vertical position toward a horizontal position, the power output decreases. Two minor peaks are also observed in power output around times t = 1.65 s and t = 2.1 s. Based on the results of the S809 airfoil simulation that showed the lift coefficient increases when water reaches the lower surface of the blade as a result of high rainfall rate or high angle of attack, the peaks in power output can also happen for the same reason. It is expected that the same pattern in power output observed in Fig. 11.6 will be repeated every 0.42 s due to the rotational symmetry of the geometry. The decrease in the horizontal wind turbine and power generation could be at least 5%.

11.5   Conclusion

1. Numerical simulation based on the coupled Lagrangian-Eulerian approach successfully captured the water film formation on a three-dimensional, horizontal-axis wind turbine, and, in turn, predicted the performance loss caused by the rain water film.

2. The lift coefficient increased with angle of attack or rotation of the turbine blades even past the stall angle compared to the single-phase case. This favorable increase in lift, however, was accompanied by an increase in the drag coefficient.
3. At lower rainfall rates, the performance of the wind turbines was highly sensitive to the rainfall rate. Once the rainfall rate was high enough to form a water film that immersed most of the surface of the blades, however, a further increase in rainfall rate did not significantly affect turbine performance.
4. The three-dimensional multiphase model showed that the rain reduces the power output of the wind turbines and makes the power output of the wind turbines oscillatory. This reduction for the NREL phase VI wind turbine was at least 5% at 7 m/s wind speed and a rainfall rate of 40 mm/h.
5. For more accurate representation of the turbulence, future work may include three-dimensional simulation of the wind turbine using large eddy simulation (LES). This requires additional computational time.

11.6 Exercises

11.6.1 Ex. 1: Polar Curve

1. Draw a drag polar curve (lift coefficient versus drag coefficient) for no-rain and 400 mm/h rainy conditions (use data from Fig. 11.4) for an S809 airfoil. What conclusions will be obtained by comparing the drag polar curve in dry and rainy conditions?

11.6.2 Ex. 2: Calculation of Power Generation

The wind speed before reaching the wind turbine is V_1, and the wind speed after the passage through the rotor plane is V_2. The rotor-swept area has a radius of R. Assume the average wind speed V through the rotor may be written: $V = (V_1 + V_2)/2$.

1. Find an expression for power extracted from the wind and its ratio to total power in the undisturbed wind streaming through exactly the same area with no rotor blocking.
2. Determine at what velocity ratio the ratio of the power extracted from the wind to the power in the undisturbed wind has its maximum values.

11.6.3   Ex. 3: Two-Dimensional Gas/Particle Flow

1. Develop two-dimensional governing equations for two-phase flow of air and fine
   sand particles in the air over an airfoil. You may use the DPM model for sand
   particles before colliding with the airfoil and single phase for flow over the airfoil.

Nomenclature

ρ	Density of the fluid
p	Pressure of the fluid
u_i	Velocity of the fluid
μ	Viscosity of the mixture
k	Turbulent kinetic energy
ω	Specific dissipation rate
$\widetilde{G_k}$	Generation of k
G_ω	Generation of ω
Y_k	Dissipation of k
Y_ω	Dissipation of ω
Γ_k	Effective diffusivity of k
Γ_ω	Effective diffusivity of ω
D_ω	Cross-diffusion term
u_{Pi}	Velocity of the parcel
m_P	Mass of the parcel
F_D	Drag coefficient
ρ_P	Densities of the droplets
g_i	Gravitational acceleration
α_q	Volume fraction of the qth phase
ρ_q	Density of the qth phase
$\boldsymbol{u}_q$	Velocity of the qth phase
S_{α_q}	Mass source term for the qth phase
F_i	Momentum source term
V	Volume of the computational cell
Δt	Time-step of the continuous phase solver
u_{ri}	Droplet's relative velocity to fluid continuous phase
u_{fi}	Velocity of the continuous fluid phase
C_L	Lift coefficient
C_D	Drag coefficient
F_{st}	Surface tension force
σ	Surface tension

References

1. Corten GP, Veldkamp HF (2002) Insects can halve wind-turbine power. Nature 412:42–43
2. Khalfallah MG, Koliub AM (2007) Effect of dust on the performance of wind turbines. Desalination 209(1-3):209–220
3. Arastoopour H (2019) The critical contribution of chemical engineering to a pathway to sustainability. Chem Eng Sci 203:247–258
4. Bragg MB, Gregorek GM, Lee JD (1986) Airfoil aerodynamics in icing conditions. J Aircraft 23(1):76–81
5. Corrigan RD, DeMiglio RD (1985) Effect of precipitation on wind turbine performance. NASA TM-86986
6. Al BC, Klumpner C, Hann DB (2011). Effect of rain on vertical axis wind turbines. International Conference on Renewable Energies and Power Quality (ICREPQ'11), Spain
7. Arastoopour H, Gidaspow D, Abbasi E (2017). Computational transport phenomena of fluid-particle systems. Mechanical Engineering Series Book, Springer, Cham, Switzerland
8. Arastoopour H (2001) Numerical simulation and experimental analysis of gas/solid flow systems: 1999 Fluor-Daniel Plenary lecture. Powder Technol 119(2):59–67
9. Abbasi E, Abbasian J, Arastoopour H (2015) CFD-PBE numerical simulation of CO_2 capture using MgO-based sorbent. Powder Technol 286:616–628
10. Valenine JR, Decker RA (1995) A Lagrangian-Eulerian scheme for flow around an airfoil in rain. Inter J Multiphase Flow 21(4):639–648
11. Yeom GS, Chang KS, Baek SW (2012) Numerical prediction of airfoil characteristics in a transonic droplet-laden air flow. Int J Heat Mass Transfer 55:453–469
12. Bezos G, Dunham RE, Gentry G (1992) Wind tunnel aerodynamic characteristics of a transport-type airfoil in simulated heavy rain environment. NASA Technical Paper, 3184
13. Thompson B, Jang J, Dion J (1995) Wing performance in moderate rain. J Aircraft 32 (5):1034–1039
14. Cai ME, Abbasi E, Arastoopour H (2013) Analysis of the performance of a wind turbine airfoil under heavy-rain conditions using a multiphase computational fluid dynamics approach. Ind Eng Chem Res 52(9):3266–3275
15. Cohan A, Arastoopour H (2016) Numerical simulation and analysis of the effect of rain and surface property on wind-turbine airfoil performance. Int J Multiphase Flow 81:46–53
16. Arastoopour H, Cohan A (2017) CFD simulation of the effect of rain on the performance of horizontal wing turbine. AIChE J 63:5375–5383
17. Simms D, Schreck S, Hand M, Fingersh (2001) NREL unsteady aerodynamics experiment in the NASA Ames wind tunnel: a comparison of predictions to measurements. NREL/TP-500-29494
18. Cohan AC (2016) A coupled Lagrangian-Eulerian multiphase model for simulation of wind turbine performance under rainy conditions. Ph.D. thesis, Illinois Institute of Technology
19. ANSYS (2014) ANSYS Fluent theory guide. ANSYS Inc.
20. Chitsomboon T, Thamthae C (2011) Adjustment of k-omega SST turbulence model for an improved prediction of stalls on wind turbine blades. World Renewable Energy Congress
21. Menter, FR (1994) Two-equation eddy-viscosity turbulence model for engineering applications. AIAA J 32(8):1598–1605
22. Brackbill JU, Kothe DB, Zemach C (1992) A continuum method for modeling surface tension. J Comput Phys 100(2):335–354

23. Hand M, Simms D, Fingersh LJ, Jager D, Larwood S, Cotrell J, Schreck S (2001) Unsteady aerodynamics experiment Phase VI: wind tunnel test configurations and available data campaigns. NREL/TP-500-29955, Golden, Colorado
24. Somers DM (1997) Design and experimental results for the S809 airfoil. National Renewable Energy Laboratory, Golden, CO
25. Sørensen, NN, Michelsen JA, Schreck S (2002) Navier-Stokes predictions of the NREL phase VI rotor in the NASA Ames 80ft x120 ft wind tunnel. Wind Energy 5(2):151–169
26. Yelmule MM, Vsj EA (2013) CFD predictions of NREL phase VI rotor experiments in NASA/AMES wind tunnel. Int J Renewable Energy Res 3(2):261–269

Chapter 12
Application of Multiphase Flow Simulation in Pharmaceutical Processes

Hamid Arastoopour, Dimitri Gidaspow, and Robert W. Lyczkowski

12.1 Introduction

Pharmaceutical processing includes several processes that are based on a multiphase system, particularly gas/solid flow and fluidization. These processes are devoted to making particles, modifying their properties, and then turning them into structured products and may include crystallization, mixing, granulation, pulverization, drying, and tableting.

Crystallization from solution is commonly used in the pharmaceutical industry as the final step in the production of active pharmaceutical ingredients (API). One of the most common methods of supersaturation generation is through the addition of an anti-solvent (which reduces solute solubility) that is often used in combination with cooling to maximize yield and to produce crystals of desired quality in terms of purity, size, morphology, and shape. Mixing is known to have a significant effect on supersaturation distribution, nucleation kinetics, and final product-size distribution [1]. Liu et al. [2] used the computational fluid dynamics (CFD) approach to simulate anti-solvent crystallization and to study the effects of agitation rate, anti-solvent feed rate, and position on mixing.

More than two-thirds of all pharmaceutical products destined for U.S. consumption are solids; thus, fluid/particle systems including fluidized beds play a significant role in the design and scale-up of pharmaceutical processes such as granulation and drying.

The granulation process is widely used in pharmaceutical industries. During the granulation process, fine powder particles form agglomerates, which are larger size granules with improved properties. Granulation is an important unit operation in the downstream tablet manufacturing process because it has a significant effect on the mechanical properties of the tableting material (e.g., hardness and dissolution rate). Sen et al. [3] presented a multi-scale hybrid model for granulation, using information from different scales of CFD, the discrete element method (DEM), and the population balance model (PBM). Granulation has been extensively modeled using the

PBM approach, which groups a lump of particles in different classes (based on size, porosity, etc.). The flow field of particles usually is calculated using the DEM. CFD calculates all forces acting on the particles due to the flow field of the fluidizing medium. The PBM can track the change in particle size and the liquid content of the particles, and calculates the new granulate/particle size distribution (PSD) at every PBM time step.

The drying process deals with removing water from solid materials. For design and scale-up of drying, understanding of mass, energy, and momentum transfer is required. The drying operation has been used extensively in chemical, pharmaceutical, and agricultural applications [4, 5]. Drying is one of the major unit operations in the manufacturing of solid pharmaceuticals. In many pharmaceutical processes, optimum design of the drying process will significantly enhance the rate and reliability of the production and, in turn, decrease the cost of the pharmaceutical products. During the pharmaceutical drying process, if the temperature and the heating rate are not maintained at the desired level, then the process could have adverse effects on the product including an increase in the cost due to loss of valuable drug materials [6]. Fluidized bed dryers are capable of reducing the drying time and can be operated close to the isothermal condition, which facilitates operational control of the bed [7–10].

In the pharmaceutical industry, fluidized beds are typically used in the batch process for protection against heat damage and increased throughput. Overall, the batch-wise operation provides a means of helping to control and maintain the uniformity of the process. In the batch drying of solids, the drying rate is controlled by two mechanisms: the constant drying rate period and the falling rate period [11, 12]. The surface moisture on the particle and water in the large pores of the particle predominantly influence the constant drying rate; whereas, the moisture, which is trapped or bound within the solids, controls the falling rate periods. Because the bound moisture is more difficult to remove at a slow rate than the surface moisture, the falling rate period approximately determines the rate of the entire drying process [13].

In recent years, the CFD approach has been used increasingly to improve the process design and scale-up of the drying process using both bubbling and spouted beds [9, 12, 14, 15]. This chapter focuses on the CFD approach for design, scale-up, and performance optimization of a pharmaceutical bubbling fluidized bed dryer based on work done by Jang and Arastoopour [16].

12.2 Model Development for Pharmaceutical Bubbling Fluidized Bed Dryer

12.2.1 Gas/Solid Flow Model

The CFD model used in this study is based on a two fluid model (TFM) with a Eulerian-Eulerian approach [16–19]. The kinetic theory of granular flow was used to describe the particulate phase flow behavior in a bubbling fluidized bed dryer and to develop a constitutive relation to close the governing equations. In this work, a set of governing equations (continuity, species, momentum, and energy) was used to describe the dynamic behavior and the heat and mass transfer phenomena of the drying process in a bubbling fluidized bed, and was solved using a commercial ANSYS CFD code [20]. The empirical correlation developed by the Syamlal-O'Brien drag model [21] was used as the drag force between phases. In this study, the continuity and momentum equations (see Chap. 2 or [19]) were solved numerically along with the following energy and species balance equations.

12.2.2 Heat and Mass Transfer Model

The conservation of energy equation for each phase can be written,

$$\frac{\partial\left(\varepsilon_g\rho_g h_g\right)}{\partial t} + \nabla\cdot\left(\varepsilon_g\rho_g\vec{v}_g h_g\right) = -\varepsilon_g\frac{\partial p_g}{\partial t} + \bar{\tau}_g$$
$$: \nabla\vec{v}_g - \nabla\cdot\vec{q}_g + Q_{sg} + \dot{m}_{sg}\Delta H_{vap} \quad (12.1)$$

$$\frac{\partial(\varepsilon_s\rho_s h_s)}{\partial t} + \nabla\cdot\left(\varepsilon_s\rho_s\vec{v}_s h_s\right) = -\varepsilon_s\frac{\partial p_s}{\partial t} + \bar{\tau}_s$$
$$: \nabla\vec{v}_s - \nabla\cdot\vec{q}_s + Q_{gs} - \dot{m}_{sg}\Delta H_{vap} \quad (12.2)$$

where h is the specific enthalpy, $\vec{q}$ is the heat flux, and Q_{sg} is the rate of heat transfer between the gas and solid phases. The first term of each equation describes the rate of change of enthalpy, the second term represents energy (enthalpy) transfer due to convection, the third term represents the energy due to the work of expansion of the void fraction, the fourth term represents energy due to shear stress dissipation, the fifth term represents the energy transfer due to conduction, the sixth term represents the rate of energy transfer between the gas and solid phases, and the last term represents energy transfer due to phase change [16].

The specific enthalpy (h) and the heat flux ($\vec{q}$) in each phase are expressed,

$$h = \int C_p dT \tag{12.3}$$

$$\vec{q} = k\nabla T \tag{12.4}$$

where k is the thermal conductivity and C_p is the heat capacity of each phase. The conservation of energy equations thus are closed using expressions for the rate of heat transfer between the gas and solid phases (Q_{sg}). The inter-phase heat exchange may be expressed as the product of the specific interfacial exchange area and the gas/particle heat transfer coefficient α_{gs} [16],

$$Q_{sg} = -Q_{gs} = \frac{6\varepsilon_s}{d_s} \alpha_{sg} (T_g - T_s) \tag{12.5}$$

In the literature, there are many empirical correlations for heat transfer coefficients in the fluidized beds. In this study, the empirical correlation introduced by Kothari [22] is used to calculate the heat transfer between the gas and particulate phases in the bubbling fluidized bed,

$$Nu = \frac{\alpha_{gs} d_s}{k_g} = 0.03 \, \text{Re}_p{}^{1/3} \tag{12.6}$$

where Nu is the Nusselt number, d_s is the particle diameter, and k_g is the thermal conductivity of the gas phase.

The latent heat due to vaporization of moisture [14] is,

$$\Delta H_{vap} = 3168 - 2.4364 \cdot T_g \tag{12.7}$$

To simulate moisture transfer from the solid phase to the gas phase, the following water species transport equations for each phase are included,

$$\frac{\partial (\varepsilon_g \rho_g Y_v)}{\partial t} + \nabla \cdot \left(\varepsilon_g \rho_g \vec{v}_g Y_v \right) = \nabla \cdot \left(D_{v,g} \rho_v \varepsilon_g \nabla Y_v \right) + \dot{m} \tag{12.8}$$

$$\frac{\partial (\varepsilon_s \rho_s X_s)}{\partial t} + \nabla \cdot \left(\varepsilon_s \rho_s \vec{v}_s X_s \right) = \nabla \cdot (D_{v,s} \rho_s \varepsilon_s \nabla X_s) - \dot{m} \tag{12.9}$$

where Y_v is the moisture content of the gas phase, X_s is the moisture content of the solid phase, $D_{v,g}$ and $D_{v,s}$ are the moisture diffusion coefficient of the gas and the solid phase, respectively, and $\dot{m}$ is the moisture mass transfer rate between the gas and solid phases.

12.2.3   Drying Rate Model

The drying rate is controlled by two mechanisms: the constant drying rate period and the falling rate period. The surface moisture on the particle and the water in the large pores of the particle predominantly influence the constant drying rate period. The moisture trapped or bound within the porous structure of the particles controls the falling rate period [23, 24].

During the constant rate period, the particles are assumed to be at wet bulb temperature with a thermodynamic equilibrium between the gas and solid phases. The drying rate is determined by the convective mass transfer of moisture through the stationary film of air surrounding the solid particles inside the granulates [24]. The driving force for convective mass transfer is the difference between the saturated moisture content of the gas phase ($Y*_i$) at the solid surface and the moisture content of the gas phase (Y_v). The expression for the mass transfer rate per unit volume for the constant rate period may be expressed,

$$\dot{m} = K_{sg} \cdot \rho_g \cdot 6/ds \cdot \left(Y_i^* - Y_v \right) \tag{12.10}$$

Many empirical correlations are available in the literature for the value of the mass transfer coefficient (K_{sg}). In this study, the empirical correlation introduced by Gunn [25] was used to calculate the mass transfer between the gas and particulate phases in a bubbling fluidized bed.

$$Sch = \frac{K_{sg}d_s}{Dv_g}$$

$$= \left(7 - 10\varepsilon_g + 5\varepsilon_g^2 \right) \left(1 + 0.7 \, \mathrm{Re}_s^{0.2} Sc^{1/3} \right)$$
$$+ \left(1.33 - 2.4\varepsilon_g + 1.2\varepsilon_g^2 \right) \mathrm{Re}_s^{0.7} Sc^{1/3} \tag{12.11}$$

$$Sc = \frac{\mu_g}{\rho_g Dv_g} \tag{12.12}$$

$$Dv_g = 2.60 \cdot 10^{-5} \cdot \left(\frac{T_s}{T_{ref}} \right)^{3/2} \tag{12.13}$$

where Sch is the Sherwood number, Sc is the Schmidt number, and A_s is the overall external particle surface area to unit volume ratio. The moisture content of the saturated drying gas at the surface of the wet particles, $Y*_i$, is calculated using the following expressions [14],

$$Y_i^* = 0.622 \frac{p_v}{760 - p_v} \tag{12.14}$$

$$p_v = 133.3 \times \exp\left(13.869 - \frac{5173}{T_s}\right) \tag{12.15}$$

When the moisture content of the solid particles, X_i, reaches a critical value of X_{cr}, the falling rate period (diffusion control) begins [13]. The following diffusion equation was used [26].

The mass transfer rate per unit volume may be expressed,

$$\dot{m} = \frac{Dv_s \cdot \pi^2}{d_s^2} \cdot \rho_s \cdot \left(X_s - X_f\right) \tag{12.16}$$

Dv_s is the moisture diffusion coefficient inside the granule particles. This value may be obtained experimentally. Due to the lack of experimental values for Dv_s, a typical constant value of 2×10^{-12} m^2/s was used, which shows the slow solid drying rate during the falling rate period. Xs is the volume-averaged moisture content of the granules and X_f is the final volume-averaged moisture content or volume averaged equilibrium moisture content of the granules.

Initial and boundary conditions are as follows. Initially, the gas velocity was set at zero throughout the entire bed. A uniform velocity profile for the gas phase was applied as an inlet condition. A value for the pressure was specified at the outlet of the fluidized bed. For the gas phase, no-slip and non-penetrating wall conditions were considered. For the solid phase, Johnson and Jackson's [27, 19] wall boundary condition was applied (see Chap. 2). Based on experimental observations in the literature, a value of restitution coefficient of 0.2 between the particle and the wall, a value of restitution coefficient of 0.9 between particles, and a specularity coefficient of 0.2 were assumed.

12.3 Numerical Simulation

Two-dimensional transient numerical simulations of the previous governing and constitutive equations along with boundary and initial conditions were performed using ANSYS computer software [20]. The results of the CFD simulation were compared with the laboratory-scale experimental data obtained at Duquesne University to validate and refine the CFD model. The modified model was then used to simulate the drying of the same material in two Abbott Laboratories kilo- and 10-kilo-scale units [16].

12.3.1   Three Different Scales of Drying Fluidized Beds

All three scales of the dryer fluidized beds were conically shaped with dimensions *summarized* in Table 12.1.

For the laboratory-scale drying experiment, the air inlet flow rate was kept at 5 m³/h at a temperature of 50 °C. The drying process was continued for up to 25 min, and the moisture content of the granule samples was measured every 5 s. Using a kilo-scale set-up, the inlet flow rate was kept at 15 m³/h at a temperature of 30 °C. The drying process was continued up to 10 min and the outlet temperature was measured every 20 s. For the 10-kilo-scale set-up, the inlet flow rate was kept at 166 m³/h at 21 °C. The drying was continued for up to 20 min and the outlet temperature was measured every 10 s.

Pharmaceutical granulated particles with a density of 1200 kg/m³, diameter of 287 μm, and initial moisture content of 0.0417 kg water/kg dried granules were used in each experiment. For each wet granulate, a critical moisture content (X_{cr}) separates the two drying rate periods. Because X_{cr} was not measured directly, it was estimated based on the laboratory-scale experimental data (0.012 kg$_{water}$/kg$_{dry\ solid)}$ as an input parameter to the CFD model [16].

12.3.2   Numerical Analysis

To avoid solution divergence, small time steps on the order of 1×10^{-4} s were chosen. Convergence was set to occur when the scaled residuals reported for all variables fell below 1×10^{-4}, except energy and species balance for which convergence was set to occur when the residuals fell below 1×10^{-6}. In order to obtain accurate and meaningful simulation results, calculated values should be independent of the grid size. Therefore, simulations for three different scale fluidized beds were carried out using different numbers of meshes and grid-independent results were obtained [16].

Table 12.1 Dimensions of fluidized bed dryers at each scale

	Height (m)	Diameter (m)	
		Top	Bottom
Laboratory scale	0.46	0.19	0.08
Kilo scale	0.56	0.30	0.14
10-Kilo scale	0.58	0.80	0.305

Table 12.2 Flow and heat and mass transfer properties (initial and inlet conditions). (This table was originally published in [16] and has been reused with permission)

Properties	Values
Particle size	287 μm
Particle density	1200 kg/m^3
Gas density	1.225 kg/m^3
Gas viscosity	1.789 × 10^{-5} kg/m/s
Inlet gas velocity	0.28 m/s
Moisture diffusivity	2 × 10^{-12} m^2/s
Inlet gas velocity	0.27 m/s (lab), 0.28 m/s (kilo), 0.53 m/s (10-l-kilo)
Initial bed height	0.12 m (lab), 0.10 m (kilo), 0.13 m (10-kilo)
Initial solid temperature	294 K
Lab-scale inlet gas temperature	293.5 K at 0–180 s (0.4083 × t + 270.68) K at 180–230 s 325 K at 230–600 s
Kilo-scale inlet gas temperature	303 K at 0–200 s (−0.014 × t + 307.25) K at 200–600 s
10-kilo-scale inlet gas temperature	293.5 K at 0–400 s (0.15 × t + 240.73) K at 400–600 s
Initial moisture content	0.041 kg$_{water}$/kg$_{dry\ solid}$
Equilibrium moisture content	0.004 kg$_{water}$/kg$_{dry\ solid}$

12.4 Simulation Results and Discussion

Table 12.2 summarizes the flow, heat, and mass transfer properties, and initial and inlet conditions used to simulate the Duquesne University (laboratory scale) and Abbott Laboratories (kilo and 10-kilo scale) drying experiments [16].

12.4.1 Gas/Solid Flow Patterns

Figure 12.1 shows the contour of the solid volume fraction of three different scale fluidized bed dryers at different times (0.6, 5.0, and 10.0 s) after the start of the simulation.

The snapshots of 5 and 10 s in the laboratory-scale dryer show bubbles passing through the middle of the bed and splitting and coalescing at the bottom of the bed. In the cases of the kilo- and 10-kilo-scale dryers, after the first bubble was broken, the large bubble split into several small bubbles. In the kilo scale and 10-kilo scale, relatively more bubbles and rather more chaotic movements of solid particles were observed. In the laboratory-scale dryer, the solid volume fraction around the wall was almost stationary in comparison with the solid volume fraction patterns in the kilo- and 10-kilo-scale fluidized bed dryers. This means that both the kilo- and 10-kilo-scale fluidized bed systems result in better mixing conditions than the

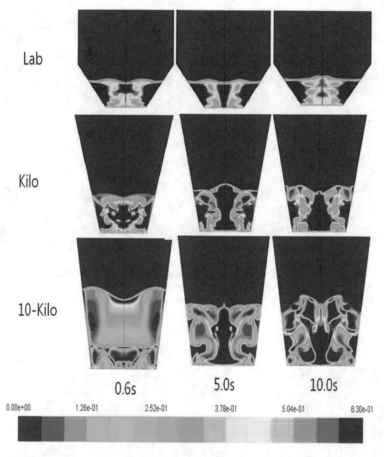

Fig. 12.1 Contour plot of the solid volume fraction of three different scale dryers at different time (0.6, 5.0, and 10.0 s); lab scale (top), kilo scale (middle), 10-kilo scale (bottom). (This figure was originally published in [16] and has been reused with permission)

laboratory-scale fluidized bed. This could be contributed to less wall effect penetration for large-scale beds. In the larger-scale dryers, more bubbles exist in the bed, which means bubbles occupy more space in the bed, which leads to an increase in the bed height.

12.4.2 Heat and Mass Transfer During the Drying Process

Figures 12.2, 12.3, and 12.4 show the simulation results of the solid volume fraction, the air and solid temperature, and the moisture content in the solid phase after 100 s of drying operation for the laboratory-, kilo-, and 10-kilo-scale dryer, respectively.

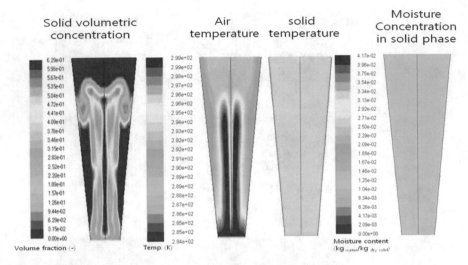

Fig. 12.2 Contour plot for solid volume fraction, air and solid temperature, and moisture content in solid phase after 100 s of drying operation for lab-scale fluidized bed dryer. (This figure was originally published in [16] and has been reused with permission)

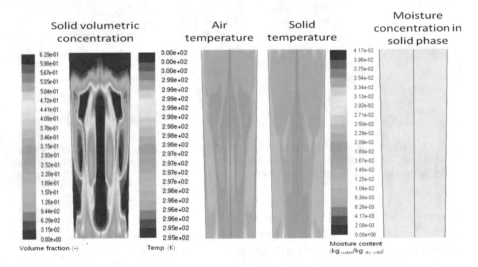

Fig. 12.3 Contour plot for solid volume fraction, air and solid temperature, and moisture content in solid phase after 100 s of drying operation for kilo-scale fluidized bed dryer. (This figure was originally published in [16] and has been reused with permission)

Figure 12.2 shows the presence of several bubbles at the center of the bed and corresponding high gas temperature in the bubbling region for the laboratory-scale process. The solid temperature and solid moisture content, however, are uniform in the entire bed due to the reasonably high degree of average mixing in the laboratory-scale fluidized bed dryer.

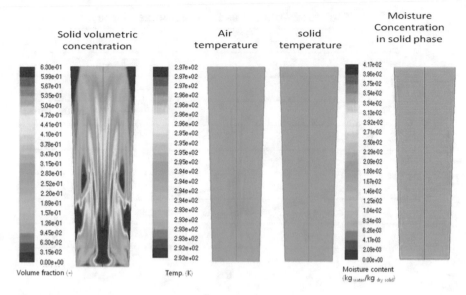

Fig. 12.4 Contour plot for solid volume fraction, air and solid temperature, and moisture content in solid phase after 100 s of drying operation for 10-kilo-scale fluidized bed dryer. (This figure was originally published in [16] and has been reused with permission)

Figures 12.3 and 12.4 show the presence of more bubbles in the kilo- and 10-kilo-scale dryer than in the laboratory-scale dryer, which means that these systems are at a higher degree of mixing than the laboratory-scale dryer. The simulation also showed that the difference between the average solid temperature and the inlet gas temperature will continuously decrease and result in more uniform temperature distribution in the bed. For example, for the constant rate drying period (e.g., 100 s of drying operation), almost uniform temperature and moisture distribution were calculated in the entire fluidized bed dryer using the CFD simulation program. Figures 12.2, 12.3, and 12.4 clearly show that gas/solid hydrodynamic behavior plays an important role in heat and mass transfer in fluidized bed dryers.

Figure 12.5 shows the moisture content in the solid phase with respect to dimensionless time (operating time/gas residence time in the bed) for three different scale dryers. Gas residence times (t_R) in the bed for each scale fluidized bed dryer was calculated as 0.55 m/s for the lab scale, 0.53 m/s for the kilo scale, and 0.3 m/s for the 10-kilo scale, respectively. Figure 12.5 clearly describes the characteristic drying curve that is separated by a constant drying rate period and falling rate period. The simulation results on three different scale dryers can be used to validate the scale-up of the drying process in fluidized bed dryers using similar dimensionless numbers such as the Reynolds number, Nusselt number, and Schmidt number.

Table 12.3 shows dimensionless groups for all three scales. According to Table 12.3, for the laboratory- and kilo-scale dryer cases, the Reynolds number (Re), Froude number (Fr), Nusselt number (Nu), and Sherwood number (Sh) are very similar, which means that the two systems have similar hydrodynamic, and heat

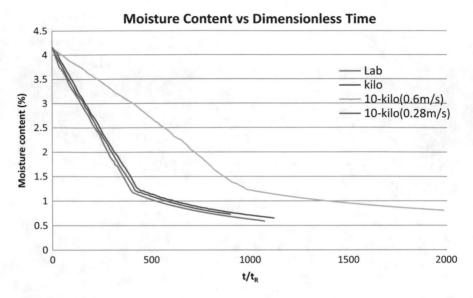

Fig. 12.5 Moisture content in solid phase as a function of dimensionless time during drying operation for the laboratory-, kilo-, and 10-kilo-scale fluidized bed dryer. (This figure was originally published in [16] and has been reused with permission)

Table 12.3 Dimensionless groups for each scale dryer. (This table was originally published in [16] and has been reused with permission

	Expression	Definition	Lab	Kilo	10-Kilo (0.6 m/s)	10-Kilo (0.28 m/s)
Re	$\frac{\rho_g d_s v_{sg}}{\mu_g}$	*Inertia force* / *Viscous force*	5.50	5.31	10.42	5.50
Fr	$\frac{v_s^2}{gh}$	*Inertia force* / *Gravitational force*	0.07	0.07	0.22	0.061
Nu	$\frac{\alpha_{sg} d_s}{k_g}$	*Convective heat transfer* / *Conductive heat transfer*	0.28	0.26	0.63	0.275
Sch	$\frac{K_{sg} d_p}{D v_g}$	*Convective MassTransfer* / *Diffusive MassTransfer*	11.34	11.24	21.68	11.34
$\frac{\dot{m}_w}{\dot{m}_g}$		*Moisture Removal Rate* / *Gas Mass Flow Rate*	0.001	0.015	0.034	0.016

and mass transfer characteristics. For the 10-kilo scale, however, dimensionless groups are different. From Fig. 12.5, except the 10-kilo scale dryer, the drying rate profile of the laboratory- and kilo-scale dryers are very similar, having similar dimensionless numbers. In order to validate if the dimensionless group analysis is a reasonable approach for the scale-up of the drying process using bubbling fluidized beds, an additional simulation was performed for the 10-kilo scale dryer using the same Reynolds number (using a lower superficial inlet gas velocity of 0.28 m/s) as the laboratory- and kilo-scale dryers.

Figure 12.5 shows that the drying curve for the 10-kilo-scale dryer using the same Reynolds number has a similar trend as the drying curve for the laboratory- and kilo-scale dryer during the constant rate period and falling rate period of the simulation.

In summary, the fact that CFD simulation results for solid particles drying as a function of dimensionless time showed the same results for three different scales with a similar dimensionless group is excellent proof that the CFD model along with a similar dimensionless group can be used as a tool to scale up the drying process from the experimental scale to both the kilo-scale and 10-kilo-scale fluidized bed dryers.

12.5   Comparison Between Simulation Results and Experimental Data

Figure 12.6 shows very good agreement between CFD simulation results and the laboratory-scale fluidized bed dryer data (Duquesne University experimental data) [16]. A proper critical moisture content value of 0.012 $kg_{water}/kg_{dry\ sol}$ was estimated. This means about 71% of the initial moisture of the granulates was on the surface of the granulates, between particles that constitute granulates, and in large surface pores (removed during the constant drying rate period). Of the remaining moisture, 19% was in the porous structure of the particles (mainly removed during the falling drying rate period) and 10% remained as the equilibrium moisture content.

Figure 12.7 shows the comparison of the outlet gas temperature simulation with the Duquesne University experimental data for 600 s of drying. Figure 12.7 shows very good agreement of the simulation results with the Duquesne University data.

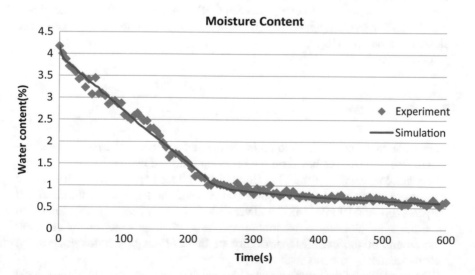

Fig. 12.6 Comparison between numerical simulation based on CFD and the Duquesne University Laboratory fluidized bed dryer experiment data. (This figure was originally published in [16] and has been reused with permission)

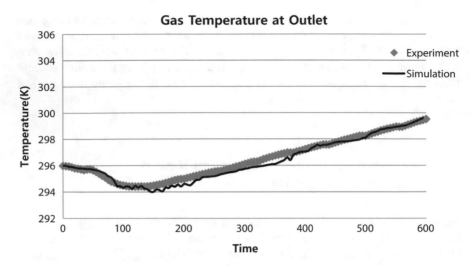

Fig. 12.7 Comparison between the numerical simulation based on CFG and the Duquesne University Laboratory fluidized bed dryer experiment data. (This figure was originally published in [16] and has been reused with permission)

The results show that the CFD model is capable of predicting the coupled heat and mass transfer phenomena during the drying process for a lab-scale fluidized bed dryer.

Because the moisture content in the kilo- and 10-kilo scales was not measured, only the outlet gas temperature was compared with the simulation results. Very good agreement was obtained, however, between CFD simulation results and the experimental data for gas temperature at the outlet of both the kilo- and 10-kilo scale fluidized bed systems [16].

12.6   Conclusion

1. The CFD model can be used to optimize the performance and assist in the scale-up of drying processes based on the bubbling fluidized bed.
2. Very good agreement of the CFD simulation with the experimental values of the outlet gas temperature of the laboratory-scale and the larger-scale fluidized bed dryers (kilo- and 10-kilo scales) and very good agreement between the simulation and the solid moisture concentration of the small-scale experiments are clear indications of the excellent capability of the CFD approach for scale-up of pharmaceutical processes such as drying.
3. CFD simulation of the drying process showed that the measurement of the critical moisture content of a single granulate is needed for simulation and design of the drying processes. The critical moisture content measurement is used as a guide to

distinguish between two drying rate periods: namely, the constant and falling rate periods.

4. CFD simulation may be used to get a more detailed understanding of pharmaceutical processes and, in turn, to assist the process management to anticipate and avoid potential operating problems. CFD simulation is also capable of identifying the key and critical factors in the design, scale-up, and operation of the pharmaceutical processes.

5. The dimensionless similarity approach is a reasonable approach for scale-up of the fluidized bed drying case.

12.7 Exercises

12.7.1 Ex. 1: Pneumatic Conveying Pharmaceutical Drying Process Modeling

Consider a drying process in which gas and pharmaceutical granulated spherical particles of density, ρ_p, and diameter, d_p, enter from the bottom into a cylindrical pneumatic conveying reactor with radius, R, and length, L. The particle volume concentration is very low such that particles move in a dispersed, suspended manner. Gas and particles flow vertically (consider axisymmetric cylindrical coordinate, z, as the vertical direction and, r, as the radial direction). Inlet (z = 0) gas and solid velocities, pressure, and solid volume fraction are V_{g1}, V_{s1}, P_1, and ϵ_{s1}, respectively. Initially, there is no particle in the riser with stationary gas at pressure, P_0. Gas enters at high temperature, T_{g1}, and particles enter at room temperature, T_0, and moisture content, X_0 (mass of water/mass of dry particle).

1. Develop transient, two-dimensional (r and z) governing equations (continuity, momentum and energy equations for each phase; and water species balance equations and the granular temperature equation for the particle phase). Assume laminar Newtonian flow for the gas phase and kinetic theory for the particle phase flows.

2. Write down all constitutive equations and all needed boundary and initial conditions for both gas and particles phases. Clearly mention any assumption that you are making and why? Assume the particle phase is incompressible and the gas phase obeys the ideal gas equation of state.

3. Is such a dilute process the most suitable process for drying? Is drying efficiency increasing or decreasing with particle size?

12.7.2   Ex. 2: Fluidized Bed Pharmaceutical Granulation Process

1. Read reference [3] and develop three-dimensional transient governing equations for a pharmaceutical granulation process in a rectangular bubbling fluidized bed system.

12.7.3   Ex. 3: Pharmaceutical Drying Using Moving Packed-Bed Process

Consider a drying process in which pharmaceutical granulated spherical particles of density ρ_p, diameter, d_p, enter from the top of a cylindrical down comer reactor (particles move downward in close to a packed bed regime) with radius, R, and length, L. Gas at a low v_{s1} enters from the bottom of the down comer. Consider transient axisymmetric cylindrical coordinate, z, as the vertical direction and, r, as the radial direction. Inlet particles velocity, pressure, and solid volume fraction are v_{s1}, P_1, and ϵ_{s1}, respectively. Initially, there is a packed bed of dry particle in the riser with stationary gas at pressure, P_0. Gas enters at high temperature, Tg_1, and particles enter at room temperature, T_0, and moisture content, X_0 (mass of water/mass of dry particle). The critical moisture content of the particles is X_c.

1. Develop governing equations (continuity, momentum, and energy equations for each phase; and water species balance equations and the granular temperature equation for the particle phase) for this drying process.
2. Write down all constitutive equations and all needed boundary and initial conditions for both the gas and particle phases. Clearly mention any assumption that you are making and why?
3. Is this a reasonable system for the drying process? What will be the effect of particle size distribution in this system?

Nomenclature

C_p	Heat capacity
D_{vg}	Moisture diffusivity of gas phase
D_{vs}	Moisture diffusivity of solid phase
d	Diameter
g	Gravity
H_{vap}	Heat of vaporization per unit mass
h	Enthalpy
K_{sg}	Mass transfer coefficient
k	Conductivity
$\dot{m}$	Mass transfer rate
Nu	Nusselt number
P	Pressure

$P_{s,fr}$	Solid frictional pressure
P_v	Vapor pressure
Qsg	Rate of heat transfer between gas and solid
q	Rate of heat flux
Re_s	Particle Reynolds number
Sc	Schmidt number
Sch	Sherwood number
t	Time
T	Temperature
v	Velocity
X_{cr}	Critical moisture content fraction
X_f	Final moisture content fraction
Xs	Moisture content in solid phase fraction
Yv	Moisture content in gas phase fraction
$Y*$	Saturated moisture fraction in gas phase

Greek Symbols

ε	Volume fraction
μ	Viscosity
ρ	Density
τ	Stress tensor (N/m^2)

Subscripts

g	Gas phase
s	Solid phase

References

1. Myerson A (2002) Handbook of industrial crystallization, 2nd edition. Butterworth Heinemann Company
2. Liu X, Hatziavramidis D, Arastoopour H, Myerson AS (2006) CFD simulations for analysis and scale-up of anti-solvent crystallization. AIChE J 52:3621–3625
3. Sen M, Barrasso D, Singh R, Ramachandran R (2014) A multi-scale hybrid CFD-DEM-PBM description of a fluid-bed granulation process. Processes 2(1):89–111, https://doi.org/10.3390/pr2010089
4. Jamaleddine TJ, Ray MB (2010) Numerical simulation of pneumatic dryers using computational fluid dynamics. Ind Eng Chem Res 49:5900–5910
5. Wu ZH, Mujumdar AS (2007) Simulation of the hydrodynamics and drying in a spouted bed dryer. Drying Technol 25:59–74
6. Levin M (2006) Pharmaceutical process scale-up, 2nd ed. Taylor & Francis, New York
7. Jang JK, Roza C, Arastoopour H (2010) CFD simulation of pharmaceutical particle drying in a bubbling fluidized bed reactor, in Fluidization XIII, p 853–860
8. Schmidt A, Renz U (2000) Numerical prediction of heat transfer in fluidized beds by a kinetic theory of granular flows. Int J Therm Sci 39:871–885
9. Rosa CA, Freire JT, Arastoopour H (2010) Numerical and experimental steady state flow and heat transfer in a continuous spouted bed. In: Kim SD, Kang Y, Lee JK, Seo YC (eds) Fluidization XIII. ECI, New York, p 669–676
10. Chandran AN, Rao SS, Varma YBG (1990) Fluidized-bed drying of solids. AIChE J 36:29–38

11. Schlunder EU (2004) Drying of porous material during the constant and the falling rate period: A critical review of existing hypotheses. Drying Technol 22:1517–1532
12. Wang HG, Yang WQ, Senior P, Raghavan RS, Duncan SR (2008) Investigation of batch fluidized-bed drying by mathematical modeling, CFD simulation and ECT measurement. AIChE J 54:427–444
13. Coumans WJ (2000) Models for drying kinetics based on drying curves of slabs. Chem Eng Process 39:53–68
14. Palancz B (1983) A mathematical-model for continuous fluidized-bed drying. Chem Eng Sci 38:1045–1059
15. Szafran RG, Kmiec A (2004) CFD modeling of heat and mass transfer in a spouted bed dryer. Ind Eng Chem Res 43:1113–1124
16. Jang J, Arastoopour H (2014) CFD simulation of a pharmaceutical bubbling bed drying process at three different scales. J Powder Technol 263:14–25
17. Jang J, Arastoopour H (2013) CFD simulation of different-scaled bubbling fluidized beds. In: Kuipers JAM, Mudde RF, Van Ommen IR, Deen NG (eds) Fluidization XI. Engineering Conference International, p 691–698
18. Arastoopour H (2001) Numerical simulation and experimental analysis of gas/solid flow systems: 1999 Fluor-Daniel plenary lecture. Powder Technol 119:59–67
19. Arastoopour H, Gidaspow D, Abbasi E (2017) Computational transport phenomena of fluid-particle systems. Mechanical Engineering Series, ISBN 978-3-319-45488-7, Springer Press
20. ANSYS Inc. (2013) ANSYS Fluent theory guide, release 15.0. Canonsburg, PA
21. Syamlal M, Rogers WA, O'Brien TJ (1993) MFIX documentation, theory guide, vol 1, NTI Service (ed), Springfield, VA
22. Kothari AK (1967) M.S., Chemical Engineering, Illinois Institute of Technology, Chicago, Illinois
23. Assari MR, Tabrizi HB, Saffar-Avval M (2007) Numerical simulation of fluid bed drying based on two-fluid model and experimental validation. Applied Therm Eng 27:422–429
24. Debaste F, Halloin V, Bossart L, Haut B (2008) New modeling approach for the prediction of yeast drying rates in fluidized beds, J Food Eng 84:335–347
25. Gunn DJ (1978) Transfer of heat or mass transfer to particles in fixed and fluidised beds. Int J Heat Mass Transfer 21:467–476
26. Crank J (1975) The mathematics of diffusion, 2nd ed. Clarendon Press, Oxford, England
27. Johnson PC, Jackson R (1987) Frictional collisional constitutive relations for antigranulocytes-materials, with application to plane shearing. J Fluid Mech 176:67–93

Chapter 13
Hydrodynamics of Fluidization with Surface Charge

Hamid Arastoopour, Dimitri Gidaspow, and Robert W. Lyczkowski

13.1 Introduction

In 1968, the first gas phase fluidized bed polymerization reactor was commercialized by Union Carbide for producing high density polyethylene (HDPE). They extended this process for producing linear low density polyethylene (LDPE) in 1975 and for polypropylene (PP) in 1985 [1, 2].

Although there are several advantages for using gas/solid fluidized bed reactors, there are also some problems [3]. The electrostatic force causes the particles to stick to the wall. This is known as wall sheeting. Hendrickson [4] has provided a thorough review of electrostatics in gas phase fluidized bed polymerization reactors. Yao et al. [5] found that an increase in relative humidity of fluidizing gas enhances the charge dissipation due to an increase in the surface conductivity of the particles. Wolny and Opalinski [6] added fine particles to large dielectric particles to neutralize the electric charge. Boland and Geldart [7] investigated the charge in fluidized beds. They found that the charge is generated around gas bubbles in the wake region and that the amount of charge increases with particle size because of enhanced interparticle contact.

Recently, Giffin and Mehrani [8] used Faraday cups to measure particle charge in bubbling and slugging fluidized bed columns filled with polyethylene particles. They found that the fines were predominantly positively charged, while the wall and dropped particles were predominantly negatively charged. Sowinski et al. [9] proposed the hypothesis that the particles adhere to the wall due to their positive charge. Rokkam et al. [3] developed the electrostatic model for fluidization of multi-size particles carrying different surface charges. Rokkam et al. [10] published a review of computational modeling of gas/solid fluidized bed polymerization reactors.

In this study, the charge-to-mass ratio of polypropylene particles in a laboratory fluidized bed has been measured and the hydrodynamics of fluidization has been computed using the kinetic theory based model. The force generated due to the charge of the particles was included in the model by solving the Poisson's equation

© Springer Nature Switzerland AG 2022
H. Arastoopour et al., *Transport Phenomena in Multiphase Systems*, Mechanical
Engineering Series, https://doi.org/10.1007/978-3-030-68578-2_13

for the voltage generated in the fluidized bed. This model explains the sheeting phenomenon described by Hendrickson [4].

Sinclair and Jackson [11] were the first to use the kinetic theory model to compute the core-annular regime for riser flow. Al-Adel et al. [12] showed that the particles move toward the wall due to the presence of surface charge. In the lamella electrosettler [13] and electrostatic separator [14] used in this study, the effect of surface charge on the voltage distribution has been neglected.

13.2   Fluidized Bed to Determine Charge-to-Mass Ratio

A two-dimensional bed was constructed from glass with an internal cross-section of 2.54 cm × 8.26 cm. The apparatus consisted of two parts: the top part, 124.46 cm tall, which contained the particles, and the bottom distributor part, 17.78 cm tall. At the base of the rectangular glass bed was a fine 304 L stainless steel wire support grid, 8 × 8 mesh. The set-up and the experimental procedure were similar to that of Kashyap et al. [15] for nanoparticles. Compressed air was conditioned prior to entering the fluidized bed. The air was blown through a filter to remove water and oil from the air stream. The air flow rate to the fluidized bed was regulated by directing the air stream through a rotameter with a manual valve. Air from the fluidized bed was discharged to the atmosphere. A schematic diagram of the fluidized bed is shown in Fig. 13.1. The fluidized bed was charged with polypropylene particles with particle size of 831 μm and particle density of 910 kg/m^3.

Bed expansion was measured as a function of applied electric field at several velocities. Figure 13.2 shows the experimental results. At low velocities, the bubbles are small and hence the bed expansion is low. The electric field decreases the bed expansion substantially, especially at high velocities.

The surface charge of particle, q_s, was determined from a steady state one-dimensional momentum balance on the particles,

$$g\varepsilon_s(\rho_s - \rho_g) + \varepsilon_s\rho_s q_s E = \beta_B v_g \qquad (13.1)$$

buoyancy + electric force = drag

As the first approximation, the electric field, E, was determined by dividing the applied voltage by the distance between the electrodes. Figure 2 in the Rokkam et al. paper [3] is in agreement with such an approximation. The voltage at all boundaries was set to zero. In this experiment, the voltages are zero at the left and right boundaries. The actual value of the electric field in the direction of gravity has to be determined iteratively. Of course, the most accurate way to determine E is to measure the voltage distribution in the bed by insertion of a stiff probe, with a metal tip connected to the electrometer used to measure the current in this study. In the one-dimensional balance, the average of velocity of the solid is zero in Eq. (13.1).

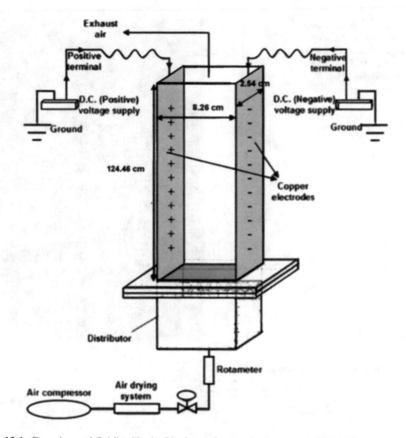

Fig. 13.1 Experimental fluidized bed with electrodes attached to two parallel walls

Table 13.1 summarizes the surface charge determination. The drag coefficient, β_B, was determined with no applied electric field. The average charge agrees with the charge per unit mass of 869 μm polyethylene particles reported in Table 1 in the Hendrickson [4] paper.

13.3   Experimental Bubbles and Bed Expansion

Typical bubble sizes are shown in Fig. 13.2. Without the application of an electric field, the bubbles are large. Either with or without the application of an electric field, at first, a small bubble forms at the bottom of the bed. The bubble then grows bigger and moves to one side of the wall. The bubble at this time has a clear nose and wake region. Finally, the bubble goes to one side of the wall.

Fig. 13.2 Volume fractions
of polypropylene particles
with no applied electric field
at gas velocity of 53 cm/s.
Left: 51.00 min, right:
51.03 min

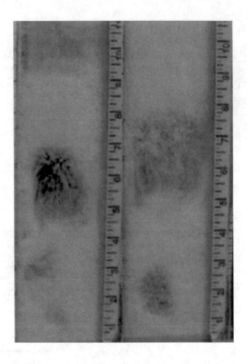

Table 13.1 Charge determination

Voltage (V)	Charge per unit volume (μC/cm^3)			
	v = 53 cm/s	v = 61 cm/s	v = 69 cm/s	v = 77 cm/s
12	0.46	0.43	0.53	0.61
14	0.47	0.48	0.58	0.64
16	0.51	0.47	0.59	0.65

Figure 13.3 shows the effect of the electric field on the bed expansion for several gas velocities. The electric force reduces the bed expansion and hence the observed bubble sizes.

13.4 Hydrodynamic Model with Surface Charge

The hydrodynamic model used in this study is similar to that described in Chap. 2 and [16, 17]. The main addition to the model is the electrical force generated by the surface charge of the particles. This approach is similar to the one-dimensional simulation of Al-Adel et al. [12], but here the Poisson's equation for voltage is solved in two dimensions. The standard drag correlation was used. The IIT computer code described in the Gidaspow and Jiradilok [18] book was used with the addition

Fig. 13.3 Effect of electric field on the bed expansion ratio (BER) at various superficial gas velocities

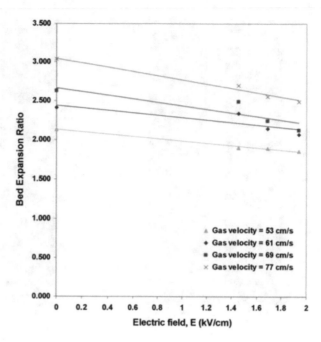

Table 13.2 Poisson's equation

$\nabla \cdot E = \frac{\rho}{\varepsilon}$
$E = -\nabla \Psi$
hence $\nabla \cdot \nabla \Psi = \nabla^2 \Psi = -\frac{\varepsilon_s \rho_s q_s}{\varepsilon_r \varepsilon_0}$
where E is the electric field (Volt/m), q_s is the specific charge (C/kg)
ε_r is dielectric constant, ε_0 is permittivity of free space
[8.8542×10^{-12} C^2/(Nm^2) and Ψ is a scalar electric potential field (Volt)
Boundary conditions for Poisson's equation are:
X-direction: Ψ = applied value
Z-direction: $\frac{\partial \Psi}{\partial z} = 0$

Table 13.3 System properties

Reactor diameter	8.75 cm
Reactor inlet diameter	8.25 cm
Reactor height	124.46 cm
Particle size	841 μm
Particle density	910 kg/m^3
Grid size	0.25 cm × 0.98 cm
Grid number ($\Delta X \times \Delta Y$)	35 (radial) × 129 (axial)
Time step	10^{-6}

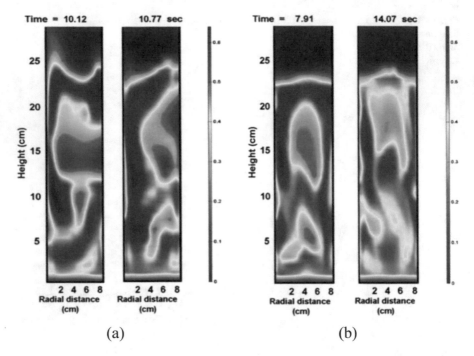

Fig. 13.4 Volume fractions of polypropylene particles: (**a**) without electric charge, and (**b**) with constant particle charge density of 0.5 µC/kg, both at gas velocity of 53 cm/s in the IIT fluidized bed simulations

of the Poisson's equation (Table 13.2) for the voltage. Table 13.3 gives the system properties for the simulation of the experiment.

Figure 13.4a shows the formation and propagation of bubbles with no surface charge of particles for the condition of the experiment. The bubbles are at the right wall. Figure 13.4b shows the computed bubbles with the measured surface charge of particles. The computed bubbles are considerably smaller than those without charge, in agreement with the smaller bed expansion. Figure 13.4b also shows that the bubbles are at the center of the bed. The bubble sizes qualitatively agree with the experiment shown in Fig. 13.2.

A qualitative comparison of the computed solid volume fraction with and without the effect of the surface charge is given in Fig. 13.5 at a gas velocity of 53 cm/s. The most significant difference is the higher solid volume fraction near the wall and the bottom of the bed. Without surface charge, the solid volume fraction has a peak at the center of the bed because the bubbles tend to form between the center of the bed and the walls.

Figure 13.6 shows the computed axial velocities with and without consideration of surface charge. Without charge, the particles move down at the center of the bed, consistent with the high solid volume fraction near the center of the bed shown in Fig. 13.5. With surface charge, however, the highest velocity is near the center of the

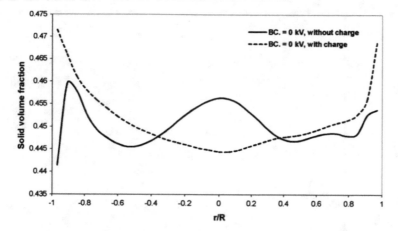

Fig. 13.5 Comparison of time-averaged solid volume fraction of polypropylene particles, with and without charge at gas velocity of 53 cm/s and height of 0.49 cm above the grid

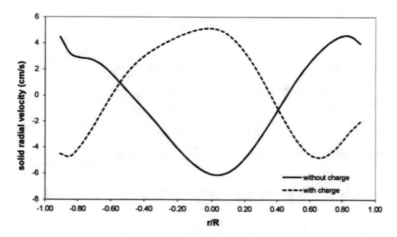

Fig. 13.6 Comparison of time-averaged axial velocity of polypropylene particles, with and without charge, at gas velocity of 53 cm/s and height of 0.49 cm above the grid

bed, consistent with the centrally located bubbles shown in Fig. 13.4b. Figure 13.7 shows the computed granular temperatures with and without charge compared to literature values. The computations of granular temperature are fully described in Gidaspow and Jiradilok [18]. The presence of the surface charge decreases the granular temperature.

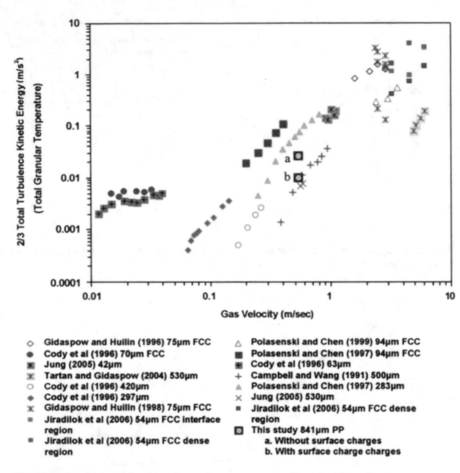

Fig. 13.7 Comparison of computed granular temperatures with and without charge

13.5   Surface Charge from Current Measurement

To measure the steady-state current produced by collision of polypropylene parti-cles, a 4.81 cm² platinum plate was attached to the left wall of the two-dimensional column 5.08 cm above the support grid. The platinum plate was connected with a Keithley 610C electrometer by a copper wire. Figure 13.8 shows the measured current density as function of gas velocity. At first, the current increases with gas velocity due to an increase in the number of collisions per unit time. The current then decreases due to decreased particle density.

The surface charge per unit mass can be calculated from the current density and wall collision frequency as shown below. The molecular wall collision frequency is

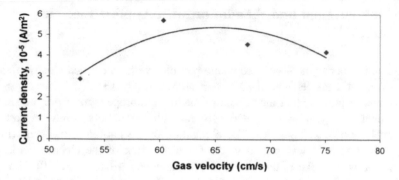

Fig. 13.8 Steady-state current produced by collision of 841 μm polypropylene particles fluidized in the two-dimensional glass fluidized bed

given on page 39 of the transport phenomena book [19]. In terms of granular temperature, θ, and the number of particles per unit volume, n,

$$\text{wall collision frequency} = \sqrt[n]{\frac{\theta}{2\pi}} \qquad (13.2)$$

hence, the current density, i, is,

$$i = \varepsilon_s \rho_s q_s \sqrt[n]{\frac{\theta}{2\pi}} \qquad (13.3)$$

where q is charge per unit mass.

For a gas velocity of 53 cm/s, Fig. 13.8 shows the current density to be 2.87×10^{-5} A/m^2. From the simulations, the granular temperature was estimated to be 0.01 (m/s)^2 and a solid volume fraction of 0.35 near the platinum plate. Equation (13.3) then provides the charge per unit mass of 2 μC/kg, compared to 0.5 μC/kg used in the simulations. To obtain better agreement, the simulations need to be repeated with the new charge. With the higher charge, it is expected that the volume fraction near the wall will increase and the granular temperature will decrease. To obtain a better value of charge, however, both the granular temperature and solid volume fraction need to be measured because, in the simulation, the electric field was computed from electrostatics. The measurement of the current reported here clearly shows that the polypropylene particles carry a charge significant to change the particle distribution in the fluidized beds.

13.6   Experiment and Simulation with Applied Electric Field

The bubbling behavior of polypropylene particles with an applied electric field of 16.9 kV/cm at a gas velocity of 53 cm/s is shown in Fig. 13.9.

The results indicate that the bubble formation and propagation from the simulation shown in Fig. 13.4a, b are consistent with the bubble behavior from the experiment shown in Fig. 13.9. The elongation of the bubble in the bed with an applied electric field was placed in the same direction of the electric field. This observation was similar to the results of Wittmann and Ademoyega [20] who gave an idealization that the bubble shapes in electrofluidized beds can be oblate ellipsoids in a three-dimensional bed and elliptical cylinders cross-section in a two-dimensional bed. In this study, the flat bubble formed near the bottom of the bed before moving to one side of the wall. The shape of the bubble changed from flat to elliptical cylinders. The bubble at the top tended to combine with the other bubble that located at their tail before bursting at the bed surface.

The computed voltages with an applied electric field are shown in Fig. 13.10. The voltage distribution shown by the dashed line in Fig. 13.10 without the effect of surface charge is linear.

The presence of surface charge produces parabolic voltage distributions for both positive and negative surface charges that are mirror images of each other. Figure 13.11 shows the computed freezing of the fluidized bed. The gas forms a

Fig. 13.9 Volume fractions of polypropylene particles with applied electric field of 16.9 kV/cm at gas velocity of 53 cm/s. Left: 5.30 min, right: 8.20 min

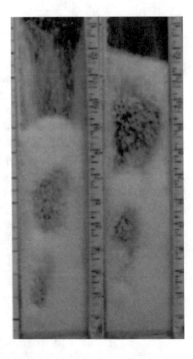

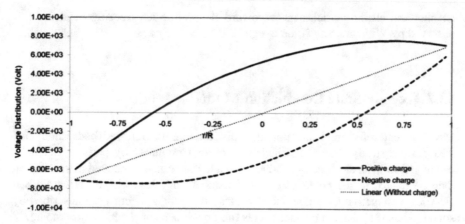

Fig. 13.10 Instantaneous potential at constant particle charge density 0.5 µC/kg and gas velocity 53 cm/s

Fig. 13.11 Computed freezing of bed due to a high applied potential. Volume fraction of polypropylene particles with 16.9 kV/cm electric field strength, at 20.0 s, gas velocity of 53 cm/s, with constant particle charge density of 2.4 µC/kg in the IIT fluidized bed simulation

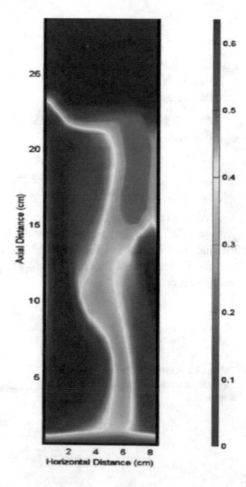

channel and all the particles move toward the wall. Such freezing behavior was observed by Wittmann and Ademoyega [20].

13.7 Steady-State Conduction Model

Measurements of current and potential distributions in fluidized beds show that frequently a steady state is reached. The current produced by particle collisions can then flow to an electrically conductive wall. The voltage generated in the bed can then be related to the potential gradient using Ohm's law with an effective bed electrical conductivity. Charge and discharge of storage batteries are described by such models [21]. Such a balance yields the one-dimensional conduction model,

$$\kappa \frac{d^2\Phi}{dx^2} = \frac{3\varepsilon_s i}{2d_p} \tag{13.4}$$

where,

i is the production of current per unit area of particle, amp/m^2
Φ is the electric potential
κ is the bed electrical conductivity

At x = 0 and at x = D, the electric potential is taken to be zero. This gives the dimensionless voltage,

$$\overline{\phi} = \frac{\phi d_p \kappa}{1.5\varepsilon_s D^2 i} \tag{13.5}$$

A comparison to the electrostatic model in Table 13.4 shows that this dimensional voltage differs from the electrostatic model by the substitution of the conductivity for the dielectric constant of the bed, and the current production for particle surface charge. The maximum dimensionless voltage is again 1/8. The maximum actual voltage becomes,

Table 13.4 One-dimensional approximation of voltage distribution	$\dfrac{\partial^2 \overline{\Psi}}{\partial \overline{x}^2} = -1$
	$\overline{\Psi} = \left(\dfrac{\epsilon_r \epsilon_0}{\varepsilon_s \rho_s q_s D^2}\right)\Psi$ and $\overline{x} = \dfrac{x}{D}$
	The maximum voltage potential is then [18],
	$\Psi_{max} = \dfrac{1}{8}\dfrac{\varepsilon_s \rho_s q_s D^2}{\epsilon_r \epsilon_0}$
	where D is the reactor diameter (m)

$$\phi_{max} = \frac{\varepsilon_s D^2 i}{16 d_p \kappa} \tag{13.6}$$

We see that a low bed conductivity produces high voltages. The use of humid air will thus decrease the electrostatic effect in the bed because the electrical conductivity of air changes with moisture content, while the dielectric constant in the electrostatic model does not.

The conduction model gives the following expression for the current measured at $x = 0$,

$$\text{current per unit wall area} = \frac{3 D \varepsilon_s i}{4 D_p} \tag{13.7}$$

It can be observed that the measured current increases with bed diameter, surface to volume ratio, and bed volume fraction. In terms of this measured current density, the maximum bed voltage becomes,

$$\phi_{max} = \frac{i_{measured} D}{4\kappa} \tag{13.8}$$

Typical values of the measured current densities in polypropylene fluidized beds are $i_{measured}$ $i_{\text{measured}} = 5 \times 10^{-6}$ amp/m^2.
For air,

$$\kappa = 10^{-14} \text{ mho/m}$$

The conductivity of polypropylene is even lower than that of air. Without particle motion, the maximum voltage would be huge. The mechanism of charge transfer in a fluidized bed, however, is due to the motion of the particles that transfer charge from one particle to another, like molecules for conduction. To obtain a theoretical value of the voltage, the kinetic theory of particles must be extended to charge transfer. Clearly, the conductivity of the fluidized bed must involve the particle motion in the bed.

13.8 Simulation of Commercial Bed (Sheeting Behavior in Commercial Polymerization Reactor)

Hendrickson [4] has reviewed the sheeting phenomenon in polymerization reactors. To compute the sheeting phenomenon in this study, uniform particle size and isothermal operation were assumed without consideration of polymerization reactions. In the study of Rokkam et al. [3], the fines were positively charged and the large particles carried a negative charge. Hence, both possibilities were explored in this study.

Fig. 13.12 Typical
commercial fluidized bed
polymerization reactor
configuration

Table 13.5 Commercial
polymerization reactor
geometry

Diameter of reaction zone	D_{rea}
Diameter of disengagement zone	1–1.4 D_{rea}
Height of reaction zone	1.9 D_{rea}
Height of disengagement zone	2.0 D_{rea}
Particle size	841 μm
Particle density	910 kg/m^3
Grid size	5 cm × 5 cm
Time step	10^{-6}

Figure 13.12 shows the geometry of a typical polymerization reactor. Tables 13.5 and 13.6 show the geometry and the parameters used in the simulation of the commercial fluidized bed.

Figure 13.13 shows the computed solid volume fractions with and without surface charge for the commercial polymerization fluidization reactor at time equal

Table 13.6 Parameters used for the commercial polymerization simulation

Description	Value
Gas density	1.54 kg/m^3
Solid density	910 kg/m^3
Solid diameter	841 μm
Restitution coefficient between particles	0.90
Restitution coefficient between particle and wall	0.60
Specularity coefficient	0.60
Simulation time	150 s
Inlet conditions	
Inlet solid velocity	0.10 m/s
Inlet gas velocity	0.55 m/s
Inlet gas temperature	334.20 K
Operating conditions	
Pressure	1.216 Pa
Temperature	345.80 K

Fig. 13.13 Computed solid volume fraction flow structure at 150 s: (**a**) with surface charge, (**b**) without surface charge

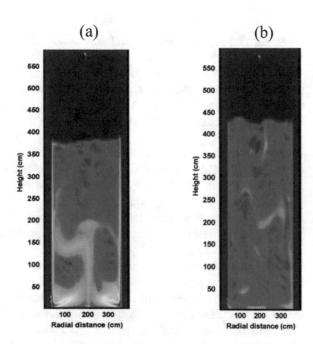

to 150 s. The surface charge was assumed to be positive and equal to 0.5 μC/kg. With the surface charge, the particle volume fractions are much higher at the walls, producing the sheeting phenomenon. With charge, there is also a higher concentration at the center near the bottom of the bed, producing down flow. Also similar to the experiment, the bed expansion with surface charge is smaller.

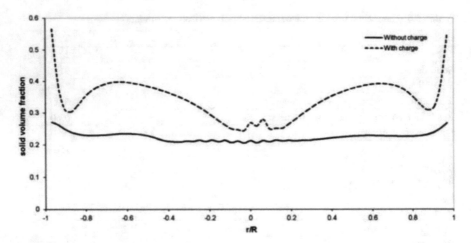

Fig. 13.14 Comparison of time-averaged solid volume fraction with and without surface charge at bed height of 15 cm

Figure 13.14 shows the computed time-averaged solid volume fractions with and without charge. With the presence of the surface charge, the volume fraction at both walls is greater than 0.5 and is near minimum packing. There is also very little motion next to the walls. Hence, the particles, which are sticky at high temperatures, agglomerate and form a sheet. Without surface charge, the solid volume fraction in the bed is almost uniform with an average value of 0.25, with an only slight increase next to the walls.

Figure 13.15a shows the instantaneous particle size distribution with a negative charge of 0.5 μC/kg. The behavior is similar to that for the positive charge in Fig. 13.13. Figure 13.15b compares the time-averaged solid volume fraction for positive and negative charges at two locations. For both positive and negative charges, the particle volume fractions at both walls are near maximum packing, producing the sheeting phenomenon. The corresponding voltage distributions are shown in Fig. 13.15c.

Clearly, the negatively charged particles repel each other and move to the walls. Similarly, positive particles move from a higher potential at the center, produced by the surface charge, to the walls maintained at zero voltage. Figure 13.16 shows the computed maximum voltage distribution at various heights compared to the one-dimensional analytical solution of the Poisson's equation. The non-uniform voltage values in the analytical solution are due to various computed solid volume fractions. Figure 13.16 also shows that the computation times of 60 and 120 s are close to the steady-state values obtained from the one-dimensional analytical solution.

Figure 13.17 shows a comparison of the electrical and gravity forces acting on the particles. The electrical force is of the order of the gravity at the walls and is zero at the center. The large electric force at the walls gives rise to the sheeting

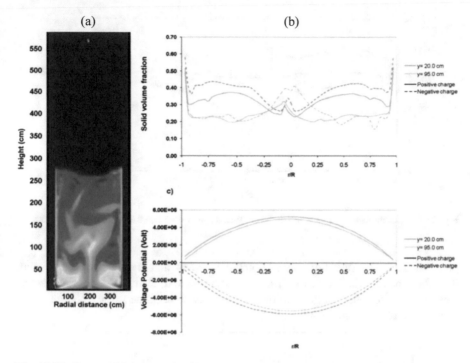

Fig. 13.15 Computed fluidization behavior with negative surface charge: (**a**) volume fraction flow structure, (**b**) time-averaged solid volume fraction, and (**c**) voltage distribution

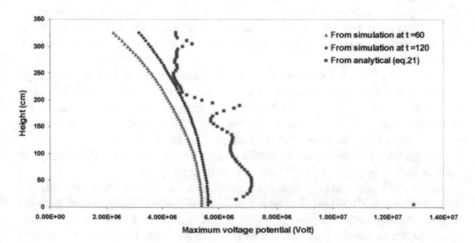

Fig. 13.16 Computed instantaneous maximum voltage potential at constant charge density of 0.5 μC/kg

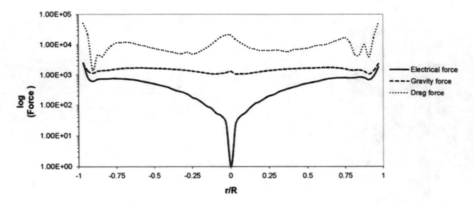

Fig. 13.17 Comparison between electric force, gravity, and drag force at bed height of 30 cm

phenomenon. The one-dimensional analysis shows that maximum voltage and, therefore, the electrical force increases with the square of the diameter of the reactor. Hence, in the commercial fluidization reactor, the effect of the surface charge of the particles becomes highly significant.

13.9   Conclusion

The surface charge-to-mass ratio for 831 μm polypropylene particles was measured in a small bubbling fluidized bed with an applied voltage on the two sides of the bed. The charge-to-mass ratio was 0.5 μC/kg, in agreement with previous measurements. The charge-to-mass ratio was determined from the effect of the electric field on the bed expansion, using a particle momentum balance, similar to the study of the effect of the electric field on the hydrodynamics of fluidized nanoparticles by Kashyap et al. [15].

The effect of the surface charge of the particles was added to a standard type kinetic theory based hydrodynamic model of fluidization. This was done by adding the electrical force to the particle momentum balance equation and by adding the Poisson's equation for voltage distribution. This model generalizes the one-dimensional approach of Al-Adel et al. [12] for the riser.

The partial differential equations with charge were added to the CFD computer code described in the book by Gidaspow and Jiradilok [18].

For the laboratory experiment, the computed solid volume fractions with and without charge in the model showed a significant difference. For example, the solid volume fraction with charge was higher at the walls. This effect is due to the electrical force created by the surface charge of the particles. The CFD computations show that the electrical force at the wall is of the order of magnitude of gravity. The electrical force is very small away from the wall. The CFD simulations show that with an applied electric field on both walls, the voltage distribution is significantly different from the linear case with no particle charge. With a high applied potential, the CFD code shows that the bed ceases to move, as observed by Wittmann and Ademoyega [20].

The CFD computations were performed for a commercial fluidized bed reactor. The reactor wall sheeting phenomena described by Hendrickson [4] were computed. The solid volume fractions were of the order of 0.5 near the walls, compared with an average solid volume fraction in the bed of about 0.25. This phenomenon is due to a high value of electrical field generated by the charge of particles. An approximate analytical solution of Poisson's equation shows that the maximum voltage varies as the square of the bed diameter. Hence, for large fluidized beds, the effect of the presence of surface charge of particles is much larger than that found in laboratory reactors.

The computed granular temperatures of polypropylene particles compare well with literature correlations. These granular temperatures were used to estimate the charge-to-mass ratio of polypropylene particles from measurements of current at four different gas velocities. The computed charge-to-mass ratio from the measurement of current is approximately -2 µC/kg. The charge-to-mass ratio obtained by Gajewski [22] for polystyrene particles flowing in the riser is close to this value at the same gas velocity and particle concentration. This value of surface charge is somewhat higher than the value used in this study to simulate the sheeting behavior of the commercial fluidized bed. Hence, the sheeting problem may have been underestimated. This novel technique of measuring the charge-to-mass ratio from the measurement of current can be applied to the commercial polymerization reactor easier than the methods used in the past.

13.10 Exercise

13.10.1 Ex. 1: Comparison Between Two Electrostatic Charging Models

1. Although electrostatic charging is still poorly understood [23], there are several models in the literature. Compare the model for electrostatic charging used in this chapter to that used by Rokkam et al. in Ref. [10].

References

1. Yang WC (2003) Handbook of fluidization and fluid-particle systems. Marcel Dekker, New York
2. Xie T, McAuley KB, Hsu JCC, Bacon DW (1994). Gas phase ethylene polymerization: production processes, polymer properties, and reactor modeling. Ind Eng Chem 33:449–479

3. Rokkam RG, Fox RO, Muhle ME (2010) Computational fluid dynamics and electrostatic modeling of polymerization fluidized-bed reactors. Powder Technol 203: 109–124
4. Hendrickson G (2006) Electrostatics and gas phase fluidized bed polymerization reactor wall sheeting. Chem Eng Sci 61:1041–1064
5. Yao L, Bi HT, Park, AH (2002) Characterization of electrostatic charges in freely bubbling fluidized beds with dielectric particles. J Electrostatics 56:183–197
6. Wolny A, Opaliñski I (1983) Electric charge neutralization by addition of fines to a fluidized bed composed of coarse dielectric particles. J Electrostatics 14:279 –289
7. Boland D, Geldart D (1971/72) Electrostatic charging in gas fluidized beds. Powder Technol 5:289–297
8. Giffin A, Mehrani P (2010) Comparison of influence of fluidization time on electrostatic charge build-up in the bubbling vs. slugging flow regimes in gas-solid fluidized beds. J Electrostatics 68:492–502
9. Sowinski A, Miller L, Mehrani P (2010) Investigation of electrostatic charge distribution in gas solid fluidized beds. Chem Eng Sci 65:2771–2781
10. Rokkam RG, Fox R, Muhle M (2010) Computational modeling of gas-solid fluidized-bed polymerization reactors, chapter 12. In: Pannala S, Syamlal M, O'Brien TJ (eds) Computational gas-solids flows and reacting systems: theory, methods and practice. Engineering Science Reference, New York
11. Sinclair JL, Jackson R (1989) Gas–particle flow in a vertical pipe with particle–particle interactions. AIChE J 35:1473–1486
12. Al-Adel MF, Saville DA, Sundaresan S (2002) The effect of static electrification on gas-solid flows in vertical risers. Ind Eng Chem Res 41:6224–6234
13. Gidaspow D, Shih YT, Bouillard J, Wasan D (1989) Hydrodynamics of a lamella electrosettler. AIChE J 35:714–724
14. Shih YT, Gidaspow D, Wasan DT (1987) Hydrodynamics of electrofluidization: separation of pyrites from coal. AIChE J 33:1322–1333
15. Kashyap M, Gidaspow D, Driscoll M (2008) Effect of electric field on the hydrodynamics of fluidized nanoparticles. Powder Technol 183:441–453
16. Jiradilok V, Gidaspow D, Damronglerd S, Koves WJ, Mostofi R (2006) Kinetic theory based CFD simulation of turbulent fluidization of FCC particles in a riser. Chem Eng Sci 61:5544–5559
17. Arastoopour H, Gidaspow D, Abbasi E (2017) Computational transport phenomena of fluid-particle systems. Mechanical engineering series, ISBN 978-3-319-45488-7, Springer Press
18. Gidaspow D, Jiradilok V (2009) Computational techniques: the multiphase CFD approach to fluidization and green energy technologies. Nova Publishers, New York
19. Bird RB, Stewart WE, Lightfoot EN (2001). Transport phenomena. John Wiley & Sons, Inc., New York
20. Wittmann CV, Ademoyega BO (1987) Hydrodynamic changes and chemical reaction in a transparent two-dimensional cross-flow electrofluidized bed. 1. experimental results, Ind Eng Chem Res 26:1586–1593
21. Gidaspow D, Baker BS (1983) A model for discharge of storage batteries. J Electrochem Soc 120:1005–1010
22. Gajewski A (1989). Measuring the charging tendency of polystyrene particles in pneumatic conveyance. J of Electrostatics 23:55–66
23. Park AH, Bi H, Grace, JR (2002) Reduction of electrostatic charges in gas-solid fluidized beds. Chem Eng Sci 57:153 –162

Appendix: A Generalization of Onsager's Multicomponent Diffusion Equations

A.1 Introduction

Onsager [1] derived the fundamental equation for multicomponent diffusion by means of a variation principle. He maximized the negative sum of the integral of the free energy plus the dissipation and obtained the fundamental equation of multicomponent diffusion that relates the gradient of the chemical potential to the sum of the friction coefficients multiplied by fluxes. Lamm [2] and others have achieved significant progress in understanding multicomponent isothermal diffusion starting with this fundamental equation. Onsager restricted his treatment to the case of vanishing volume average velocity. He stated that such a situation could be achieved by neglecting volume changes due to mixing. Onsager also neglected the kinetic energy of diffusion. DeGroot and Mazur [3] present a form of such kinetic energy for a multicomponent mixture. Neglect of such a kinetic energy produces generalizations of Fick's law that have the same fault as the usual Fick's law. Contrary to physical reality, concentration pulses propagate at infinite rates. The treatment below permits the addition of the kinetic energy usually neglected.

A.2 Onsager's Dissipation Function

Onsager's [1] dissipation function, F, with sign changed is,

$$F = -\frac{1}{2} \sum_{i=1}^{N} \sum_{k=1}^{N} R_{ik} J_i^0 J_k^0 \tag{A.1}$$

The fluxes of i-th component, J_i^0, are referred to the zero bulk velocity of the fluid and for later convenience expressed in mass units by,

© Springer Nature Switzerland AG 2022
H. Arastoopour et al., *Transport Phenomena in Multiphase Systems*, Mechanical Engineering Series, https://doi.org/10.1007/978-3-030-68578-2

$$J_i^0 = \rho_i v_i \tag{A.2}$$

where ρ_i is the i-th concentration, say grains per cubic liter, and v_i is the i-th species velocity. The coefficients R_{ik} have become known as the friction coefficients. The origin of Eq. (A.1) [4] is the assumption that the forces X_i,

$$X_i = \sum_{k=1}^{N} R_{ik} J_k^0 \tag{A.3}$$

can be expressed as a linear combination of the fluxes. Entropy production [4] is the reciprocal absolute temperature times the dissipation function, consistent with the understanding that dissipation produces heat.

$$2F \geq 0 \tag{A.4}$$

and,

$$R_{ik} = R_{ki} \tag{A.5}$$

noting that it is the sum

$$R_{ik} J_i J_k + R_{ki} J_k J_i = (R_{ik} + R_{ki}) J_i J_k \tag{A.6}$$

that enters into the dissipation function.

Since there is no dissipation when all species velocities are equal,

$$-F(\rho_i v_i, \rho_i v_i) = \frac{1}{2} \sum_{i=1}^{N} \rho_i v_i^2 \sum_{k=1}^{N} R_{ik} \rho_k = 0 \tag{A.7}$$

This yields the important relation between friction coefficients (Onsager's Eq. 7b, [1]),

$$\sum_{k=1}^{N} R_{ik} \rho_k = 0 \tag{A.8}$$

For N diffusing components, Eq. (A.8) gives N relations between friction coefficients. Together with the reciprocal relations given by Eq. (A.5), as noted by Onsager, there are then $\frac{1}{2}(N^2 - N)$ independent friction coefficients. Usually Eq. (A.8) is used to eliminate the diagonal friction coefficients, R_{ii}. For a 2×2 case, the dissipation function, F, can be easily shown to give,

$$F_{2\times2} = \frac{1}{2}R_{12}\rho_1\rho_2(v_1 - v_2)^2 \tag{A.9}$$

Equation (A.9) was obtained using the definition of F given by Eq. (A.1), the reciprocal relation, and Eq. (A.8). To obtain a positive sign for R_{12}, the definition of F was taken to be the negative of one first given by Onsager [1]. Lamm [2] already noted the need for this slight change. Equation (A.9) shows that dissipation is caused by relative motion only.

The variation of F can be shown to be,

$$-\delta F = \sum_{i=1}^{N}\sum_{k=1}^{N} R_{ik}J_k\delta J_i \tag{A.10}$$

It is easy to see this for a 2×2 system.

$$-F(J_i) = \frac{1}{2}R_{11}J_1^2 + R_{12}J_1J_2 + \frac{1}{2}R_{22}J_2^2 \tag{A.11}$$

$$-F(J_i + \Delta J_i) = \frac{1}{2}R_{11}(J_1 + \Delta J_1)^2 + R_{12}(J_1 + \Delta J_1)(J_2 + \Delta J_2)$$
$$+ \frac{1}{2}R_{22}(J_2 + \Delta J_2)^2 \tag{A.12}$$

The difference becomes,

$$-F = R_{11}J_1\delta J_1 + R_{12}J_2\delta J_1 + R_{12}J_1\delta J_2 + R_{22}\delta J_2 + R_{22}\Delta J_2$$
$$+ \text{terms of } O(\Delta J_i)^2 \tag{A.13}$$

The variation of F is the expression on the right hand side of Eq. (A.13) as the second order terms vanish,

$$-\delta F = R_{11}J_1\delta J_1 + R_{12}J_2\delta J_1 + R_{12}J_1\delta J_2 + R_{22}\delta J_2 \tag{A.14}$$

Onsager's dissipation function can also be written in terms of fluxes relative to a mean velocity,

$$J_i = \rho_i(v_i - v) \tag{A.15}$$

where v is defined,

$$\rho v = \sum_{i=1}^{n} \rho_i v_i \tag{A.16}$$

yielding the usual restraint for diffusion of n species,

$$\sum_{i=1}^{n} J_i = 0 \tag{A.17}$$

To obtain the dissipation relative to the weight average velocity defined above, note that using the relation between friction factors, Eq. (A.8), F can be written,

$$-F = \frac{1}{2} \sum_{i=1}^{N} \sum_{k=1}^{N} R_{ik} J_i^0 J_k^0 - \frac{1}{2} v \sum_{i=1}^{N} J_i^0 \sum_{k=1}^{N} R_{ik} \rho_k$$

$$- \frac{1}{2} v \sum_{k=1}^{N} J_k^0 \sum_{i=1}^{N} R_{ik} \rho_k + \frac{1}{2} v^2 \sum_{k=1}^{N} \rho_k \sum_{i=1}^{N} R_{ik} \rho_i \tag{A.18}$$

since the last three sums are each zero. Rearrangement of Eq. (A.18) gives,

$$-F = \frac{1}{2} \sum_{i=1}^{N} \sum_{k=1}^{N} R_{ik} \rho_i (v_i - v) \rho_k (v_k - v) \tag{A.19}$$

A similar result is obtained using molar fluxes, $C_i(v_i - v^*)$.

A.3   Entropy with Kinetic Energy of Diffusion

In a manner identical to that presented by deGroot and Mazur [3], decompose the internal energy per unit mass of mixture, U^* into the usual internal energy U plus contributions due to the kinetic energies of the components relative to the weight average velocity,

$$U^* = U(S, V, \omega_1, \omega_2, \ldots, \omega_n) + \sum_{i=1}^{n} \frac{1}{2} \omega_i \left(\vec{v}_i - \vec{v} \right)^2 \tag{A.20}$$

Equations (A.20) and (A.32) of deGroot and Mazur [3] are identical. Equation (A.20) is now differentiated convectively, where,

$$\frac{d}{dt} = \frac{\partial}{\partial t} + \vec{v} \cdot \nabla \tag{A.21}$$

The result is deGroot and Mazur's Eq (36) [3],

$$T\frac{dS}{dt} = \frac{dU^*}{dt} + P\frac{dV}{dt} - \sum_{i=1}^{n}\left[\mu_i + \frac{1}{2}\left(\vec{v}_i - \vec{v}\right)^2\right]\frac{d\omega_i}{dt} - \rho^{-1}\sum_{i=1}^{n}\vec{J}_i$$
$$\cdot\frac{d\left(\vec{v}_i - \vec{v}\right)}{dt} \tag{A.22}$$

where the chemical potential is,

$$\mu_i = \left|\frac{\partial U}{\partial \omega_i}\right| \tag{A.23}$$

To obtain an expression for entropy production, replace dU^*/dt by its equivalent in terms of an energy balance, and $d\omega_i/dt$ must also be replaced by a relation involving relative fluxes, using mass balances.

A.4   Conservation of Species

The conservation of a species equation is,

$$\frac{\partial \rho_i}{\partial t} + \nabla \cdot \rho_i\vec{v_i} = r_iM_i \tag{A.24}$$

where r_i is the molar production of species i by chemical reaction. Using the definition of $\vec{J}_i$ it can be written,

$$\frac{d\rho_i}{dt} + \rho_i\nabla \cdot \vec{v}_i + \nabla \cdot \vec{J}_i = r_iM_i \tag{A.25}$$

Then using the identity,

$$\frac{d\rho_i}{dt} = \rho\frac{d\omega_i}{dt} + \omega_i\frac{d\rho}{dt} \tag{A.26}$$

and the sum of Eq. (A.25) over i,

$$\frac{d\rho}{\partial t} = \rho\nabla \cdot \vec{v} \tag{A.27}$$

In Eq. (A.26), the species balance can be written,

$$\rho \frac{d\omega_i}{dt} = -\nabla \cdot \vec{J}_i + r_i M_i \tag{A.28}$$

A.5 Mixture Energy Balance

An energy balance on a system of mass m, moving with the weight average velocity v can be written,

$$\frac{d\vec{U}^*}{dt} = \frac{dQ}{dt} - \frac{dW}{dt} \tag{A.29}$$

where

$$\vec{U}^* = \iiint\limits_{V(t)} U^* \rho dV \tag{A.30}$$

and where dQ/dt is the rate of heat entering the system, plus its production by dissipation, and dW/dt is the work done by the system per unit time. The rate of work done by the system to be considered is the work of expansion and the work done by the external forces $\vec{f}_i$ acting on component i. In terms of a mean pressure P over a small volume of the system V, the work becomes,

$$\frac{dW}{dt} = P \frac{d}{dt} \iiint\limits_{V(t)} dV - \iiint\limits_{V(t)} \sum_{i=1}^{n} \vec{J}_i \cdot \vec{f}_i dV \tag{A.31}$$

The heat entering the system consists of the conduction flux and a multicomponent convection flux considered by Merk [5] and Aris [6]. The latter effect arises due to an excess of mass flow into the system by diffusion. In other words, the system considered is an open system in thermodynamic terminology. With the mass flow, there is associated an energy flow and PV work. Therefore, the rate of heat transfer across a closed system surface S is,

$$-\frac{dQ_S}{dt} = \oiint_{S(t)} \vec{q} \cdot \vec{n} dS + \oiint_{S(t)} \sum_{i=1}^{n} h_i \vec{J}_i \cdot \vec{n} dS \tag{A.32}$$

where q is the rate of heat flux out of the system by conduction and h_i is the specific enthalpy of species i.

The transformation of the internal plus kinetic diffusion energies, U^* into heat by means of Onsager's dissipation function, F, must be incorporated because the balance is made using U^*. This dissipation can be included into heat Q_D,

$$\frac{dQ_D}{dt} = \iiint\limits_{V(t)} F dV \tag{A.33}$$

Then the total rate of heat production and net rate of outflow is,

$$\frac{dQ}{dt} = \frac{dQ_D}{dt} + \frac{dQ_S}{dt} \tag{A.34}$$

For the purpose of the energy balance, it does not matter whether the dissipation is considered as heat, work, or as an energy generation. Truesdell [7] treats it as the "rate of working of diffusing drags."

Application of the divergence theorem to Eq. (A.32), the Reynolds transport theorem to $d\vec{U}^* /dt$ and to $P\frac{d}{dt}\int dV$, and the usual contradiction argument applied to the arbitrary system volume V yield the differential form of the energy equation. Rates of chemical reactions, r_i are set to zero,

$$\frac{\partial(U^*\rho)}{\partial t} + \nabla \cdot U^*\rho\vec{v} + P\nabla \cdot \vec{v} + \nabla \cdot q + \nabla \cdot \sum_{i=1}^{n} h_i\vec{J}_i - \sum_{i=1}^{n}\vec{J}_i \cdot \vec{f}_i - F$$
$$= 0 \tag{A.35}$$

Using the continuity equation with zero r_i, the energy equation, Eq. (A.35), can be rewritten in convective form,

$$\frac{\partial U^*}{\partial t} = -V\nabla \cdot J_q - P\frac{dV}{dt} + V\sum_{i=1}^{n}\vec{J}_i \cdot \vec{f}_i + FV \tag{A.36}$$

where the specific volume, V, is the reciprocal of ρ in Eq. (A.27), and where the total heat flux $\vec{J}_q$ is defined,

$$\vec{J}_q = \vec{q} + \sum_{i=1}^{n} h_i\vec{J}_i \tag{A.37}$$

Equation (A.36) is an extended form of deGroot and Mazur's Eq. (39) [3]. The chief difference is that Eq. (A.36) contains the kinetic energies of diffusion as shown by Eq. (A.20) and the corresponding dissipation function. There is also no viscous dissipation which is included in deGroot and Mazur's equation. The heat flux J_q contains the multicomponent diffusion effects deleted in the latter, however.

Equation (A.36) is in the form needed to eliminate dU^*/dt in the entropy expression, Eq. (A.22).

A.6   Entropy Production

Substitution of the energy balance, Eq. (A.36), and the species balance, Eq. (A.28), into Eq. (A.22) gives an expression for the rate of change of the mixture entropy,

$$
\rho T \frac{dS}{dt} = -\nabla \cdot \vec{J}_q + \sum_{i=1}^{n} \vec{J}_i \cdot \vec{f}_i + F + \sum_{i=1}^{n} \left[\mu_i + \frac{1}{2}\left(\vec{v}_i - \vec{v}\right)^2 \right] \nabla \cdot \vec{J}_i
$$

$$
- \sum_{i=1}^{n} \vec{J}_i \cdot \frac{d\left(\vec{v} - \vec{v}\right)}{dt} \tag{A.38}
$$

To obtain the equation in a form in which the entropy production can be recognized, the first and fourth terms on the right side of Eq. (A.38) are put into a form of flux times a force using the identities,

$$
\nabla \cdot \frac{\vec{J}_q}{T} = \vec{J}_q \nabla \frac{1}{T} + \frac{1}{T} \nabla \cdot \vec{J}_q \approx -\frac{1}{T^2} \vec{J}_q \cdot \nabla T + \frac{1}{T} \nabla \cdot \vec{J}_q \tag{A.39}
$$

$$
\nabla \cdot \frac{\mu_i^*}{T} \vec{J}_i = \frac{\mu_i^*}{T} \nabla \cdot \vec{J}_i + \vec{J}_i \cdot \nabla \frac{\mu_i^*}{T} \tag{A.40}
$$

This gives an equation similar to that given by deGroot and Mazur [3],

$$
\rho \frac{dS}{dt} = -\nabla \cdot \left(\frac{\vec{J}_q - \sum_{i=1}^{n} \mu_i^* \vec{J}_i}{T} \right) - \frac{1}{T^2} \vec{J}_q \cdot \nabla T
$$

$$
- \sum_{i=1}^{n} \vec{J}_i \cdot \left[\nabla \frac{\mu_i + \frac{1}{2}\left(\vec{v}_i - \vec{v}\right)^2}{T} + \frac{1}{T}\frac{d\left(\vec{v}_i - \vec{v}\right)}{dt} - \frac{\vec{f}_i}{T} - \sum_{k=1}^{N} \frac{1}{2T} R_{ik} \vec{J}_k \right] \tag{A.41}
$$

Using the continuity equation, pdS/dt can be written in conservative form. Then this term and the first term in Eq. (A.41) can be written,

$$
\frac{\partial(\rho S)}{\partial t} + \nabla \cdot (\rho S v) + \nabla \cdot \frac{\vec{q}}{T} + \sum_{i=1}^{n} \nabla \cdot \vec{J}_i \left[S_i - \frac{1}{2}(v_i - v)^2 T^{-1} \right] = \sigma \tag{A.42}
$$

where S_i is the partial specific entropy of component i.

The second and third terms on the right-hand side of Eq. (A.41) are the entropy production σ,

$$
\sigma = -\frac{1}{T^2}\vec{J}_q \cdot \nabla T - \sum_{i=1}^{n}\vec{J}_i
$$

$$
\cdot \left[\nabla \frac{\mu_i + \frac{1}{2}\left(\vec{v}_i - v\right)^2}{T} + \frac{1}{T}\frac{d(v_i - v)}{dt} - \frac{f_i}{T} - \sum_{k=1}^{n}\frac{1}{2T}R_{ik}\vec{J}_k\right]
\tag{A.42a}
$$

A.7   Isothermal Equations of Motion

Under isothermal conditions, the variation of the entropy production is obtained from Eq. (A.41),

$$
T\delta\sigma = -\sum_{i=1}^{n}\delta\vec{J}_i\left[\nabla\mu_i + \frac{\partial\left(\vec{v}_i - \vec{v}\right)}{\partial t} + \vec{v}_i\nabla\cdot\left(\vec{v}_i - \vec{v}\right) - \vec{f}_i - \sum_{k=1}^{N}R_{ik}\vec{J}_k\right]
$$

$$
\tag{A.43}
$$

For a stationary state, the variation of the entropy production was set to zero. For arbitrary variations $\delta\vec{J}_i$, the equations of motion (component momentum balances) were obtained,

$$
\frac{\partial\left(\vec{v}_i - \vec{v}\right)}{\partial t} + \vec{v}_i\nabla\cdot\left(\vec{v}_i - \vec{v}\right) + \nabla\mu_i = \sum_{k=1}^{N}R_{ik}\vec{J}_k + \vec{f}
\tag{A.44}
$$

where $i = 1, 2, 3, \ldots n$ components. Equation (A.44) differs from the famous Onsager's Eq. (14) [1] principally by the presence of the acceleration terms on the left side of Eq. (A.44), which can be written as relative acceleration,

$$
\text{relative acceleration} = \frac{d\left(\vec{v}_i - \vec{v}\right)}{dt^i}
\tag{A.45}
$$

The component momentum balances reduce themselves to the fundamental multicomponent equation used by Lamm [1]. To obtain his Eq. (A.1), acceleration and body forces are neglected in Eq. (A.44). Further, the friction coefficient sum on the right-hand side of Eq. (A.44) is rearranged using Eq. (A.8). Thus, elimination of the diagonal terms yields the simplified momentum balances for each component i,

$$-\nabla \mu_i = \sum_{k=1}^{n} R_{ik} \rho_k \left(\vec{v}_i - \vec{v}_k \right) \tag{A.46}$$

The advantage of the form Eq. (A.46) is that the diagonal terms R_{ii} do not appear since the fluxes are relative to the component for which a momentum balance is made. Lamm's coefficient ϕ_{ij} is related to R_{ij} by means of the equation,

$$\phi_{ij} = \rho_i \rho_j R_{ij} \tag{A.47}$$

Equation (A.46) is usually the starting point for modern studies of multicomponent diffusion. Its generalization is Eq. (A.44). It can be used to simplify the entropy production in Eq. (A.41). Using Eq. (A.44), the entropy production becomes after substituting in the expression for σ into Eq. (A.41),

$$T\sigma = -\sum_{i=1}^{N} \sum_{k=1}^{n} \frac{1}{2} \left(R_{ik} \vec{J}_i \right) \cdot \vec{J}_k = F \tag{A.48}$$

Thus, heat production is via friction coefficients only, as should occur in practice. The other terms in Eq. (A.42) are the accumulation and the entropy flow terms.

The sum of the component momentum equations should yield the mixture momentum equation, that is, Newton's second law of motion. To obtain such a result, Eq. (A.44) is multiplied by ρ_i and summed over i. The Gibbs-Duhem relation at constant temperature,

$$\sum_{i=1}^{n} \rho_i \nabla \mu_i = \nabla P \tag{A.49}$$

is used. Furthermore Eq. (A.8) shows that the double sum involving the friction coefficients, R_{ik}, vanishes. This leaves the equation,

$$\sum_{i=1}^{n} \rho_i \frac{d\left(\vec{v}_i - \vec{v} \right)}{dt^i} + \nabla P = \sum_{i=1}^{n} \vec{f}_i \rho_i \tag{A.50}$$

The continuity equation, Eq. (A.24), is multiplied by $\left(\vec{v}_i - \vec{v} \right)$ to give, with zero r_i,

$$\left(\vec{v}_i - \vec{v} \right) \frac{\partial \rho_i}{\partial t} + \left(\vec{v}_i - \vec{v} \right) \nabla \cdot \left(\rho_i \vec{v}_i \right) = 0 \tag{A.51}$$

Addition of Eq. (A.51) summed over i to Eq. (A.50) yields the balance in conservative form,

$$\sum_{i=1}^{n} \frac{\partial}{\partial t}\left[\rho_i\left(\vec{v}_i - \vec{v}\right)\right] + \nabla \rho_i \vec{v}_i \left(\vec{v}_i - \vec{v}\right) + \nabla P = \sum_{i=1}^{n} f_i \rho_i \qquad (A.52)$$

The first sum in Eq. (A.52) vanishes in view of the definition of $\vec{v}$. The second sum can be rearranged using the identity that relates motion with the weight average velocity to that with individual components,

$$\sum_{i=1}^{n} \rho_i \frac{d\vec{v}_i}{dt^i} = \rho \frac{d\vec{v}}{dt} + \nabla \sum_{i=1}^{n} \left[\rho_i \vec{v}_i \cdot \left(\vec{v}_i - \vec{v}\right)\right] \qquad (A.53)$$

Identities of type Eq. (A.53) also hold for enthalpies, energies, etc. Using Eq. (A.53), Eq. (A.52) becomes the momentum balance,

$$\sum_{i=1}^{n} \rho_i \frac{d\vec{v}_i}{dt^i} - \rho \frac{d\vec{v}}{dt} + \nabla P = \sum_{i=1}^{n} \vec{f}_i \rho_i \qquad (A.54)$$

Without the term $\rho d\vec{v}/dt$, Eq. (A.54) is the mixture momentum balance with respect to a stationary frame of reference for inviscid flow. For example, it is Muller's Eq. (2.12) [8]. To see this, one uses Muller's identity,

$$\sum_{i=1}^{n} \rho_i \frac{d\vec{v}_i}{dt^i} = \rho \frac{d\vec{v}}{dt} + \nabla \sum_{i=1}^{n} \rho_i \left(\vec{v}_i - \vec{v}\right) \cdot \left(\vec{v}_i - \vec{v}\right) \qquad (A.55)$$

The reason that Eq. (A.54) lacked the term $\rho d\vec{v}/dt$ is because the system chosen was in the frame of reference moving with this velocity, v.

To determine motion of n components in a fluid, one also has to give an expression for the chemical potential, μ_i, as a function of the weight fractions, ω_i. The isothermal motion is then determined by the 3n momentum balances, Eq. (A.44), and by the n conservation of species balances, Eq. (A.24) with zero r_i. There are 4n variables: n vector quantities, v_i in three dimensions, and n weight fractions, ω_i. These quantities are determined by the system of 4n first order partial differential equations. In one dimension, the set of 2n equations can be solved by the method of characteristics. It is expected that the characteristics, that is, the directions along which the signals propagate, will be real and distinct. Then, given some initial distribution of weight fractions and velocities and properly assigned boundary conditions determined by the sign of characteristics (see Courant and Hilbert [9]), the problem will be well posed. Progress of motion and concentrations can then be marched out in time using the method of characteristics. When friction dominates, it is expected that the hyperbolic system will degenerate into a parabolic system, that is, one governed by Fick's type of diffusion with no inertia.

A.8   Fick's Law: Zero Acceleration

For zero acceleration and in the absence of external forces, the equation of motion, Eq. (A.44), for component "1" in a binary mixture is simply the usual balance between the gradient of chemical potential and the friction caused by relative motion. Equation (A.46) converted to molar units using $\vec{\bar{\mu}}_1 = M_1\mu_1$, $\rho_2 = M_2C_2$, and $\overline{R}_{12} = M_1M_2R_{12}$ becomes,

$$-C_1\nabla\bar{\mu}_1 = \overline{R}_{12}C_1C_2\left(\vec{v}_1 - \vec{v}_2\right) \tag{A.56}$$

The binary diffusivity, D, is usually [10] defined by,

$$C_1\left(\vec{v}_1 - \vec{v}^*\right) = -D_{12}C\nabla y_1 \tag{A.57}$$

For a binary mixture, the flux relative to the mean velocity can also be written,

$$C_1\left(\vec{v}_1 - \vec{v}^*\right) = C_1y_2\left(\vec{v}_1 - \vec{v}_2\right) = \frac{C_1C_2}{C}\left(\vec{v}_1 - \vec{v}_2\right) \tag{A.58}$$

using the definition of the mean velocity. To obtain an expression for D_{12}, a relation of the chemical potential in terms of mole fractions is needed,

$$\bar{\mu}_i = \bar{\mu}_i(y_1, y_2, T), \quad i = 1, 2 \tag{A.59}$$

Such a relation is usually written in terms of activities, a, or in terms of activity coefficients γ. For component one, it is usually written,

$$\bar{\mu}_i = \bar{\mu}_1^0 + RT \ln a_1 \tag{A.60}$$

and

$$a_1 = \gamma_1 y_1 \tag{A.61}$$

Differentiation of Eq. (A.59) gives,

$$\nabla\mu_1 = \frac{\partial\bar{\mu}_i}{\partial y_i}\nabla y_i \tag{A.62}$$

The stability condition of thermostatics [11] requires,

$$\frac{\partial \bar{\mu}_i}{\partial y_i} > 0 \qquad \text{(A.63)}$$

for an n-component system.

An expression for diffusivity can now be easily obtained using Eqs. (A.58), (A.57), (A.62), and (A.56),

$$C_1 C_2 \left(\vec{v}_1 - \vec{v}_2 \right) = -D_{12} C^2 \nabla y_1 = -\frac{C_1}{\bar{R}_{12}} \frac{\partial \bar{\mu}_i}{\partial y_i} \nabla y_1 \qquad \text{(A.64)}$$

Equation (A.64) shows that,

$$-D_{12} = \frac{y_1}{C\bar{R}_{12}} \frac{\partial \bar{\mu}_i}{\partial y_i} \qquad \text{(A.65)}$$

Since $\bar{R}_{12}$ was positive and the change of chemical potential with mole fraction is positive, D_{12} is positive, as is well-known. Conventionally, it is written in terms of a thermodynamic coefficient,

$$B_{ij} = 1 + \frac{\partial \ell n \gamma_i}{\partial \ell n y_i} \qquad \text{(A.66)}$$

which becomes one for an ideal solution. Using Eqs. (A.60), (A.61), (A.62), and (A.66),

$$\frac{\partial \bar{\mu}_1}{\partial y_1} = \frac{RT}{y_1} B_{11} \qquad \text{(A.67)}$$

Then D_{12} becomes,

$$D_{12} = \frac{RT}{CM_1 M_2 R_{12}} B_{11} \qquad \text{(A.68)}$$

This is Lamm's Eq. (42) [2], where, in his notation, $\phi_{12} = \rho_1 \rho_2 R_{12}$. Since the friction coefficient R_{12} is still an unknown, its predictive capability is limited. Nevertheless, assuming a constant friction coefficient, it shows that, for ideal gases, diffusivity is inversely proportional to pressure and to the square of the absolute temperature. Such a pressure dependence is usually observed, while the temperature dependence is usually closer to the 1.75 power indicating a variation of the friction coefficient with temperature. The link to thermodynamic properties is in B_{11}.

Einstein's formula Eq. (33) [12] is,

$$D_{12} = \frac{RT}{N} \frac{1}{6\pi\eta\rho} \tag{A.69}$$

where η is the viscosity of the solvent and ρ is the radius of the diffusing molecule obtained from Eq. (A.68) by assuming that friction is due to viscous motion of N spherical molecules per mole, each of which is far enough apart and moves slowly enough to obey Stokes' law. In relating the friction coefficient R_{12} to the Stokes' friction coefficient, B_{11}, and y_2, the mole fraction of solvent is set to unity.

For the case of forced diffusion, neglecting acceleration, Eq. (A.44) for a binary system gives,

$$\rho_1 \nabla \mu_1 = -R_{12}\rho_1\rho_2 \left(\vec{v}_1 - \vec{v}_2 \right) + \rho_1 \vec{f}_1 \tag{A.70}$$

or

$$\rho_1 \left(\vec{v}_1 - \vec{v} \right) = -\frac{\omega_1}{R_{12}} \frac{\partial \mu_1}{\partial \omega_1} \nabla \omega_1 + \frac{\omega_1}{R_{12}} \vec{f}_1 \tag{A.71}$$

The reciprocal of R_{12} on the right-hand side of Eq. (A.71) is usually called the mobility.

A.9   Perfect Gas Mixture

The component momentum balances, Eq. (A.44), reduce themselves to the Maxwell-Stefan hydrodynamic equations for diffusion [7] for the case of a perfect gas mixture whose mass average velocity is zero. This follows from the expression of the chemical potential for a perfect gas mixture,

$$\bar{\mu}_1 = \bar{\mu}_1^0 + RT\ell nP_i = M_i\mu_1 \tag{A.72}$$

and the ideal gas law,

$$P_i = C_iRT \tag{A.73}$$

where P_i is the partial pressure of component i, together with the relationship for concentrations, C_i,

$$\rho_i = C_iM_i \tag{A.74}$$

Thus, for ideal gases, the equations of motion relative to a fixed reference for constituent i are,

$$\rho_i \vec{a}_i + \nabla P_i = \sum_{k=1}^{n} R_{ik} \rho_i \rho_k \left(\vec{v}_k - \vec{v}_k \right) + \rho_i \vec{f}_i \qquad (A.75)$$

Equation (A.75) is Truesdell's Eq. (3) [7]. Then the momentum balance for component i and the corresponding mass balance for one-dimensional propagation can be written in matrix form,

$$\begin{pmatrix} \dfrac{\partial v_i}{\partial t} \\[2mm] \dfrac{\partial C_i}{\partial t} \end{pmatrix} + \begin{pmatrix} v_i & \dfrac{RTB_{11}}{C_1 M_1} \\[2mm] C_i & v_i \end{pmatrix} \begin{pmatrix} \dfrac{\partial v_i}{\partial x} \\[2mm] \dfrac{\partial C_i}{\partial x} \end{pmatrix} = \begin{pmatrix} M_2 R_{12} C_2 (v_2 - v_1) \\[2mm] qr_1 \end{pmatrix} \qquad (A.76)$$

The characteristic determinant of this system is,

$$\begin{vmatrix} v_i - \lambda & \dfrac{RT}{\rho_i} \\[2mm] C_i & v_i - \lambda \end{vmatrix} = 0 \qquad (A.77)$$

The characteristics roots are real and equal to,

$$\lambda = v_i \pm \sqrt{RT/M_i} \qquad (A.78)$$

The system of differential equations is hyperbolic. Until the signal arrives at a given location, the change at that location is determined by the initial data. Thus, the well-known paradox associated with the use of Fick's law of diffusion does not occur [13–15]. However, Eq. (A.76) makes it clear that the disturbances propagate very fast for the case of an ideal gas mixture. They propagate with the isothermal sound speed which is of the same order of magnitude as the usual adiabatic speed of sound. Therefore in this case, the effect of including inertia is not expected to be very large except for short times or for rapid processes. The isothermal assumption made in deriving the equations of motion may be lifted to give a perhaps physically more realistic description of diffusion in gases.

A.10   Binary Diffusion with Inertia

Consider diffusion in a constant molar density mixture with zero average velocity. In such a simplified situation for a binary system, one needs only one equation of motion and one mass balance. In one dimension, Eqs. (A.44) and (A.24) then become for component 1,

$$\begin{pmatrix} \dfrac{\partial v_1}{\partial t} \\[2mm] \dfrac{\partial C_1}{\partial t} \end{pmatrix} + \begin{pmatrix} v_1 & \dfrac{RTB_{11}}{C_1 M_1} \\[2mm] C_1 & v_i \end{pmatrix} \begin{pmatrix} \dfrac{\partial v_1}{\partial x} \\[2mm] \dfrac{\partial C_1}{\partial x} \end{pmatrix} = \begin{pmatrix} M_2 R_{12} C_2 (v_2 - v_1) \\[2mm] r_1 \end{pmatrix} \tag{A.79}$$

The characteristic roots are now given by,

$$\lambda = v_1 \pm \sqrt{RTB_{11}/M_1} = v_1 \pm \sqrt{y_1 \dfrac{\partial \bar{\mu}_1}{\partial y_1}} \tag{A.80}$$

In view of the stability condition of thermostatics, Eq. (A.63), the characteristics are again real. The system of partial differential equations describing diffusion is hyperbolic. However, unlike in the case of perfect gas mixtures, the concentration pulse may in this case propagate quite slowly. This is clear from Eq. (A.80) applied to a mixture with an activity coefficient substantially below one.

Consider a step change of concentration of component one at a boundary. Until the concentration signal arrives at some position in the mixture, the behavior at that point is determined by the initial data. Only after the arrival of that signal does the boundary data affect the behavior of the mixture at the said point. The usual parabolic diffusion equations incorrectly predict an instantaneous response to a step change at the boundary. The more general hyperbolic equations derived in this appendix remove this well-known paradox.

A.11　Inertial Correction of Fick's Law

For low fluxes and zero average molar velocity, a simple correction to Fick's second law of diffusion can be obtained. Dropping the velocity square term in Eq. (A.79) and combining it with the continuity equation, the momentum equation for species "1" becomes,

$$\frac{\partial (C_1 v_1)}{\partial t} + \frac{RTB_{11}}{M_1} + \frac{\partial C_1}{\partial x} = -R_{12} M_2 CC_1 (v_1 - v^*) \tag{A.81}$$

(inertial) (driving force) (friction)

where Eq. (A.58) was used to change $(v_1 - v_2)$ into $(v_1 - v^*)$. As stated above, next v^* is set equal to zero, as will be the case for equimolal counter-diffusion. Then, the molar flux for component "1," $N_1 = C_1 v_1$ is given by Eq. (A.81),

$$N_1 = -\frac{RTB_{11}}{M_1 M_2 C R_{12}} \frac{\partial C_1}{\partial x} - \frac{1}{R_{12} M_2 C} \frac{\partial N_1}{\partial t} \tag{A.82}$$

which reads,

molar flux = (diffusivity) × (concentration gradient) − (relaxation time)
× (inertia of component one)

In terms of a binary type diffusion coefficient D_{12} given by Eq. (A.68), the expression for flux can be written,

$$N_1 = -D_{12} \frac{\partial C_1}{\partial x} - \frac{D_{12}}{\left(\frac{RTB_{11}}{M_1}\right) C} \frac{\partial N_1}{\partial t} \tag{A.83}$$

The inertial correction thus involves a ratio of binary diffusivity to the square of the propagation velocity. Its effect can thus be readily assessed in terms of known quantities. It will be clearly important for rapid transients. Substitution of Eq. (A.83) into the mass balance,

$$\frac{\partial C_1}{\partial t} + \frac{\partial N_1}{\partial x} = 0 \tag{A.84}$$

and elimination of the second order partial of N_1 using Eq. (A.84), constant D_{12}, and propagation velocities yields the hyperbolic equation for diffusion,

$$\frac{\partial C_1}{\partial t} = D_{12} \frac{\partial^2 C_1}{\partial x^2} - \frac{D_{12}}{\left(\frac{RTB_{11}}{M_1}\right)} \frac{\partial^2 C_1}{\partial t^2} \tag{A.85}$$

In the above form, such an equation has been discussed for mass transfer in capillary-porous bodies [13] and for heat conduction [14]. In the latter reference, Baumeister and Hamill discuss the physically unreasonable behavior given by the error function solution and the physically more correct representation obtained from the solution of Eq. (A.85). The error function solution predicts instantaneous response at all points in the medium and infinite flux at the boundary in the limit of zero time. Inertial correction removes these paradoxes.

Bowen [16] has derived component momentum balance equations of the type given by Eq. (A.44) with component viscosities included. He has interpreted these momentum balances in terms of a diffusion theory. However, he has not considered the inertial correction to Fick's law discussed here.

Nomenclature

a_1 Activity of component 1
$\vec{a}_1$ Acceleration, Eq. (A.75)

B_{ij}	Thermodynamic coefficient, Eq. (A.66)
C_i	Molar concentration of species i
C	Molar density of the mixture
D_{12}	Diffusivity of component in a binary dissipation
F	Function, Eq. (A.1)
f_i	External force acting on component i
h_i	Partial specific enthalpy of component i
J_i	Flux of component i relative to weight average velocity
J_i^0	Flux of component i, Eq. (A.2)
J_q	Heat flux, Eq. (A.37)
M	Molecular weight
m	Mass
N	Avogadro's number, Eq. (A.69)
N_1	Molar flux of component 1
n	Number of components
P	Pressure
P_i	Partial pressure of i-th component
Q	Heat into the system
q	Net rate of heat outflow
R	Gas law constant
R_{ij}	Onsager-type friction coefficient
$\overline{R}_{ij}$	Onsager-type friction coefficient in molar units
r_i	Molar rate of production of species i
S	Mixture entropy per gram
t	Time
T	Absolute temperature
U	Internal energy of the mixture per gram
U^*	Internal plus kinetic energy per gram
v_i	Velocity of i-th component
v	Weight average velocity of the mixture
v^*	Molar average velocity of the mixture
V	Volume
W	Work done by the system
X_i	i-th force
x	Spatial coordinate
y_i	Mole fraction of i-th component

Greek Symbols

γ	Activity coefficient
δ	Variation
η	Viscosity
λ	Eigenvalue
μ_i	Chemical potential of component i

$\bar{\mu}_i$	Molar chemical potential of component i
ρ	Radius of molecule or density
σ	Entropy production per unit volume
$\phi_{ij} = \rho_i \rho_j R_{ij}$	Lamm's friction coefficient
ω_i	Weight fraction

References

1. Onsager L (1945) Theories and problems of liquid diffusion. Ann NY Acad Sci 46: 241–265
2. Lamm O (1964) Studies in the kinematics of isothermal diffusion: a macro-dynamical theory of multicomponent fluid diffusion. Adv Chem Phys VI:291–313
3. deGroot SR, Mazur P (1962) Non-equilibrium thermodynamics. North Holland Publishing Co., Amsterdam
4. Onsager L (1931) Reciprocal relations in irreversible processes, I. Phys Rev 37:405–426
5. Merk HJ (1959) The macroscopic equations for simultaneous heat and mass transfer in isotropic, continuous and closed systems. Appl Sci Res A 8:73–99
6. Aris R (1962) Vectors, tensors, and the equations of fluid mechanics. Prentice Hall, Englewood Cliffs, NJ
7. Truesdell CI (1962) Mechanical basis of diffusion. J Chem Phys 37:2336–2344
8. Muller I (1968) A thermodynamic theory of mixtures of fluids. AVRMA 28:1–39
9. Courant R, Hilbert D (1962) Methods of mathematical physics. partial differential equations, Vol. II. Interscience Publishers, New York
10. Bird RB, Stewart, WE, Lightfoot EN (1962) Transport phenomena. John Wiley & Sons, New York
11. Callen HB (1960) Thermodynamics. John Wiley & Sons, New York
12. Einstein Albert (1956) Investigations on the theory of the Brownian movement. Dover Publications, New York
13. Luikov AG (1956) Application of the methods of thermodynamics of irreversible processes to the investigation of heat and mass transfer. J Eng Phys 9(3):189–202
14. Baumeister, KJ, Hamill TD (1969) Hyperbolic heat conduction equation: a solution for the Sermi-infinite body problem. ASME J Heat Transfer 543-548
15. Morse PM, Feshbach H (1953) Methods of theoretical physics, vol. 1. McGraw-Hill, New York, p 865
16. Bowen RM (1976) Theory of mixtures. In: Eringen AC (ed) Continuum physics, vol III. Academic Press, New York, pp 1–127

Index

Printed in the United States
by Baker & Taylor Publisher Services